Effect of Dietary Antioxidants in Chronic Disease Prevention

Effect of Dietary Antioxidants in Chronic Disease Prevention

Guest Editor

Baojun Xu

MDPI

Basel • Beijing • Wuhan • Barcelona • Belgrade • Novi Sad • Cluj • Manchester

Guest Editor
Baojun Xu
Department of Life Sciences
Beijing Normal-Hong Kong
Baptist University
Zhuhai
China

Editorial Office
MDPI AG
Grosspeteranlage 5
4052 Basel, Switzerland

This is a reprint of the Special Issue, published open access by the journal *Antioxidants* (ISSN 2076-3921), freely accessible at: https://www.mdpi.com/journal/antioxidants/special_issues/Connection_Oxidative_Stress_Chronic_Dietary_Antioxidants.

For citation purposes, cite each article independently as indicated on the article page online and as indicated below:

Lastname, A.A.; Lastname, B.B. Article Title. *Journal Name* **Year**, *Volume Number*, Page Range.

ISBN 978-3-7258-6222-1 (Hbk)
ISBN 978-3-7258-6223-8 (PDF)
https://doi.org/10.3390/books978-3-7258-6223-8

Contents

About the Editor

Baojun Xu

Baojun Xu is a Chair Professor at Beijing Normal-Hong Kong Baptist University (BNBU, a full English teaching university in China), a Fellow of the Royal Society of Chemistry, a Zhuhai Scholar Distinguished Professor, Department Head of the Department of Life Sciences, Program Director of the Food Science and Technology Program, and author of over 430 peer-reviewed papers. Dr. Xu received a Ph.D. in Food Science from Chungnam National University, South Korea. He conducted postdoctoral research work at North Dakota State University (NDSU), Purdue University, and the Gerald P. Murphy Cancer Foundation in the USA from 2005 to 2009. He carried out short-term visiting research at NDSU in 2012 and the University of Georgia in 2014, followed by visiting research during his sabbatical leave (7 months) at Pennsylvania State University in the USA in 2016. Dr. Xu is serving as Associate Editor-in-Chief of *Food Science and Human Wellness*, Associate Editor of *Food Research International*, Associate Editor of *Food Frontiers*, and an Editorial Board Member of around 10 international journals. He received the inaugural President's Award for Outstanding Research at the UIC in 2016 and the President's Award for Outstanding Service at the UIC in 2020. Dr. Xu has been listed as one of the world's top 2% of scientists by Stanford University for the past six consecutive years and has been listed as the Best Scientist in the world in the field of Biology and Biochemistry by Research.com in 2023 and 2024. Prof. Xu was named a Highly Ranked Scholar (top 0.05%) by ScholarGPSTM, ranking at #8 in Food Science and Technology in the world and #14 in Agricultural and Natural Sciences.

antioxidants

Editorial

Effect of Dietary Antioxidants in Chronic Disease Prevention

Baojun Xu

Food Science and Technology Program, Department of Life Sciences, Beijing Normal-Hong Kong Baptist University, Zhuhai 519087, China; baojunxu@uic.edu.cn

Chronic diseases are a major global public health challenge, with increasing incidence and mortality rates [1]. Oxidative stress, characterized by an imbalance between oxidants and antioxidants, leading to cellular damage, is a key mechanism in their development [2]. Low levels of reactive oxygen species are necessary for many processes such as intracellular signal transduction, metabolism, immune and hypoxic responses, and transcriptional regulation. However, excessive reactive oxygen species (ROS) may be pathological and contribute to the development and progression of chronic diseases [3]. Dietary antioxidants, found in fruits, vegetables, nuts, and whole grains, can neutralize free radicals and inhibit oxidative stress [4], thus playing a crucial role in chronic disease prevention [5,6]. This Special Issue delves into the latest research findings of dietary antioxidants in chronic disease prevention, including eight research articles and two reviews, which provides insights for future research and dietary guidelines.

Huang et al. studied a multicomponent dietary supplement's impact on tear secretion and ocular surface inflammation in dry eye syndrome (DES) patients (Contribution 1). The supplement (45 mg/day eicosapentaenoic acid, 30 mg/day docosahexaenoic acid, 30 mg/day lutein, and 1.8 mg/day zeaxanthin) significantly increased tear secretion and decreased the ocular surface disease index score after 12 weeks. Inflammatory markers IL-6 and IL-8 also decreased.

Yang et al. found that Lycii Radicis Cortex (LRC) can inhibit anaerobic bacterial proliferation and inflammatory cell infiltration in rat gingival tissues, showing anti-inflammatory effects (Contribution 2). Additionally, LRC reduces malondialdehyde levels and inducible nitric oxide synthase activity, demonstrating antioxidant properties. In addition, it can effectively prevent connective tissue degradation. LRC also decreased the receptor activator of NF-κB ligand/osteoprotegerin ratio and the number and area of osteoclasts on the alveolar bone surface, thereby inhibiting alveolar bone loss.

In a cohort study, Bonfiglio et al. found that an intake of merely 165 mg of flavonoids per day exerts a protective effect against MASLD, reducing the risk of this condition (Contribution 3).

Kim et al. demonstrated through in vitro experiments using LPS-stimulated BV2 microglial cells that *Quercus acuta* Thunb. (QA) effectively mitigates microglia-mediated neuroinflammatory responses by inhibiting NF-κB and MAPK signaling pathways and activating the Nrf2/HO-1 pathway (Contribution 4).

Sandoval et al. explored in C57BL/6 mice if oral β-carotene mitigates P4502E1 (CYP2E1) expression in ethanol-exposed subjects (Contribution 5). Their findings imply β-carotene might amplify liver damage from both low and high alcohol doses. Thus, factors like alcohol volume, exposure duration, damage regulation, and signaling pathways involved in the consumption of both alcohol and antioxidants must all be weighed.

Park et al. found that Menthae Herba (MH) reduces ROS and NF-κB-mediated inflammatory signaling pathways while upregulating CREB/Nrf2/HO-1-related antioxidant signaling in microglial cells (Contribution 6).

Zhang et al. investigated the synergistic anti-inflammatory effects of proanthocyanidin (LGSP) and camellia oil (CO) in vitro (Contribution 7). The results indicated that the combined treatment of LGSP (20 μg/mL) and CO (1 mg/mL) synergistically suppressed the production of NO, TNF-, IL-6, and ROS. The synergistic effect was attributed to their suppression of the activation of nuclear factor-κB (NF-κB) and mitogen-activated protein kinase (MAPK) signaling pathways.

Yu et al. investigated the protective effect of *Mallotus oblongifolius* polyphenols (MOP) on ethanol-induced gastric mucosal injury in rats. The results showed that MOP could increase the expression of antioxidant enzymes and decrease the expression oxidative enzymes to prevent ethanol-induced acute gastric mucosal injury. These effects may be related to the inhibition of p38/ERK/JNK phosphorylation and the activation of the Nrf2 signaling pathway by MOP (Contribution 8).

The review of You et al. recorded oxidative stress and dietary antioxidants in head and neck cancer (HNSCC) (Contribution 9). The review summarized the role of oxidative stress in carcinogenesis, particularly focusing on the three major risk factors for HNSCC, including smoking, alcohol consumption, and high-risk human papillomavirus (HR-HPV) infection. Additionally, this review points out that how nine dietary antioxidants, such as vitamin C, vitamin E, carotenoids, epigallocatechin-3-gallate (EGCG), and curcumin, mitigate ROS, influence cancer-related signaling pathways, and modulate the tumor microenvironment in the development and progression of HNSCC.

Review conducted by Zhang et al., summarizes how genistein exerts therapeutic effects by inhibiting oxidative stress. Additionally, the review focuses on the mechanisms by which genistein may combat five common diseases induced by oxidative stress, including Parkinson's disease, Alzheimer's disease, diabetes, cardiovascular disease, and cancer. Moreover, it evaluates strategies for enhancing the water solubility and bioavailability of genistein (Contribution 10).

This Special Issue comprises a total of 10 articles, encompassing two clinical cohort studies, three in vivo experimental studies, three in vitro experimental studies, and two reviews. These studies collectively demonstrate the significant potential of dietary antioxidants in mitigating chronic diseases through their anti-inflammatory and antioxidant properties, which highlight the importance of targeted nutritional interventions in chronic disease prevention and management. Future research should focus on elucidating the deeper mechanisms and conducting larger-scale clinical trials to further validate these promising results and optimize dietary recommendations.

Funding: This research received no external funding.

Institutional Review Board Statement: There are humans involved in this study, it is an editorial.

Informed Consent Statement: It is not applicable to this editorial.

Data Availability Statement: There are no new data for this editorial.

Conflicts of Interest: The author declares no conflicts of interest.

List of Contributions:

1. Huang, S.C.; Lei, Y.P.; Hsiao, M.C.; Hsieh, Y.K.; Tang, Q.P.; Chen, C.; Hsu, M.Y. Multicomponent Dietary Supplementation: Impact on Tear Secretion and Ocular Surface Inflammation in Dry Eye Syndrome Patients. *Antioxidants* **2025**, *14*, 103. https://doi.org/10.3390/antiox14010103.

2. Yang, J.; Song, H.; Lee, J.; Chung, H.; Kwon, Y.S.; Jegal, K.H.; Kim, J.K.; Ku, S.K. The Effect of the Root Bark of *Lycium chinense* (Lycii Radicis Cortex) on Experimental Periodontitis and Alveolar Bone Loss in Sprague-Dawley Rats. *Antioxidants* **2024**, *13*, 1332. https://doi.org/10.3390/antiox13111332.
3. Bonfiglio, C.; Tatoli, R.; Donghia, R.; Guido, D.; Giannelli, G. Exploratory Role of Flavonoids on Metabolic Dysfunction-Associated Steatotic Liver Disease (MASLD) in a South Italian Cohort. *Antioxidants* **2024**, *13*, 1286. https://doi.org/10.3390/antiox13111286.
4. Kim, J.K.; Yang, H.J.; Go, Y. *Quercus acuta* Thunb. Suppresses LPS-Induced Neuroinflammation in BV2 Microglial Cells via Regulating MAPK/NF-κB and Nrf2/HO-1 Pathway. *Antioxidants* **2022**, *11*, 1851. https://doi.org/10.3390/antiox11101851.
5. Sandoval, C.; Mella, L.; Godoy, K.; Adeli, K.; Farías, J. β-Carotene Increases Activity of Cytochrome P450 2E1 during Ethanol Consumption. *Antioxidants* **2022**, *11*, 1033. https://doi.org/10.3390/antiox11051033.
6. Park, Y.J.; Yang, H.J.; Li, W.; Oh, Y.C.; Go, Y. Menthae Herba Attenuates Neuroinflammation by Regulating CREB/Nrf2/HO-1 Pathway in BV2 Microglial Cells. *Antioxidants* **2022**, *11*, 649. https://doi.org/10.3390/antiox11040649.
7. Zhang, L.; Chen, J.; Liang, R.; Liu, C.; Chen, M.; Chen, J. Synergistic Anti-Inflammatory Effects of Lipophilic Grape Seed Proanthocyanidin and Camellia Oil Combination in LPS-Stimulated RAW264.7 Cells. *Antioxidants* **2022**, *11*, 289. https://doi.org/10.3390/antiox11020289.
8. Yu, S.; Duan, Z.; Li, P.; Wang, S.; Guo, L.; Xia, G.; Xie, H. Protective Effect of Polyphenols Purified from Mallotus oblongfolius on Ethanol-Induced Gastric Mucosal Injury by Regulating Nrf2 and MAPKs Pathways. *Antioxidants* **2022**, *11*, 2452. https://doi.org/10.3390/antiox11122452.
9. You, A.J.; Park, J.; Shin, J.M.; Kim, T.H. Oxidative Stress and Dietary Antioxidants in Head and Neck Cancer. *Antioxidants* **2025**, *14*, 508. https://doi.org/10.3390/antiox14050508.
10. Zhang, K.; Wang, J.; Xu, B. Critical Review on Molecular Mechanisms for Genistein's Beneficial Effects on Health Through Oxidative Stress Reduction. *Antioxidants* **2025**, *14*, 904. https://doi.org/10.3390/antiox14080904.

References

1. Greenberg, H.; Pi-Sunyer, F.X. Preventing preventable chronic disease: An essential goal. *Prog. Cardiovasc. Dis.* **2019**, *62*, 303–305. [CrossRef] [PubMed]
2. Sies, H. Oxidative stress: A concept in redox biology and medicine. *Redox Biol.* **2015**, *4*, 180–183. [CrossRef] [PubMed]
3. Kishi, S.; Nagasu, H.; Kidokoro, K.; Kashihara, N. Oxidative stress and the role of redox signalling in chronic kidney disease. *Nat. Rev. Nephrol.* **2024**, *20*, 101–119. [CrossRef] [PubMed]
4. Rajaram, S.; Jones, J.; Lee, G.J. Plant-based dietary patterns, plant foods, and age-related cognitive decline. *Adv. Nutr.* **2019**, *10*, S422–S436. [CrossRef] [PubMed]
5. Qu, G.; Chen, J.; Guo, X. The beneficial and deleterious role of dietary polyphenols on chronic degenerative diseases by regulating gene expression. *Biosci. Trends* **2019**, *12*, 526–536. [CrossRef] [PubMed]
6. Khemka, S.; Reddy, A.; Garcia, R.I.; Jacobs, M.; Reddy, R.P.; Roghani, A.K.; Pattoor, V.; Basu, T.; Sehar, U.; Reddy, P.H. Role of diet and exercise in aging, Alzheimer's disease, and other chronic diseases. *Ageing Res. Rev.* **2023**, *91*, 102091. [CrossRef] [PubMed]

antioxidants

MDPI

Article

Multicomponent Dietary Supplementation: Impact on Tear Secretion and Ocular Surface Inflammation in Dry Eye Syndrome Patients

Shih-Chien Huang [1,2], Yen-Ping Lei [3], Min-Chien Hsiao [4], Yu-Kai Hsieh [5], Quei-Ping Tang [6], Connie Chen [7,8,9] and Min-Yen Hsu [7,10,*]

1 Department of Nutrition, Chung Shan Medical University, Taichung City 402, Taiwan; schuang@csmu.edu.tw
2 Department of Nutrition, Chung Shan Medical University Hospital, Taichung City 402, Taiwan
3 Department of Nursing, National Yang Ming Chiao Tung University, Hsinchu City 300, Taiwan; yplei@nycu.edu.tw
4 Department of Medical Education, Changhua Christian Hospital, Changhua City 500, Taiwan; 184984@cch.org.tw
5 School of Medicine, Taipei Medical University, Taipei City 110, Taiwan; b101111038@tmu.edu.tw
6 Department of Nutrition, Wei Gong Memorial Hospital, Toufen City 351, Taiwan; 046346@tool.caaumed.org.tw
7 Department of Ophthalmology, Chung Shan Medical University Hospital, Taichung City 402, Taiwan; cconnie7@csmu.edu.tw
8 Department of Optometry, Chung Shan Medical University, Taichung City 402, Taiwan
9 Institute of Optometry, Chung Shan Medical University, Taichung City 402, Taiwan
10 School of Medicine, Chung Shan Medical University, Taichung City 402, Taiwan
* Correspondence: cshy1769@csh.org.tw

Abstract: Dry eye syndrome (DES) is a prevalent ocular condition characterized by tear film instability, inflammation, and discomfort, affecting millions worldwide. DES is related to oxidative stress imbalance and ocular surface inflammation, which are important factors in the development of the condition. Recent studies have demonstrated that fish oil, lutein, and zeaxanthin possess anti-inflammatory and antioxidant properties. This study investigated the efficacy of a multicomponent dietary supplement in improving tear secretion and mitigating ocular surface inflammation in patients with DES. It was an open-label intervention trial. In total, 52 participants were randomly assigned to control ($n = 23$) and supplement (45 mg/day eicosapentaenoic acid, 30 mg/day docosahexaenoic acid, 30 mg/day lutein, and 1.8 mg/day zeaxanthin; $n = 29$) groups for 12 weeks. The participants were evaluated using Schirmer's test and the ocular surface disease index (OSDI) as ocular surface parameters. Moreover, blood or tear oxidative stress, antioxidant capacities, and tear inflammatory indicators were measured at weeks 0 and 12. The results indicated a significant increase in tear secretion and a significant reduction in OSDI scores in the supplement group. Additionally, inflammatory markers, such as interleukin (IL)-6 and IL-8, significantly decreased after the intervention. However, the OSDI of the supplement group significantly improved by 6.60 points ($\beta = -6.60$, $p = 0.01$). These findings support the potential of targeted nutritional supplementation as a safe and effective strategy for alleviating DES symptoms, offering an alternative to conventional treatments that exclusively focus on symptom management. This study highlights the role of specific nutrients in modulating tear production and inflammation, thereby providing a foundation for dietary approaches to DES treatment. Future research should explore the long-term benefits of such interventions and their impact on overall ocular health.

Keywords: dry eye syndrome; inflammation; oxidative stress; fish oil; lutein; zeaxanthin

1. Introduction

Dry eye syndrome (DES), a multifactorial disease of the ocular surface, is primarily characterized by tear film instability. This instability leads to increased tear film osmolarity and ocular surface inflammation or damage, resulting in discomfort [1]. DES significantly impacts patients' quality of life [2]. As of 2021, the global prevalence of DES was estimated at 29.5%. Notably, Africa exhibited the highest regional prevalence rate (47.9%), followed by West Asia (29.0%); East Asia (19.4%); and Europe, South America, and Oceania (13.7–14.9%). Taiwan, Japan, and South Korea have reported rates of approximately 30.0% [3]. In Taiwan, approximately 25% of the population suffers from DES, and the incidence is increasing annually [4].

DES potentially arises from a multitude of single or multiple risk factors, encompassing personal attributes (e.g., age, post-menopausal status, and contact lens use), environmental conditions (e.g., prolonged eye fixation without adequate blinking), clinical diseases (e.g., autoimmune and chronic diseases), medications (e.g., psychiatric drugs, antiviral medications, beta blockers, and diuretics), and ocular surgeries (e.g., laser and cataract operations). These factors have been identified and discussed in various studies [5–7].

Inflammation is a key mechanism underlying the pathogenesis of DES. Research has demonstrated that in DES, epithelial cells release relatively high concentrations of chemokines and cytokines, notably tumor necrosis factor-alpha (TNF-α), interleukin (IL)-1β, IL-6, and IL-8 [8,9]. Among these, IL-6 and TNF-α have been identified as critical inflammatory markers in DES, contributing to lacrimal gland cell apoptosis and reduced tear production [10]. Patients with DES may develop an adaptive immune response leading to chronic inflammation [11]. This cytokine-mediated process affects sensory neurons and cytokine alterations, resulting in decreased lacrimal secretion and associated ocular surface discomfort [12,13].

Oxidative stress is another crucial mechanism contributing to the exacerbation of DES. Prolonged exposure of the ocular surface to factors such as ultraviolet radiation, air pollution, hormonal changes, and bacterial infections potentially induces the overproduction of reactive oxygen species (ROS). When antioxidant capabilities are insufficient or imbalanced, ROS can directly damage the ocular glands, leading to epithelial and goblet cell apoptosis or harm to corneal nerves, thereby diminishing tear secretion quality and tear film stability [14]. ROS also trigger inflammatory responses, rendering their overproduction a key factor in the DES cycle [15].

Clinically, artificial tears are commonly used to manage DES, aiming to increase moisture and prevent tear evaporation; however, they merely alleviate symptoms temporarily [16]. Researchers suggest that intense pulsed light therapy significantly improves DES symptoms; nevertheless, it is considered unsafe for upper-eyelid treatment, requires annual or biennial sessions, and is costly [17]. Autologous serum eye drops have recently emerged, and they are made from the patient's own blood and cellular components to mimic the biologically active nutrients in natural tears, such as vitamins A and C, lysozyme, and immunoglobulins [18,19]. However, they are not suitable for patients with autoimmune disease, are prone to microbial contamination, and require refrigeration [20]. Additionally, the Food and Drug Administration has approved anti-inflammatory drugs such as cyclosporine A and lifitegrast 5% for DES; while they can improve symptoms, their long-term use may cause side effects such as conjunctival hyperemia, secretions, foreign body sensation, and decreased vision [16,21]. Consequently, dietary supplements for eye health have actively been explored and are considered safer for long-term use [22].

An in vitro study found omega-3 fatty acids, particularly eicosapentaenoic acid (EPA) and docosahexaenoic acid (DHA), to significantly enhance cell viability and reduce inflammation in corneal epithelial cells [23]. Numerous studies have indicated that high doses of fish oil, specifically EPA (1500–2000 mg) plus DHA (1000–1050 mg) daily for 3–6 months, potentially improve tear film break-up time, tear secretion, and the ocular surface disease index (OSDI) [24,25]. In contrast, low-dose fish oil supplementation often uses a compound formula to enhance efficacy. A study administered a high-dose fish oil compound (EPA 1050 mg + DHA 127.5 mg/day), which also included vitamins (A, C, and E), zinc, magnesium, copper, selenium, tyrosine, and cysteine, to patients with DES for 3 months. This regimen significantly increased tear secretion and tear film break-up time compared with the control [24]. Another study administered a fish oil compound (EPA 45 mg + DHA 700 mg/day), which also contained vitamins (A, C, and E), zinc, copper, and selenium, to patients with DES for 3 months, resulting in a significant reduction in inflammatory factors IL-1β, IL-6, and IL-8 [26].

Lutein and zeaxanthin, widely recognized as beneficial for eye health, can absorb 40–90% of incident blue light, thus protecting the retina from photodamage and reducing light scattering [27]. Their antioxidant properties help scavenge free radicals and enhance overall antioxidative capacity, preventing oxidative damage [28,29]. Furthermore, their anti-inflammatory properties are significant, as inflammation is a key pathogenic mechanism underlying several ocular diseases. Lutein and zeaxanthin can prevent oxidative stress-induced cytokine increase and regulate the expression of inflammation-related genes [30]. Research has demonstrated that administering lutein (20 mg/day) and zeaxanthin (2 mg/day) to patients with DES for 3 months significantly improves tear film break-up time and tear meniscus height [22]. Another study providing lutein (20 mg/day) and zeaxanthin (4 mg/day) for 8 weeks resulted in significant improvements in tear film break-up time, tear secretion, the OSDI, tear osmolarity, and matrix metallopeptidase-9 (MMP-9) levels [31]. These benefits may be attributed to lutein's ability to inhibit IL-6 secretion by epithelial cells through the nuclear factor kappa B signaling pathway, highlighting its anti-inflammatory potential [32].

The incidence of DES has been increasing annually and is affecting younger populations [4]. However, the long-term health effects of high-dose fish oil supplementation are concerning. Moreover, the efficacy of dietary supplements containing a combination of fish oil, lutein, and zeaxanthin in ameliorating inflammation and oxidative stress and enhancing antioxidant capacity is yet to be elucidated. Therefore, this study aimed to investigate the effects of 12-week supplementation with a compound containing fish oil, lutein, and zeaxanthin on symptoms, oxidative stress, antioxidant capacity, and inflammation in patients with DES.

2. Methods

2.1. Study Design and Sample Size Determination

This study was an open-label intervention trial. The participants were randomly assigned to control and supplement (45 mg/day EPA, 30 mg/day DHA, 30 mg/day lutein, and 1.8 mg/day zeaxanthin) (Far East Bio-Tec Co., Ltd., Taipei City, Taiwan) groups. Supplementation was administered for 12 weeks. Each participant had two visits (weeks 0 and 12) throughout the study period. To assure compliance with the intervention, the participants were asked to return their bottles for capsule counts. In addition, the study co-executor made phone calls to remind each participant to take the capsules every week during the intervention period. A participant would be excluded if their compliance was <80%. In a previous study, the supplement group exhibited a significant increase in tear break-up time (TBUT) of 4.6 s ($p < 0.05$) compared with the control group [33]. We subsequently calculated the sample size based on the detection of a significant 4.6 s increase between two groups with a power of 80% and a two-sided test with an α value of 0.05. A total of 42 patients were required to match the calculation criteria. The final recruitment number was 52 patients, exceeding our original calculation.

2.2. Participants

In total, 60 participants were initially enrolled at baseline from the Ophthalmology Outpatient Clinic of Chung Shan Medical University Hospital, Taiwan. The inclusion criteria were as follows: (1) age between 20 and 80 years and (2) the presence of moderate or severe DES. The exclusion criteria were as follows: (1) ophthalmic surgery within the preceding 3 months, (2) autoimmune disease, (3) ocular allergy, (4) the consumption of fish oil or antioxidant supplements within the preceding 3 months, and (5) pregnancy or lactation. The participant enrollment process is illustrated in Figure 1. A total of 8 participants withdrew owing to an inability to attend follow-up appointments and personal reasons, resulting in a final sample of 52 participants. The total dropout rate was 13.3%. This study was approved by the Institutional Review Board (IRB) of Chung Shan Medical University Hospital (IRB CSMUH No. CSI-20199). Each patient signed an informed consent form prior to participating, and the informed consent process was consistent with the principles of the Declaration of Helsinki.

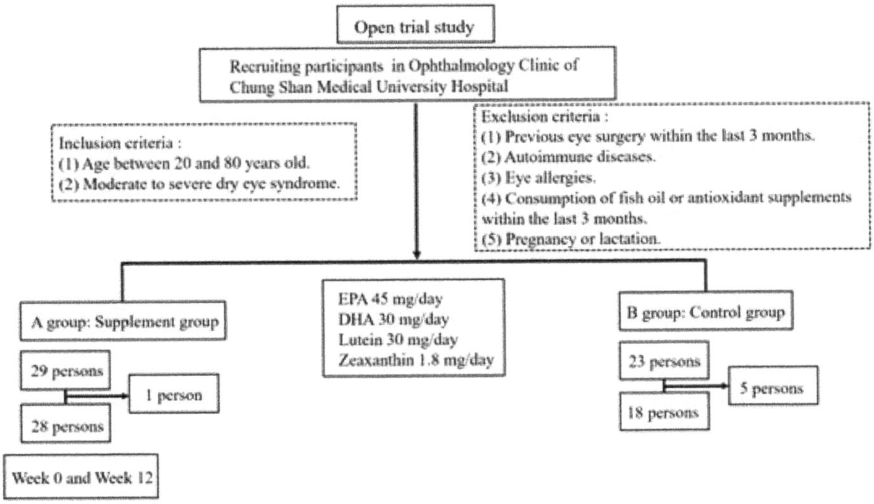

Figure 1. Clinical trial flowchart.

2.3. Data Collection and Measurements

Data regarding each participant's age, gender, self-reported DES symptoms, ophthalmic condition inquiries (high myopia: −5.00 D or more; high astigmatism: −1.50 D or more; past eye surgeries), medication usage, and lifestyle habits were collected. Additionally, the OSDI questionnaire was administered.

2.4. Schirmer Test

The Schirmer test is a common clinical method used to assess tear secretion. The procedure was repeated twice using filter paper strips placed inside the lower eyelid. After 5 min, each strip was removed, and the length of the moist section was measured to ascertain the amount of moisture absorbed. The conversion volume has been described in a previous study [34], as shown in Figure 2.

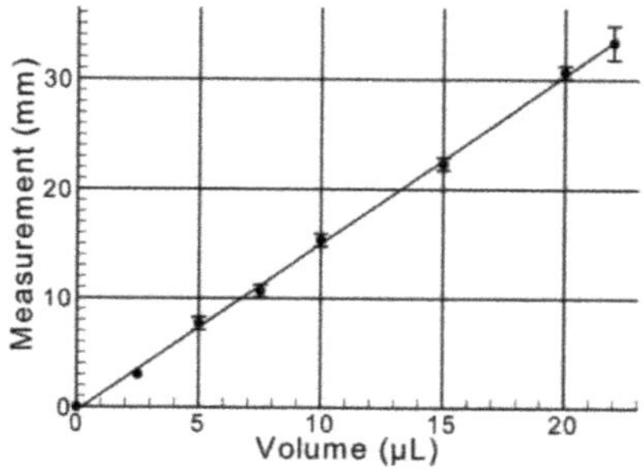

Figure 2. Graphical representation of tear volume measurements using Schirmer strips.

2.5. Ocular Surface Analyzer (OSA)

Dry eye analysis was performed using a non-invasive SBM Sistemi OSA to examine the tear film and meibomian glands. This non-contact method avoids errors resulting from contact irritation or fluorescein use. The analyzer automatically focuses and records data for quantitative analysis. Researchers have demonstrated that OSAs accurately assess tear film function, providing valuable insights into the clinical diagnosis of DES [34]. The parameters measured included non-invasive TBUT (NIBUT), lipid layer thickness (LLT), tear meniscus height (TMH), and meibomian gland loss (MG loss %).

2.6. Dietary Assessment

To ensure that the patients were maintaining their usual dietary intake, they had to complete 24 h dietary recalls at weeks 0 and 12. Nutritional composition was calculated using Nutritional composition was calculated using Nutritionist Pro™ software v7.9 (Nutrition Butler Enhanced Edition, E-Kitchen Business Corporation, Taichung City, Taiwan, 2002), and the nutrient database was based on the Taiwan food composition table (Department of Health, 2021).

2.7. Blood and Tear Collection

Fasting blood samples were drawn on the designated day. Blood specimens were collected in vacutainer tubes (Becton Dickinson, Rutherford, NJ, USA). Tear samples were collected using filter paper strips and diluted in 330 μL of phosphate-buffered saline. The original concentration of the tears was calculated based on the dilution [35,36].

2.8. Oxidative Stress and Total Antioxidant Capacity Evaluation Using Blood and Tear Specimens

To evaluate oxidative stress and antioxidant capacity, plasma and tear malondialdehyde (MDA) levels were assessed as indicators of oxidative stress. The plasma MDA concentration was measured along with thiobarbituric acid reactive substances at excitation and emission wavelengths of 515 and 555 nm, respectively, using a fluorescence spectrophotometer [37]. Tear MDA levels were measured using a commercial competitive enzyme-linked immunosorbent assay kit (Arigo, Hsinchu, Taiwan) [38]. Plasma and tear

Trolox equivalent antioxidant capacity (TEAC) was analyzed using a previously described method [39].

2.9. Multiplex Cytokine Analysis

Tear samples were tested using the Bio-Rad Bio-Plex® Pro Human Cytokine 27-Plex Assay on the Bio-Rad MAGPIX™ Multiplex Reader for the following inflammatory cytokines: IL-1β, IL-2, IL-6, IL-8, IL-17A, and TNF-α. Data were acquired using the Bio-Plex Array Reader System 200 (Bio-Rad, Hercules, CA, USA). The lower limits of quantitation (LOQs) for each cytokine are as follows: IL-1β: 0.00029 pg/μL; IL-2: 0.00129 pg/μL; IL-6: 0.00038 pg/μL; IL-8: 0.00085 pg/μL; IL-17α: 0.00244 pg/μL; and TNF-α: 0.00333 pg/μL.

2.10. Statistical Analysis

All data were analyzed using SigmaPlot software (version 12.5; Systat Software, San Jose, CA, USA) and presented on an intention-to-treat basis. A Shapiro–Wilk test was performed to determine normality. Between-group differences in demographic characteristics and experimental data were evaluated for significance using Student's *t*-test or the Mann–Whitney rank-sum test. The χ^2 or Fisher's exact test was used to analyze categorical variables. Paired comparisons were performed using the paired t-test or Wilcoxon signed-rank test to examine changes in outcomes before and after the intervention. Partial Spearman's correlation analysis was used to assess the association of changes in blood and tear oxidative stress, antioxidant capacity, and inflammatory factors with DES, after adjusting for potential confounders. Multiple linear regression analyses, with or without dietary supplementation as the dependent variable, were used to examine the association of changes in blood and tear oxidative stress, antioxidant capacity, and cytokines with DES, after adjusting for potential confounders. Statistical significance was set at $p < 0.05$.

3. Results

Participant Demographics and Self-Assessment of Dry Eye Symptoms

This study enrolled 52 participants (29 participants in the supplement group and 23 in the control group). Their mean age was 58.0 years, with a range of 39–80 years. The compliance of the supplement group was 96.6 ± 5.4%, and no participants reflected any tolerance after taking the supplement. No significant differences in age, sex, or the proportion of participants with severe myopia or severe astigmatism were noted between the groups. However, the supplement group had a significantly greater proportion of participants who had undergone previous eye-related surgery than the control group. The self-assessment of dry eye symptoms revealed that dry eyes (80.8%) constituted the greatest proportion of symptoms, followed by eye fatigue (78.8%) and blurred or worsened vision (75.0%). No significant differences in various self-reported dry eye symptoms occurred between the two groups (Table 1).

The use of medication and lifestyle habits of participants before and after the intervention are shown in Table 2. Most participants did not use anti-inflammatory drugs. Regarding lifestyle habits, a significant difference in the average daily sleep duration was observed between the two groups, both before and after the intervention. However, no significant differences in other lifestyle habits were found between the groups.

Table 1. Characteristics, visual outcome experiences, and dry eye questionnaires of study groups at baseline.

Variables [1]	Control Group (n = 23)	Supplement Group (n = 29)	p-Value
Age (y)	56.48 ± 7.81	59.59 ± 10.33	0.24
Gender (M/F)	2/23	8/29	0.16
Age of menopause	50.23 ± 4.44	51.71 ± 6.06	0.15
Menopause (n, %)	12 (52%)	14 (48%)	0.1
High myopia (n, %)	7 (30%)	7 (24%)	0.85
High astigmatism (n, %)	4 (17%)	10 (34%)	0.29
Eye surgery experience (n, %)	6 (26%) *	19 (66%)	0.01
Symptoms of dry eye			
Dryness (n, %)	20 (87%)	22 (76%)	0.48
Fatigue (n, %)	20 (87%)	21 (72%)	0.31
Foreign body sensation (n, %)	14 (61%)	15 (52%)	0.71
Photophobia (n, %)	13 (57%)	17 (59%)	0.90
Itching (n, %)	11 (48%)	14 (48%)	0.81
Burning sensation (n, %)	4 (17%)	4 (14%)	1.00
Blurred or poor vision (n, %)	17 (74%)	22 (76%)	0.87
Pain (n, %)	5 (22%)	7 (24%)	0.90

[1] Values are presented as mean ± SD. * Values are significantly different between control and supplement groups at baseline.

Table 2. Lifestyle habits of subjects from study groups at weeks 0 and 12.

Variables [1]	Control Group (n = 23)		Supplement Group (n = 29)		Week 0	Week 12
	Week 0	Week 12	Week 0	Week 12	p-Value	
Anti-inflammatory drugs (n, %)						
No	20 (87%)	19 (90%)	28 (97%)	28 (100%)	0.31	0.18
Yes	3 (13%)	2 (10%)	1 (3%)	0 (0%)		
Frequency of using artificial tears (times, %)						
No	6 (26%)	5 (24%)	11 (38%)	11 (39%)		
1–2	5 (22%)	6 (29%)	5 (17%)	5 (18%)		
3–4	11 (48%)	10 (47%)	8 (28%)	8 (29%)	0.39	0.22
5–6	1 (4%)	0 (0%)	4 (14%)	2 (7%)		
>6	0 (0%)	0 (0%)	1 (3%)	2 (7%)		
Frequency of using video display units (hour) (n, %)						
0	0 (0%)	1 (5%)	0 (0%)	0 (0%)		
2	5 (22%)	4 (19%)	4 (14%)	4 (15%)		
2–4	4 (17%)	3 (14%)	8 (28%)	6 (21%)	0.14	0.45
4–6	5 (22%)	4 (19%)	1 (3%)	2 (7%)		
>6	9 (39%)	9 (43%)	16 (55%)	16 (57%)		
Frequency of reading (hour) (n, %)						
0	11 (48%)	14 (66%)	14 (49%)	15 (54%)		
2	11 (48%)	6 (29%)	13 (45%)	12 (43%)		
2–4	0 (0%)	1 (5%)	1 (3%)	1 (3%)	0.84	0.59
4–6	1 (4%)	0 (0%)	1 (3%)	0 (0%)		
>6	0 (0%)	0 (0%)	0 (0%)	0 (0%)		
Stay up late frequency (n, %)						
Never	7 (30%)	7 (33%)	10 (34%)	10 (36%)		
Seldom	7 (30%)	8 (38%)	11 (38%)	12 (43%)		
Sometimes	5 (22%)	4 (19%)	4 (14%)	1 (3%)	0.85	0.32
Often	3 (13%)	0 (0%)	2 (7%)	0 (0%)		
Always	1 (5%)	2 (10%)	2 (7%)	5 (18%)		
Average sleep time (hour) (n, %)						
<6	1 (4%)	2 (10%)	10 (35%)	5 (18%)		
6–8	17 (74%)	15 (71%)	16 (55%)	23 (82%)	0.03	0.04
>8	5 (22%)	4 (19%)	3 (10%)	0 (0%)		
Contact lens frequency (n, %)						
Never	23 (100%)	21 (100%)	27 (94%)	26 (93%)		
Seldom	0 (0%)	0 (0%)	1 (3%)	0 (0%)		
Sometimes	0 (0%)	0 (0%)	1 (3%)	2 (7%)	0.44	0.50
Often	0 (0%)	0 (0%)	0 (0%)	0 (0%)		
Always	0 (0%)	0 (0%)	0 (0%)	0 (0%)		

[1] Values are presented as number with percentage.

Table 3 shows the results of DES-related indicators before and after the intervention. After the intervention, the tear secretion volume significantly increased by 30% in the left eye of the supplement group. Furthermore, the OSDI not only improved compared with that before the intervention but was also significantly lower than that of the control group. Post-intervention NIBUT significantly decreased in both the control and supplement groups (right eye) compared with that before the intervention. The degree of MG loss in the right eye of the supplement group significantly increased compared with that before the intervention.

Table 3. Ocular surface parameters of dry eye in study groups at weeks 0 and 12.

Variables [1]	Control Group (n = 23)		Supplement Group (n = 29)	
	Week 0	Week 12	Week 0	Week 12
Schirmer test (mm)				
Right eye	6.17 ± 4.74	6.67 ± 5.94	5.93 ± 5.99	5.75 ± 4.32
Left eye	5.65 ± 5.11	5.38 ± 4.90	4.62 ± 4.79 [b]	6.00 ± 4.55 [a]
OSDI (score/100)	34.78 ± 12.32	33.43 ± 20.15 *	33.26 ± 16.04 [a]	18.83 ± 15.19 [b]
NIBUT (sec)				
Right eye	9.79 ± 2.94 [a]	7.79 ± 2.21 [b]	10.27 ± 4.67 [a]	8.47 ± 1.96 [b]
Left eye	9.07 ± 2.44	8.21 ± 2.15	10.63 ± 4.94	9.02 ± 3.00
TMH (mm)				
Right eye	0.21 ± 0.12	0.18 ± 0.07	0.18 ± 0.05	0.19 ± 0.07
Left eye	0.21 ± 0.08	0.19 ± 0.12	0.21 ± 0.06	0.20 ± 0.08
MG loss (%)				
Right eye	54.26 ± 15.45	52.81 ± 21.11	48.07 ± 18.97 [b]	58.18 ± 20.83 [a]
Left eye	50.22 ± 20.36	47.90 ± 15.14	48.07 ± 16.00	48.89 ± 17.05
LLT (0 of 6) right eye				
0 (n, %)	10 (43%)	13 (62%)	13 (45%)	15 (54%)
1 (n, %)	9 (39%)	6 (29%)	9 (31%)	8 (29%)
2 (n, %)	2 (9%)	2 (10%)	5 (17%)	5 (18%)
3 (n, %)	1 (4%)	0 (0%)	2 (7%)	0 (0%)
4 (n, %)	1 (4%)	0 (0%)	0 (0%)	0 (0%)
5 (n, %)	0 (0%)	0 (0%)	0 (0%)	0 (0%)
6 (n, %)	0 (0%)	0 (0%)	0 (0%)	0 (0%)
LLT (0 of 6) left eye				
0 (n, %)	9 (39%)	11 (52%)	8 (28%)	16 (57%)
1 (n, %)	8 (35%)	8 (38%)	15 (52%)	9 (32%)
2 (n, %)	5 (22%)	0 (0%)	5 (17%)	3 (11%)
3 (n, %)	1 (4%)	2 (10%)	1 (3%)	0 (0%)
4 (n, %)	0 (0%)	0 (0%)	0 (0%)	0 (0%)
5 (n, %)	0 (0%)	0 (0%)	0 (0%)	0 (0%)
6 (n, %)	0 (0%)	0 (0%)	0 (0%)	0 (0%)

[1] Values are presented as mean ± SD and number with percentage; OSDI, ocular surface disease index; NIBUT, non-invasive tear break-up time; TMH, tear meniscus height; MG, meibomian gland; LLT, lipid layer thickness. [a, b] Values are significantly different between weeks 0 and 12 within the group; $p < 0.05$. * Values are significantly different between control and supplement groups at week 0 or 12; $p < 0.05$.

No significant change in calorie intake occurred in either group before and after the intervention. Regarding macronutrients, protein intake significantly increased in the control group after the intervention, while a significant increase in fat intake occurred in the supplement group. Additionally, the control group exhibited a significantly higher lipid intake and lower carbohydrate intake than the supplement group, and fat intake in both groups ranged from 31% to 39% (Table 4).

No significant differences in tear TEAC levels were noted between the two groups or from week 0 to week 12. Regarding oxidative stress, plasma MDA levels significantly decreased, whereas tear fluid MDA levels significantly increased in the control group at week 12. In contrast, in the supplement group, no significant changes in oxidative stress manifested in either blood or tears from week 0 to week 12. After 12 weeks, the levels of tear cytokines, namely, IL-6 and IL-8, significantly decreased compared with those at week 0, and IL-6 also significantly decreased compared with that in the control group at week 12. Additionally, the post-intervention IL-17A concentration significantly increased in the control group at week 12 (Table 5).

Table 4. Daily nutrient intakes of subjects from study groups at weeks 0 and 12.

Variables [1]	Control Group (n = 23)		Supplement Group (n = 29)	
	Week 0	Week 12	Week 0	Week 12
Energy (Kcal/day)	1339.65 ± 352.82	1295.38 ± 399.70	1444.42 ± 489.84	1329.96 ± 465.37
Protein (g/day)	55.69 ± 32.76	65.87 ± 32.76	62.63 ± 32.76	56.97 ± 32.76
Protein (% total energy)	17.00 ± 4.00 [b]	20.00 ± 5.00 [a]	17.00 ± 5.00	17.00 ± 5.00
Carbohydrate (g/day)	147.07 ± 32.76	143.06 ± 32.76 *	188.15 ± 32.76 [a]	149.21 ± 32.76 [b]
Carbohydrate (% total energy)	44.00 ± 12.00 *	43.00 ± 12.00	52.00 ± 14.00	46.00 ± 12.00
Lipid (g/day)	54.41 ± 32.76	53.82 ± 32.76	50.66 ± 32.76	58.30 ± 32.76
Lipid (% total energy)	37.00 ± 11.00 *	39.00 ± 10.00	31.00 ± 10.00 [b]	39.00 ± 10.00 [a]
Saturated fatty acid (mg)	5294.55 ± 4932.79	4302.55 ± 3190.90	4516.58 ± 3700.62	7791.71 ± 7475.99
MUFA (mg)	5338.82 ± 4891.44	4985.13 ± 3283.96	4983.83 ± 4712.27	7572.83 ± 7478.71
PUFA (mg)	4793.86 ± 7681.52	7047.41 ± 5990.41	4981.86 ± 5060.43	4820.06 ± 6107.76
ALA (18:3) (mg)	494.98 ± 886.48	711.82 ± 672.75 *	441.21 ± 503.39	416.16 ± 628.13
EPA (20:5) (mg)	4.35 ± 7.64 [b]	86.50 ± 259.19 [a]	8.68 ± 15.72	22.05 ± 55.70
DHA (22:6) (mg)	14.66 ± 17.03 [b]	50.26 ± 77.75 [a]	19.87 ± 25.92	27.40 ± 48.41
S-EPA (20:5) (mg)	4.35 ± 7.64 [b]	86.50 ± 259.19 [a,*]	8.68 ± 15.72 [b]	67.05 ± 55.70 [a]
S-DHA (22:6) (mg)	14.66 ± 17.03 [b]	50.26 ± 77.75 [a,*]	19.87 ± 25.92 [b]	57.40 ± 48.41 [a]

[1] Values are presented as mean ± SD. MUFA, monounsaturated fatty acid; PUFA, polyunsaturated fatty acid; ALA, α-linoleic acid; EPA, eicosapentaenoic acid; DHA, docosahexaenoic acid; S-EPA, includes dietary and supplement of eicosapentaenoic acid; S-DHA, includes dietary and supplement of docosahexaenoic acid. [a, b] Values are significantly different between weeks 0 and 12 within the group; $p < 0.05$. * Values are significantly different between control and supplement groups at week 0 or 12; $p < 0.05$.

Table 5. Antioxidant capacity, oxidative stress, and cytokines levels in study groups at weeks 0 and 12.

Variables [1]	Control Group (n = 23)		Supplement Group (n = 29)	
	Week 0	Week 12	Week 0	Week 12
Antioxidant capacities				
Plasma TEAC (µmol/L)	4489.67 ± 280.44 ^{a} (4531.05)	4217.58 ± 233.51 ^{b} (4204.20)	4306.85 ± 397.63 (4345.32)	4315.88 ± 283.65 (4229.81)
Tear TEAC (µmol/µL)	11,242.65 ± 8775.13 (8281.86)	9772.83 ± 8014.59 (7924.94)	10,864.55 ± 6119.1 (10920.00)	8910.12 ± 4084.9 (8182.44)
Oxidative stress				
Plasma MDA (µmol/L)	1.02 ± 0.22 ^{a} (1.02)	0.87 ± 0.19 ^{b} (0.94)	1.10 ± 0.20 (1.06)	1.06 ± 0.26 (0.98)
Tear MDA (µmol/µL)	0.88 ± 0.99 ^{b} (0.51)	1.70 ± 1.45 ^{a} (1.28)	1.21 ± 1.22 (0.73)	1.37 ± 1.22 (1.21)
Cytokines				
IL 1β (pg/µL)	0.02 ± 0.05 (0.00)	0.02 ± 0.03 (0.01)	0.01 ± 0.01 (0.01)	0.01 ± 0.02 (0.01)
IL 2 (pg/µL)	0.01 ± 0.01 (0.01)	0.02 ± 0.02 (0.01)	0.02 ± 0.02 (0.01)	0.02 ± 0.02 (0.01)
IL 6 (pg/µL)	0.23 ± 0.60 (0.07)	0.26 ± 0.59 * (0.10)	0.17 ± 0.16 ^{a} (0.13)	0.10 ± 0.20 ^{b} (0.06)
IL 8 (pg/µL)	2.27 ± 3.35 * (0.49)	2.02 ± 2.90 (0.76)	3.75 ± 3.83 ^{a} (2.80)	1.57 ± 1.98 ^{b} (0.48)
IL 17A (pg/µL)	0.02 ± 0.01 ^{b} (0.01)	0.05 ± 0.06 ^{a} (0.02)	0.03 ± 0.04 (0.01)	0.03 ± 0.02 (0.02)
TNF-α (pg/µL)	0.11 ± 0.17 (0.05)	0.13 ± 0.18 (0.07)	0.08 ± 0.07 (0.05)	0.11 ± 0.1 (0.06)

[1] Values are presented as mean ± SD (median). TEAC, Trolox equivalent antioxidant capacity. MDA, malondialdehyde. IL, interleukin. TNF-α, tumor necrosis factor-α. [a, b] Values are significantly different between weeks 0 and 12 within the group; $p < 0.05$. * Values are significantly different between control and supplement groups at week 0 or 12; $p < 0.05$.

We subsequently performed partial Spearman's correlation coefficient analyses to assess the association of changes in oxidative stress indicators, antioxidant activity, and inflammatory responses with changes in DES indicators, after adjusting for age, gender, and menopausal status (Table 6). Tear secretion changes negatively correlated with changes in tear TEAC and MDA levels (right eye). Plasma TEAC changes negatively correlated with changes in the meniscus height (left eye) and OSDI of tears. In addition, the results indicated that alterations in tear secretion were inversely related to IL-1β (right eye) and IL-2, IL-8, and TNF-α (both eyes) levels. IL-17A changes were negatively associated with NIBUT (right eye) and TMH (left eye). Furthermore, the degree of MG loss exhibited a positive relationship with IL-6 (left eye) and IL-8 (left eye).

To ascertain whether supplement intake influenced dry eye-related indicators, oxidative stress indicators, antioxidant activity, and inflammatory responses, we conducted multiple linear regression analyses. Supplement intake was the dependent variable, and the models were adjusted for age, gender, and menopausal status. The results are presented in Table 7.

The findings revealed a significant improvement in the OSDI and an elevation in plasma TEAC levels following supplementation. After adjusting for potential confounders, an association was established between supplementation and OSDI improvement; however, the effect on plasma antioxidant capacity was not evident.

Table 6. The correlation between the changes in ocular surface parameters of dry eye and changes in antioxidant capacity, oxidative stress, and cytokines levels [1].

Variables [3]	Δ Schirmer Test		Δ NIBUT (sec)		Δ LLT (0 of 6)		Δ TMH (mm)		Δ MG Loss (%)		Δ OSDI
	Eye										
	Right	Left	Right	Left	Right	Left	Right	Left	Right	Left	
	r [2]										
Antioxidant capacities											
Δ Plasma TEAC (μmol/L)	−0.03	0.02	−0.01	0.03	−0.14	−0.18	−0.14	−0.32 *	0.06	0.67	−0.30 *
Δ Tear TEAC (μmol/μL)	−0.72 *	−0.64 *	0.22	−0.17	−0.08	−0.18	0.17	−0.10	−0.14	0.23	−0.12
Oxidative stress											
Δ Plasma MDA (μmol/L)	0.18	−0.12	0.16	−0.05	0.01	0.07	0.02	0.22	0.05	−0.14	−0.04
Δ Tear MDA (μmol/μL)	−0.33 *	−0.23	0.28	−0.14	0.10	0.08	−0.06	0.06	0.20	−0.01	0.09
Cytokines											
Δ IL 1β (pg/μL)	−0.43 *	−0.25	−0.03	−0.06	−0.09	0.10	0.08	−0.08	−0.14	0.28	−0.06
Δ IL 2 (pg/μL)	−0.50 *	−0.48 *	0.12	−0.07	0.01	0.14	−0.10	−0.01	−0.13	−0.01	−0.06
Δ IL 6 (pg/μL)	−0.20	−0.26	0.16	0.05	0.09	0.23	0.09	0.12	0.01	0.44 *	−0.03
Δ IL 8 (pg/μL)	−0.33 *	−0.29 *	0.28	0.02	0.04	0.18	0.04	0.11	0.04	0.45 *	−0.16
Δ IL 17A (pg/μL)	−0.29	−0.27	−0.40 *	−0.26	−0.34	−0.29	−0.13	−0.41 *	−0.11	−0.02	0.05
Δ TNF-α (pg/μL)	−0.06 *	−0.56 *	0.30	0.04	−0.10	0.01	0.17	−0.11	−0.13	0.21	−0.22

[1] NIBUT, non-invasive tear break-up time; LLT, lipid layer thickness; TMH, tear meniscus height; MG, meibomian gland. OSDI, ocular surface disease index. TEAC, Trolox equivalent antioxidant capacity. MDA, malondialdehyde. IL, interleukin. TNF-α, tumor necrosis factor-α. [2] r, correlation coefficient. [3] Variables are changes at weeks 0 and 12 [Δ (week 12-0)] in both groups. * Values are significant correlations between the changes in all parameters, $p < 0.05$.

Table 7. Multiple linear regression analysis of treatment with the changes in ocular surface parameters, oxidative stress, antioxidant capacities, and inflammatory indicators of dry eye in patents with dry eye syndrome.

Dependent Variables [3]	Eye	Model 1 [1]		Model 2 [1]	
		β [2]	SE [2]	β [2]	SE [2]
Δ Schirmer test (mm)	Right	−1.10	1.94	−1.20	1.97
	Left	1.10	1.19	1.00	1.21
Δ NIBUT (sec)	Right	0.05	1.26	0.04	1.28
	Left	−0.89	1.07	−0.85	1.09
Δ LLT (0 of 6)	Right	0.19	0.21	0.18	0.21
	Left	−0.13	0.19	−0.11	0.19
Δ TMH (mm)	Right	0.02	0.03	0.03	0.03
	Left	0.01	0.03	0.01	0.03
Δ MG loss (%)	Right	11.14	0.04	10.63	5.69
	Left	3.05	6.20	2.55	6.28
Δ OSDI (score/100)		−6.38 *	2.49	−6.60 *	2.52
Δ Plasma TEAC (μM)		239.32 *	114.57	221.40	114.48
Δ Tear TEAC (pg/μL)		−21.93	2383.89	14.60	2426.32
Δ Plasma MDA (μM)		0.05	0.07	0.06	0.07
Δ Tear MDA (pg/μL)		−0.56	0.51	−0.61	0.53
Δ IL 1β (pg/μL)		0.01	0.01	0.01	0.01
Δ IL 2 (pg/μL)		−0.01	0.01	−0.01	0.01
Δ IL 6 (pg/μL)		−0.08	0.07	−0.01	0.07
Δ IL 8 (pg/μL)		0.25	1.93	0.22	1.96
Δ IL 17A (pg/μL)		−0.03	0.02	−0.03	0.02
Δ TNF-a (pg/μL)		0.01	0.05	0.01	0.05

[1] Model 1 was not adjusted; model 2 was adjusted for age, gender, and menopause. [2] β, regression; SE, standard error, * $p < 0.05$. NIBUT, non-invasive tear break-up time; LLT, lipid layer thickness; TMH, tear meniscus height; MG, meibomian gland. OSDI, ocular surface disease index. TEAC, Trolox equivalent antioxidant capacity. MDA, malondialdehyde. IL, interleukin, TNF-α, tumor necrosis factor-α. [3] Variables are changes at weeks 0 and 12 [Δ (week 12-0)] in both groups.

4. Discussion

After 12-week supplementation, the supplement group exhibited a significant increase in tear secretion and a significant reduction in tear inflammatory markers. This observation is consistent with the crucial role of inflammation in the pathogenesis of DES. Ocular surface epithelial cells are exposed to oxidative stress [40,41], and the innate immune response is triggered, leading to the release of TNF-α, IL-1β, and IL-6, which induce inflammation [42]. IL-6 stimulates the synthesis and release of acute-phase proteins, leading to the secretion of the pro-inflammatory cytokine IL-17, which induces apoptosis in lacrimal gland cells, thereby reducing tear production. Numerous clinical studies have confirmed a negative correlation between tear secretion and IL-6 and TNF-α [43,44]. Furthermore, inflammation or prolonged dryness of the ocular surface triggers a significant release of IL-8, which attracts neutrophil migration [10,45]. A previous study confirmed that increased tear film inflammation elevates ROS, leading to reduced tear secretion [46]. This study observed significantly negative correlations among changes in tear secretion, changes in tear MDA and TEAC levels, and inflammatory markers.

EPA and DHA have been shown to inhibit oxidative stress and mitigate the production of inflammatory mediators [47]. Lutein and zeaxanthin exhibit ROS scavenging properties [48]. Studies indicate that lutein can decrease the sensitivity of cell membranes to oxidative damage [49] and improve inflammation [50]. Clinical research has demonstrated that a 3-month supplementation regimen can significantly reduce tear IL-6 levels compared with the control [26] or increase tear secretion [24]. A 12-week supplementation regimen involving a combination of lutein, zeaxanthin, and curcumin significantly increased tear secretion in patients with DES [22]. Our findings are consistent with those obtained in the above studies, indicating that compound supplements significantly ameliorate inflammation. Notably, IL-17A significantly increased in the control group.

Additionally, we observed a significant increase in oxidative stress in the tears of the control group after 12 weeks. This finding is consistent with previous research that highlights how the disruption of redox homeostasis in DES contributes to the creation of a localized oxidative environment within the tear film [14]. Furthermore, both plasma antioxidant capacity and oxidative stress were significantly reduced in the control group. This paradoxical observation may reflect the complex role oxidative stress plays in the clinical progression of DES. The increase in oxidative stress could lead to the depletion of antioxidant defenses, as the body attempts to counteract the damaging effects of oxidative stress. Such depletion may represent an adaptive response to chronic oxidative stress, where cellular protective mechanisms engage antioxidants to neutralize harmful free radicals. As Liguori et al. [51] suggest, oxidative metabolism produces ROS, which further exacerbate oxidative stress. This feedback mechanism, although initially protective, may ultimately result in a reduction in both antioxidant capacity and oxidative stress markers in plasma. Moreover, additional factors such as lifestyle changes, environmental exposures, and dietary influences may further modulate oxidative and antioxidant systems, contributing to the complexity of the response. Future research should explore these variables and their intricate interactions to provide a more comprehensive understanding of the mechanisms driving DES progression and its associated oxidative stress. However, no changes in oxidative stress and antioxidant capacity indicators were observed in the supplement group, possibly because of the compound supplement providing sufficient antioxidant capacity to balance oxidants and antioxidants, thereby attenuating the potential for further deterioration. Despite the relatively low dosage of the compound supplement in this study compared with that in previous research, it effectively reduced inflammatory markers and increased tear secretion. Notably, previous studies have failed to verify a relationship between DES severity and IL-1β concentration; nonetheless, this study established a correlation between increased tear secretion and decreased IL-1β levels.

After 12-week supplementation, the supplement group displayed a significant OSDI decrease of 6.6 points compared with the control group, corroborating the antioxidant and anti-inflammatory properties of fish oil, lutein, and zeaxanthin. DHA depletion is reportedly caused by H_2O_2, which also enhances lipid oxidation in retinal pigment epithelial cells; moreover, lutein and zeaxanthin enhance cellular defense against oxidative stress [52]. Studies have revealed that antioxidant supplementation may improve plasma antioxidant capacity, increase TBUT, and enhance tear secretion, thereby alleviating dry eye discomfort in patients with DES [53–55]. Similarly, this study observed a negative correlation between OSDI changes and plasma antioxidant capacity. A previous study involving a 12-week regimen of lutein (20 mg/day) and zeaxanthin (4 mg/day), including vitamin D and curcumin, yielded a 13-point reduction in the OSDI [31]. Another study involving EPA (360 mg/day) and DHA (240 mg/day) supplementation for 12 weeks demonstrated a significant 9.4-point reduction in the OSDI [56]. Our results align with these findings, showing a significant 14.6-point reduction in the OSDI, despite the lower dosages than those in previous studies.

DES owing to meibomian gland dysfunction is a clinical problem encountered in ophthalmology [57], and increased ROS levels can directly damage meibomian glands, leading to apoptosis [58], further reduced LLT, and increased tear evaporation, affecting TMH [59]. Previous studies have shown that the anti-inflammatory and antioxidant properties of fish oil, lutein, and zeaxanthin supplements can reduce eyelid inflammation [33]. We also observed a significantly positive correlation between MG loss and inflammation in tears. Several studies have demonstrated that varying dosages of EPA (1680–2000 mg/day) + DHA (560–1000 mg/day) for 12 weeks to 1 year significantly increase TBUT [60–62]. However, we found a significant increase in MG loss, a significant decrease in TBUT, and no significant changes in LLT and TMH in the supplement group. Meibomian gland function is easily influenced by external factors, such as screen time, sleep duration, environment, and ocular surgery [33]. Numerous studies confirm that ocular surgery reduces TBUT [63–65].

In this study, the supplement group had a significantly greater proportion of participants with a history of ocular surgery (66%) than the control group (26%). Additionally, the supplement group participants reported significantly less average daily sleep, potentially contributing to the observed lack of improvement. Moreover, the significant decrease in TBUT in the control group after 12 weeks might have been related to the significant increase in tear IL-17A levels; furthermore, we found a negative correlation between TBUT and IL-17A. IL-17A leads to ocular surface epithelial and tear function impairment [66–68] and stimulates MMPs, causing ocular surface damage and affecting tear film stability [69,70].

The strength of this study was its ability to successfully overcome challenges in collecting tear samples from patients with DES. This enabled a thorough investigation into tear antioxidant capacity, oxidative stress, and cytokine profiles. Notwithstanding, this study also had certain limitations. First, it was an open-label trial. This decision was based on previous studies indicating that using oil as a placebo may improve the tear film lipid layer and reduce ocular discomfort, thus impeding the elucidation of the effects of supplementation. To address this confounding factor, an open-label trial design was selected. Second, the supplement group had a significantly higher proportion of participants with a history of ocular surgery and a reduced sleep duration, which may have contributed to the observed increase in meibomian gland loss and the decrease in tear break-up time, potentially affecting the supplementation's overall efficacy. Finally, this study involved a relatively small sample. Larger sample sizes will aid future studies in comprehensively elucidating the antioxidant effects of multicomponent supplements.

5. Conclusions

In conclusion, this study suggests that a 12-week supplementation regimen comprising EPA, DHA, lutein, and zeaxanthin has the ability to alleviate dry eye symptoms, enhance tear secretion, and mitigate the tear inflammatory response in patients with DES. However, a larger study is required to confirm these findings and provide more definitive recommendations for DES treatment.

Author Contributions: Conceptualization, S.-C.H. and M.-Y.H.; methodology, S.-C.H., Y.-P.L. and Q.-P.T.; software, S.-C.H., Y.-P.L. and Q.-P.T.; validation, S.-C.H., Y.-P.L. and Q.-P.T.; formal analysis, S.-C.H., Y.-P.L., Q.-P.T. and M.-Y.H.; investigation, S.-C.H., Y.-P.L. and Q.-P.T.; resources, S.-C.H., C.C. and M.-Y.H.; data curation, S.-C.H., Y.-P.L. and Q.-P.T.; writing—original draft preparation, S.-C.H., M.-C.H., Y.-K.H., Q.-P.T. and M.-Y.H.; writing—review and editing, S.-C.H., Y.-P.L. and M.-Y.H.; visualization, S.-C.H., Y.-P.L. and M.-Y.H.; supervision, S.-C.H., Y.-P.L. and M.-Y.H.; project administration, S.-C.H., Y.-P.L., Q.-P.T. and M.-Y.H.; funding acquisition, S.-C.H. and M.-Y.H. All authors have read and agreed to the published version of the manuscript.

Funding: This research was supported by the Taiwan National Science and Technology Council (No. 111-2314-B-040 -026 -MY2 and 113-2314-B-040 -009 -). The funders had no role in the study design, data collection, or analysis.

Institutional Review Board Statement: This study was conducted in accordance with the Declaration of Helsinki and approved by the Institutional Review Board of Chung Shan Medical University Hospital, Taichung, Taiwan (IRB CSMUH No. CSI-20199).

Informed Consent Statement: Informed consent was obtained from all subjects involved in the study.

Data Availability Statement: Data collected from human participants, as described in Table 1, are not available for confidentiality reasons.

Conflicts of Interest: The authors declare no conflicts of interest.

References

1. Walter, K. What Is Dry Eye Disease? *JAMA* **2022**, *328*, 84. [CrossRef] [PubMed]
2. Kojima, T.; Dogru, M.; Kawashima, M.; Nakamura, S.; Tsubota, K. Advances in the diagnosis and treatment of dry eye. *Prog. Retin. Eye Res.* **2020**, *78*, 100842. [CrossRef]
3. Papas, E.B. The global prevalence of dry eye disease: A Bayesian view. *Ophthalmic Physiol. Opt.* **2021**, *41*, 1254–1266. [CrossRef] [PubMed]
4. Kuo, Y.K.; Lin, I.C.; Chien, L.N.; Lin, T.Y.; How, Y.T.; Chen, K.H.; Dusting, G.J.; Tseng, C.L. Dry Eye Disease: A Review of Epidemiology in Taiwan, and Its Clinical Treatment and Merits. *J. Clin. Med.* **2019**, *8*, 1227. [CrossRef] [PubMed]
5. Gomes, J.A.P.; Azar, D.T.; Baudouin, C.; Efron, N.; Hirayama, M.; Horwath-Winter, J.; Kim, T.; Mehta, J.S.; Messmer, E.M.; Pepose, J.S.; et al. TFOS DEWS II iatrogenic report. *Ocul. Surf.* **2017**, *15*, 511–538. [CrossRef]
6. Milner, M.S.; Beckman, K.A.; Luchs, J.I.; Allen, Q.B.; Awdeh, R.M.; Berdahl, J.; Boland, T.S.; Buznego, C.; Gira, J.P.; Goldberg, D.F.; et al. Dysfunctional tear syndrome, dry eye disease and associated tear film disorders—New strategies for diagnosis and treatment. *Curr. Opin. Ophthalmol.* **2017**, *28* (Suppl. S1), 3–47. [CrossRef]
7. Abbott, K.; Hanson, K.S.; Lally, J. Prevalence of dry eye disease in the low vision population at the University of Colorado. *J. Optom.* **2024**, *17*, 100501. [CrossRef]
8. Blanco-Vázquez, M.; Vázquez, A.; Fernández, I.; Novo-Diez, A.; Martínez-Plaza, E.; García-Vázquez, C.; González-García, M.J.; Sobas, E.M.; Calonge, M.; Enríquez-de-Salamanca, A. Inflammation-related molecules in tears of patients with chronic ocular pain and dry eye disease. *Exp. Eye Res.* **2022**, *219*, 109057. [CrossRef]
9. Xia, Y.; Zhang, Y.; Du, Y.; Wang, Z.; Cheng, L.; Du, Z. Comprehensive dry eye therapy, overcoming ocular surface barrier and combating inflammation; oxidation; and mitochondrial damage. *J. Nanobiotechnol.* **2024**, *22*, 233. [CrossRef]
10. Lee, H.; Chung, B.; Kim, K.S.; Seo, K.Y.; Choi, B.J.; Kim, T.I. Effects of topical loteprednol etabonate on tear cytokines and clinical outcomes in moderate and severe meibomian gland dysfunction, Randomized clinical trial. *Am. J. Ophthalmol.* **2014**, *158*, 1172–1183. [CrossRef]
11. Park, Y.; Hwang, H.B.; Kim, H.S. Observation of Influence of Cataract Surgery on the Ocular Surface. *PLoS ONE* **2016**, *11*, e0152460. [CrossRef] [PubMed]
12. Landsend, E.C.S.; Utheim, Ø.A.; Pedersen, H.R.; Aass, H.C.D.; Lagali, N.; Dartt, D.A.; Baraas, R.C.; Utheim, T.P. The Level of Inflammatory Tear Cytokines is Elevated in Congenital Aniridia and Associated with Meibomian Gland Dysfunction. *Investig. Opthalmol. Vis. Sci.* **2018**, *59*, 2197–2204. [CrossRef] [PubMed]
13. Mrugacz, M.; Ostrowska, L.; Bryl, A.; Szulc, A.; Zelazowska-Rutkowska, B.; Mrugacz, G. Pro-inflammatory cytokines associated with clinical severity of dry eye disease of patients with depression. *Adv. Med. Sci.* **2017**, *62*, 338–344. [CrossRef]
14. Deng, R.; Hua, X.; Li, J.; Chi, W.; Zhang, Z.; Lu, F.; Zhang, L.; Pflugfelder, S.C.; Li, D.Q. Oxidative stress markers induced by hyperosmolarity in primary human corneal epithelial cells. *PLoS ONE* **2015**, *10*, e0126561. [CrossRef] [PubMed]
15. Seen, S.; Tong, L. Dry eye disease and oxidative stress. *Acta Ophthalmol.* **2018**, *96*, e412–e420. [CrossRef] [PubMed]
16. Messmer, E.M. The pathophysiology; diagnosis, and treatment of dry eye disease. *Dtsch. Aerzteblatt Int.* **2015**, *112*, 71–82. [CrossRef]
17. Toyos, R.; McGill, W.; Briscoe, D. Intense pulsed light treatment for dry eye disease due to meibomian gland dysfunction; a 3-year retrospective study. *Photomed. Laser Surg.* **2015**, *33*, 41–46. [CrossRef]
18. Pan, Q.; Angelina, A.; Marrone, M.; Stark, W.J.; Akpek, E.K. Autologous serum eye drops for dry eye. *Cochrane Database Syst. Rev.* **2017**, *2*, CD009327. [CrossRef]
19. Noble, B.A.; Loh, R.S.; MacLennan, S.; Pesudovs, K.; Reynolds, A.; Bridges, L.R.; Burr, J.; Stewart, O.; Quereshi, S. Comparison of autologous serum eye drops with conventional therapy in a randomised controlled crossover trial for ocular surface disease. *Br. J. Ophthalmol.* **2004**, *88*, 647–652. [CrossRef]

20. Gus, P.I.; Marinho, D.; Zelanis, S.; Belló-Klein, A.; Locatelli, C.; Nicola, F.; Kunzler, A.L.; Fernandes, T.R.; Carraro, C.C.; Barbosa, L. A Case-Control Study on the Oxidative Balance of 50% Autologous Serum Eye Drops. *Oxidative Med. Cell. Longev.* **2016**, *2016*, 9780193. [CrossRef]
21. O'Neil, E.C.; Henderson, M.; Massaro-Giordano, M.; Bunya, V.Y. Advances in dry eye disease treatment. *Curr. Opin. Ophthalmol.* **2019**, *30*, 166–178. [CrossRef] [PubMed]
22. Kan, J.; Wang, M.; Liu, Y.; Liu, H.; Chen, L.; Zhang, X.; Huang, C.; Liu, B.Y.; Gu, Z.; Du, J. A novel botanical formula improves eye fatigue and dry eye: A randomized, double-blind, placebo-controlled study. *Am. J. Clin. Nutr.* **2020**, *112*, 334–342. [CrossRef] [PubMed]
23. Alharbi, A. Exploring the Therapeutic Potential of Omega-3 Fatty Acid Supplementation in Dry Eye Syndrome: An In vitro Investigation. *J. Pharm. Bioallied Sci.* **2024**, *16* (Suppl. S3), S2673–S2675. [CrossRef] [PubMed]
24. Gatell-Tortajada, J. Oral supplementation with a nutraceutical formulation containing omega-3 fatty acids, vitamins, minerals, and antioxidants in a large series of patients with dry eye symptoms, results of a prospective study. *Clin. Interv. Aging* **2016**, *11*, 571–578. [CrossRef]
25. Oydanich, M.; Maguire, M.G.; Pistilli, M.; Hamrah, P.; Greiner, J.V.; Lin, M.C.; Asbell, P.A.; Dry Eye Assessment and Management Study Research Group. Effects of Omega-3 Supplementation on Exploratory Outcomes in the Dry Eye Assessment and Management Study. *Ophthalmology* **2020**, *127*, 136–138. [CrossRef]
26. Pinazo-Durán, M.D.; Galbis-Estrada, C.; Pons-Vázquez, S.; Cantú-Dibildox, J.; Marco-Ramírez, C.; Benítez-del-Castillo, J. Effects of a nutraceutical formulation based on the combination of antioxidants and ω-3 essential fatty acids in the expression of inflammation and immune response mediators in tears from patients with dry eye disorders. *Clin. Interv. Aging* **2013**, *8*, 139–148. [CrossRef]
27. Barker, F.M., 2nd; Snodderly, D.M.; Johnson, E.J.; Schalch, W.; Koepcke, W.; Gerss, J.; Neuringer, M. Nutritional manipulation of primate retinas, V: Effects of lutein, zeaxanthin, and n-3 fatty acids on retinal sensitivity to blue-light-induced damage. *Investig. Ophthalmol. Vis. Sci.* **2011**, *52*, 3934–3942. [CrossRef]
28. Gao, N.; Gao, X.; Du, M.; Xiang, Y.; Zuo, H.; Huang, R.; Wan, W.; Hu, K. Lutein protects senescent ciliary muscle against oxidative stress through the Keap1/Nrf2/ARE pathway. *Phytomedicine* **2024**, *134*, 155982. [CrossRef]
29. Morita, S.; Sueyasu, T.; Tokuda, H.; Kaneda, Y.; Izumo, T.; Nakao, Y. Lutein and zeaxanthin reduce neuronal cell damage caused by lipid peroxidation. *Biochem. Biophys. Rep.* **2024**, *40*, 101835. [CrossRef]
30. Bian, Q.; Gao, S.; Zhou, J.; Qin, J.; Taylor, A.; Johnson, E.J.; Tang, G.; Sparrow, J.R.; Gierhart, D.; Shang, F. Lutein and zeaxanthin supplementation reduces photooxidative damage and modulates the expression of inflammation-related genes in retinal pigment epithelial cells. *Free Radic. Biol. Med.* **2012**, *53*, 1298–1307. [CrossRef]
31. Radkar, P.; Lakshmanan, P.S.; Mary, J.J.; Chaudhary, S.; Durairaj, S.K. A Novel Multi-Ingredient Supplement Reduces Inflammation of the Eye and Improves Production and Quality of Tears in Humans. *Ophthalmol. Ther.* **2021**, *10*, 581–599. [CrossRef] [PubMed]
32. Chao, S.C.; Nien, C.W.; Iacob, C.; Hu, D.N.; Huang, S.C.; Lin, H.Y. Effects of Lutein on Hyperosmoticity-Induced Upregulation of IL-6 in Cultured Corneal Epithelial Cells and Its Relevant Signal Pathways. *J. Ophthalmol.* **2016**, *2016*, 8341439. [CrossRef] [PubMed]
33. Oleñik, A.; Jiménez-Alfaro, I.; Alejandre-Alba, N.; Mahillo-Fernández, I. A randomized, double-masked study to evaluate the effect of omega-3 fatty acids supplementation in meibomian gland dysfunction. *Clin. Interv. Aging* **2013**, *8*, 1133–1138. [CrossRef]
34. VanDerMeid, K.R.; Su, S.P.; Krenzer, K.L.; Ward, K.W.; Zhang, J.Z. A method to extract cytokines and matrix metalloproteinases from Schirmer strips and analyze using Luminex. *Mol. Vis.* **2011**, *17*, 1056–1063.
35. Sánchez-González, M.C.; Capote-Puente, R.; García-Romera, M.C.; De-Hita-Cantalejo, C.; Bautista-Llamas, M.J.; Silva-Viguera, C.; Sánchez-González, J.M. Dry eye disease and tear film assessment through a novel non-invasive ocular surface analyzer. *OSA Protocol. Front. Med.* **2022**, *9*, 938484. [CrossRef]
36. VanDerMeid, K.R.; Su, S.P.; Ward, K.W.; Zhang, J.Z. Correlation of tear inflammatory cytokines and matrix metalloproteinases with four dry eye diagnostic tests. *Investig. Opthalmol. Vis. Sci.* **2012**, *53*, 1512–1518. [CrossRef]
37. Lapenna, D.; Ciofani, G.; Pierdomenico, S.D.; Giamberardino, M.A.; Cuccurullo, F. Reaction conditions affecting the relationship between thiobarbituric acid reactivity and lipid peroxides in human plasma. *Free Radic. Biol. Med.* **2001**, *31*, 331–335. [CrossRef]
38. Clark, D.; Sheppard, J.; Brady, T.C. A randomized double-masked phase 2a trial to evaluate activity and safety of topical ocular reproxalap, a novel RASP inhibitor, in dry eye disease. *J. Ocul. Pharmacol. Ther.* **2021**, *37*, 193–199. [CrossRef]
39. Arnao, M.B.; Cano, A.; Hernández-Ruiz, J.; García-Cánovas, F.; Acosta, M. Inhibition by L-ascorbic acid and other antioxidants of the 2.2′-azino-bis (3-ethylbenzthiazoline-6-sulfonic acid) oxidation catalyzed by peroxidase, a new approach for determining total antioxidant status of foods. *Anal. Biochem.* **1996**, *236*, 255–261. [CrossRef]
40. Jee, D.; Park, S.H.; Kim, M.S.; Kim, E.C. Antioxidant and inflammatory cytokine in tears of patients with dry eye syndrome treated with preservative-free versus preserved eye drops. *Investig. Ophthalmol. Vis. Sci.* **2014**, *55*, 5081–5089. [CrossRef]

41. Ren, Y.; Lu, H.; Reinach, P.S.; Zheng, Q.; Li, J.; Tan, Q.; Zhu, H.; Chen, W. Hyperosmolarity-induced AQP5 upregulation promotes inflammation and cell death via JNK1/2 Activation in human corneal epithelial cells. *Sci. Rep.* **2017**, *7*, 4727. [CrossRef] [PubMed]
42. Stern, M.E.; Schaumburg, C.S.; Siemasko, K.F.; Gao, J.; Wheeler, L.A.; Grupe, D.A.; De Paiva, C.S.; Calder, V.L.; Calonge, M.; Niederkorn, J.Y.; et al. Autoantibodies contribute to the immunopathogenesis of experimental dry eye disease. *Investig. Ophthalmol. Vis. Sci.* **2012**, *3*, 2062–2075. [CrossRef] [PubMed]
43. Tong, L.; Beuerman, R.; Simonyi, S.; Hollander, D.A.; Stern, M.E. Effects of Punctal Occlusion on Clinical Signs and Symptoms and on Tear Cytokine Levels in Patients with Dry Eye. *Ocul. Surf.* **2016**, *14*, 233–241. [CrossRef]
44. McDonnell, P.J.; Pflugfelder, S.C.; Stern, M.E.; Hardten, D.R.; Conway, T.; Villanueva, L.; Hollander, D.A. Study design and baseline findings from the progression of ocular findings (PROOF) natural history study of dry eye. *BMC Ophthalmol.* **2017**, *17*, 265. [CrossRef]
45. López-Miguel, A.; Tesón, M.; Martín-Montañez, V.; Enríquez-de-Salamanca, A.; Stern, M.E.; González-García, M.J.; Calonge, M. Clinical and Molecular Inflammatory Response in Sjögren Syndrome-Associated Dry Eye Patients Under Desiccating Stress. *Am. J. Ophthalmol.* **2016**, *161*, 133–141.e412. [CrossRef]
46. Choi, W.; Lian, C.; Ying, L.; Kim, G.E.; You, I.C.; Park, S.H.; Yoon, K.C. Expression of Lipid Peroxidation Markers in the Tear Film and Ocular Surface of Patients with Non-Sjogren Syndrome, Potential Biomarkers for Dry Eye Disease. *Curr. Eye Res.* **2016**, *41*, 1143–1149. [CrossRef]
47. Djuricic, I.; Calder, P.C. Beneficial Outcomes of Omega-6 and Omega-3 Polyunsaturated Fatty Acids on Human Health: An Update for 2021. *Nutrients* **2021**, *13*, 2421. [CrossRef]
48. Kumar, P.; Banik, S.P.; Ohia, S.E.; Moriyama, H.; Chakraborty, S.; Wang, C.K.; Song, Y.S.; Goel, A.; Bagchi, M.; Bagchi, D. Current Insights on the Photoprotective Mechanism of the Macular Carotenoids, Lutein and Zeaxanthin: Safety, Efficacy and Bio-Delivery. *J. Am. Nutr. Assoc.* **2024**, *43*, 505–518. [CrossRef]
49. Havaux, M.; Dall'osto, L.; Bassi, R. Zeaxanthin has enhanced antioxidant capacity with respect to all other xanthophylls in Arabidopsis leaves and functions independent of binding to PSII antennae. *Plant Physiol.* **2007**, *145*, 1506–1520. [CrossRef]
50. Andrade, A.S.; Salomon, T.B.; Behling, C.S.; Mahl, C.D.; Hackenhaar, F.S.; Putti, J.; Benfato, M.S. Alpha-lipoic acid restores tear production in an animal model of dry eye. *Exp. Eye Res.* **2014**, *120*, 1–9. [CrossRef] [PubMed]
51. Liguori, I.; Russo, G.; Curcio, F.; Bulli, G.; Aran, L.; Della-Morte, D.; Gargiulo, G.; Testa, G.; Cacciatore, F.; Bonaduce, D.; et al. Oxidative stress, aging, and diseases. *Clin. Interv. Aging* **2018**, *13*, 757–772. [CrossRef] [PubMed]
52. Leung, H.H.; Galano, J.M.; Crauste, C.; Durand, T.; Lee, J.C. Combination of Lutein and Zeaxanthin, and DHA Regulated Polyunsaturated Fatty Acid Oxidation in H2O2-Stressed Retinal Cells. *Neurochem. Res.* **2020**, *45*, 1007–1019. [CrossRef] [PubMed]
53. Dogru, M.; Matsumoto, Y.; Yamamoto, Y.; Goto, E.; Saiki, M.; Shimazaki, J.; Takebayashi, T.; Tsubota, K. Lactoferrin in Sjögren's syndrome. *Ophthalmology* **2007**, *114*, 2366–2367. [CrossRef] [PubMed]
54. Peponis, V.; Bonovas, S.; Kapranou, A.; Peponi, E.; Filioussi, K.; Magkou, C.; Sitaras, N.M. Conjunctival and tear film changes after vitamin C and E administration in non-insulin dependent diabetes mellitus. *Med. Sci. Monit.* **2004**, *10*, CR213–CR217.
55. Peponis, V.; Papathanasiou, M.; Kapranou, A.; Magkou, C.; Tyligada, A.; Melidonis, A.; Drosos, T.; Sitaras, N.M. Protective role of oral antioxidant supplementation in ocular surface of diabetic patients. *Br. J. Ophthalmol.* **2002**, *86*, 1369–1373. [CrossRef]
56. Kangari, H.; Eftekhari, M.H.; Sardari, S.; Hashemi, H.; Salamzadeh, J.; Ghassemi-Broumand, M.; Khabazkhoob, M. Short-term consumption of oral omega-3 and dry eye syndrome. *Ophthalmology* **2013**, *120*, 2191–2196. [CrossRef]
57. Geerling, G.; Tauber, J.; Baudouin, C.; Goto, E.; Matsumoto, Y.; O'Brien, T.; Rolando, M.; Tsubota, K.; Nichols, K.K. The international workshop on meibomian gland dysfunction, report of the subcommittee on management and treatment of meibomian gland dysfunction. *Investig. Ophthalmol. Vis. Sci.* **2011**, *52*, 2050–2064. [CrossRef]
58. Dogru, M.; Kojima, T.; Simsek, C.; Tsubota, K. Potential Role of Oxidative Stress in Ocular Surface Inflammation and Dry Eye Disease. *Investig. Ophthalmol. Vis. Sci.* **2018**, *59*, DES163–DES168. [CrossRef]
59. Arita, R.; Morishige, N.; Fujii, T.; Fukuoka, S.; Chung, J.L.; Seo, K.Y.; Itoh, K. Tear Interferometric Patterns Reflect Clinical Tear Dynamics in Dry Eye Patients. *Investig. Ophthalmol. Vis. Sci.* **2016**, *57*, 3928–3934. [CrossRef]
60. Malhotra, C.; Singh, S.; Chakma, P.; Jain, A.K. Effect of oral omega-3 Fatty Acid supplementation on contrast sensitivity in patients with moderate meibomian gland dysfunction, a prospective placebo-controlled study. *Cornea* **2015**, *34*, 37–43. [CrossRef]
61. Epitropoulos, A.T.; Donnenfeld, E.D.; Shah, Z.A.; Holland, E.J.; Gross, M.; Faulkner, W.J.; Matossian, C.; Lane, S.S.; Toyos, M.; Bucci, F.A., Jr.; et al. Effect of Oral Re-esterified Omega-3 Nutritional Supplementation on Dry Eyes. *Cornea* **2016**, *35*, 1185–1191. [CrossRef] [PubMed]
62. Dry Eye Assessment and Management Study Research Group; Asbell, P.A.; Maguire, M.G.; Pistilli, M.; Ying, G.S.; Szczotka-Flynn, L.B.; Hardten, D.R.; Lin, M.C.; Shtein, R.M. n-3 Fatty Acid Supplementation for the Treatment of Dry Eye Disease. *N. Engl. J. Med.* **2018**, *378*, 1681–1690. [PubMed]

63. Denoyer, A.; Landman, E.; Trinh, L.; Faure, J.F.; Auclin, F.; Baudouin, C. Dry eye disease after refractive surgery, comparative outcomes of small incision lenticule extraction versus LASIK. *Ophthalmology* **2015**, *122*, 669–676. [CrossRef] [PubMed]
64. Wang, B.; Naidu, R.K.; Chu, R.; Dai, J.; Qu, X.; Zhou, H. Dry Eye Disease following Refractive Surgery: A 12-Month Follow-Up of SMILE Versus FS-LASIK in High Myopia. *J. Ophthalmol.* **2015**, *2015*, 132417. [CrossRef]
65. Sutu, C.; Fukuoka, H.; Afshari, N.A. Mechanisms and management of dry eye in cataract surgery patients. *Curr. Opin. Ophthalmol.* **2016**, *27*, 24–30. [CrossRef]
66. El Annan, J.; Chauhan, S.K.; Ecoiffier, T.; Zhang, Q.; Saban, D.R.; Dana, R. Characterization of effector T cells in dry eye disease. *Investig. Ophthalmol. Vis. Sci.* **2009**, *50*, 3802–3807. [CrossRef]
67. Chauhan, S.K.; El Annan, J.; Ecoiffier, T.; Goyal, S.; Zhang, Q.; Saban, D.R.; Dana, R. Autoimmunity in dry eye is due to resistance of Th17 to Treg suppression. *J. Immunol.* **2009**, *182*, 1247–1252. [CrossRef]
68. Chauhan, S.K.; Jin, Y.; Goyal, S.; Lee, H.S.; Fuchsluger, T.A.; Lee, H.K.; Dana, R. A novel pro-lymphangiogenic function for Th17/IL-17. *Blood* **2011**, *118*, 4630–4634. [CrossRef]
69. Schaumburg, C.S.; Siemasko, K.F.; De Paiva, C.S.; Wheeler, L.A.; Niederkorn, J.Y.; Pflugfelder, S.C.; Stern, M.E. Ocular surface APCs are necessary for autoreactive T cell-mediated experimental autoimmune lacrimal keratoconjunctivitis. *J. Immunol.* **2011**, *187*, 3653–3662. [CrossRef]
70. García-Posadas, L.; Hodges, R.R.; Li, D.; Shatos, M.A.; Storr-Paulsen, T.; Diebold, Y.; Dartt, D.A. Interaction of IFN-γ with cholinergic agonists to modulate rat and human goblet cell function. *Mucosal Immunol.* **2016**, *9*, 206–217. [CrossRef]

antioxidants

MDPI

Article

The Effect of the Root Bark of *Lycium chinense* (Lycii Radicis Cortex) on Experimental Periodontitis and Alveolar Bone Loss in Sprague-Dawley Rats

Jinwon Yang [1,†], Hyosun Song [2,†], Jeongjun Lee [3], Hunsuk Chung [3], Young-Sam Kwon [2], Kyung-Hwan Jegal [4], Jae-Kwang Kim [5,*] and Sae-Kwang Ku [1,*]

1 Department of Anatomy and Histology, College of Korean Medicine, Daegu Haany University, Gyeongsan 38610, Republic of Korea; yangjinwon@dhu.ac.kr
2 Department of Veterinary Surgery, College of Veterinary Medicine, Kyungpook National University, Daegu 41566, Republic of Korea; legendx0070@knu.ac.kr (H.S.); kwon@knu.ac.kr (Y.-S.K.)
3 GAPI BIO Co., Ltd., Hwaseong 18622, Republic of Korea; orglab@gapibio.co.kr (J.L.); hunsukchung@dongbangchem.co.kr (H.C.)
4 Department of Korean Medical Classics, College of Korean Medicine, Daegu Haany University, Gyeongsan 38610, Republic of Korea; jegalkh@dhu.ac.kr
5 Department of Physiology, College of Korean Medicine, Daegu Haany University, Gyeongsan 38610, Republic of Korea
* Correspondence: kim-jk@dhu.ac.kr (J.-K.K.); gucci200@dhu.ac.kr (S.-K.K.)
† These authors contributed equally to this work.

Abstract: Lycii Radicis Cortex (LRC), the dried root bark of *Lycium chinese* Mill., has traditionally been used as a medicinal herb in East Asia to treat fever and hyperhidrosis. In the present study, we investigated the effects of LRC extract on ligation-induced experimental periodontitis and associated alveolar bone loss in rats. Twenty-four hours after ligation placement, LRC was orally administered once daily for 10 days. Firstly, LRC administration inhibited anaerobic bacterial proliferation and inflammatory cell infiltration in gingival tissues. Additionally, LRC exhibited anti-inflammatory effects by reducing the expression of inflammatory mediators, including prostaglandin E_2, interleukin-1β, and tumor necrosis factor-α. LRC treatment also downregulated mRNA expression of these inflammatory mediators in lipopolysaccharide-stimulated RAW 264.7 cells by inhibiting the mitogen-activated protein kinases and nuclear factor-κB (NF-κB) signaling pathways. Furthermore, LRC showed an antioxidant effect by decreasing the malondialdehyde level and inducible nitric oxide synthase activity in gingival tissues. Moreover, LRC effectively prevented the connective tissue degradation by inhibiting matrix metalloproteinase-8 expression and the loss of collagen-occupied areas in gingival tissues. LRC also decreased the receptor activator of NF-κB ligand/osteoprotegerin (RANKL/OPG) ratio, as well as the number and occupied areas of osteoclasts on the alveolar bone surface, thereby inhibiting alveolar bone loss. In summary, these findings suggest that LRC is a promising medicinal herb for alleviating periodontitis and related alveolar bone loss through its antimicrobial, anti-inflammatory, and antioxidant properties.

Keywords: periodontitis; alveolar bone loss; anti-inflammatory; antioxidant; Lycii Radicis Cortex

1. Introduction

Periodontitis is an inflammatory disease triggered by an imbalance in the oral microbiota and an abnormal host immune response, which ultimately leads to the destruction of the connective tissues and bone that support teeth [1]. With its high prevalence exceeding 10% among the global adult population, periodontitis presents a significant public health concern by diminishing quality of life through tooth loss and pain [2,3]. Particularly, the progression of attachment loss and tooth loss has been linked to socio-economic factors, such as low educational attainment and income levels, highlighting the necessity for developing simple and accessible preventive measures [4]. Moreover, epidemiological studies have highlighted the relationship between periodontitis and systemic diseases such as diabetes and cardiovascular diseases, in the terms of prevalence, progression, and severity [5,6]. As interest in the prevention and treatment of periodontitis continues to grow, research on natural products as adjuncts to overcome the limitations of conventional therapies such as antibiotics, scaling and root planning, is becoming increasingly active [7,8].

Inadequate oral hygiene facilitates the formation and persistence of bacterial biofilm, commonly known as dental plaque. Additionally, factors including individual susceptibility, host immune responses, and behavioral risks such as smoking disrupt the balance between resident microbes and the host's immune system, triggering an inflammatory response in the gingival tissue [9]. The excessive release of proinflammatory mediators in gingival tissues leads to the infiltration of immune cells including polymorphonuclear neutrophils (PMNs), promoting their phagocytic activity and bacterial ingestion [10]. Gingivitis, the inflammation of gingiva, can further progress to periodontitis which is a chronic inflammatory condition that causes the destruction of connective tissue and alveolar bone. While the presence of neutrophils is essential for pathogen elimination, their hyperactivation can contribute to the destruction of periodontal connective tissue by releasing proteolytic enzymes such as matrix metalloproteinases (MMPs) [11]. Furthermore, the overproduction of cytokines, chemokines, and receptor activator of nuclear factor-kB ligand (RANKL) stimulates osteoclast formation, eventually contributing to alveolar bone loss [12].

Recently, oxidative stress has increasingly been recognized as a key factor in the development of periodontitis. When a pathogenic biofilm triggers an inflammatory response, neutrophils become the predominant immune cells in periodontal tissue and are the primary source of reactive oxygen species (ROS) in periodontitis [13]. Normally, ROS production by phagocytes is essential for the elimination of microbes. However, chronic and excessive production of ROS by neutrophils induces lipid peroxidation, generating metabolites such as malondialdehyde (MDA) and 4-hydroxynonenal [14]. This oxidative imbalance directly and indirectly promotes the persistence of inflammation and the destruction of periodontal tissue [15,16].

Lycii Radicis Cortex (LRC), the dried root bark of *Lycium chinense* Mill. or *L. barbarum* L. has long been used as a medicinal herb in East Asia to treat symptoms such as fever and hyperhidrosis. In traditional Chinese medicine (TCM), LRC has been employed to treat Yin-deficiency-heat syndrome, which results from heat generated internally due to Yin deficiency, often associated with prolonged illness and exhaustion [17]. According to TCM principles, periodontitis is categorized as a condition related to stomach-heat and kidney-yin deficiency syndromes [18–20], which is related to usages of LRC. In Korea, the "Kwangjebikeup", a book published in 1790 and intended to provide remedies for rural communities with limited access to medical care, documents the use of a decoction of LRC for treating bleeding gums or gingivitis. Among the two source species of LRC, *L. chinense* Mill. is widely cultivated in various regions of Korea for use as functional food or medicinal herb [21]. Previous studies have reported the biological properties of LRC derived from *L. chinense* Mill. against muscle atrophy, glioblastomas, gastric ulcer [22–24]. In particular, numerous experimental studies have been conducted to evaluate its anti-osteoporosis effects on ovariectomy- and steroid-induced osteoporosis models [25–27]. However, the effect of LRC on periodontal diseases has not been investigated. Therefore, we investigated the protective effects of LRC (from *L. chinense* Mill.) on ligation-induced periodontitis in rats and explored the molecular mechanisms of its anti-inflammatory effect in lipopolysaccharide (LPS)-stimulated RAW 264.7 cells in the present study.

2. Materials and Methods

2.1. Preparation of LRC

The LRC extract powder was manufactured and supplied by Nutracore (Suwon, Republic of Korea) as described in the supplementary information (Batch No. LC-G221114).

Some specimens of LRC extract (Code No. LRC2022Ku01) were deposited in the herbarium of the Medical Research center for Herbal Convergence on Liver Disease, Daegu Haany University (Gyeongsan, Republic of Korea). Kukoamine B was identified as a marker component using high-performance liquid chromatographic (HPLC) analysis (Supplementary Figure S1). Quantification of the peak area indicated that LRC contains 23 mg/g of kukoamine B.

2.2. Methods for In Vitro Experiments

2.2.1. Cell Culture

RAW 264.7 cells were obtained from American Type Culture Collection (ATCC; Rockville, MD, USA). The cells were cultured in Dulbecco's modified Eagle's medium (DMEM; HyClone Laboratories, Logan, UT, USA) supplemented with 10% fetal bovine serum (Lonza, Walkersville, MD, USA) and 1% Antibiotic-Antimycotic solution (HyClone Laboratories), and were maintained in a CO_2 incubator at 37°C with 5% CO_2 under a humidified atmosphere.

2.2.2. Measurement of Nitric Oxide (NO) Production

RAW 264.7 cells were pretreated with either LRC (0.3–3 mg/mL) or dexamethasone (1 μM, Sigma-Aldrich, St. Louis, MO, USA) for 1 h, followed by exposure to 0.3 μg/mL of LPS (Sigma-Aldrich) for 18 h. After the treatment period, 100 μL of conditioned media was mixed with an equal volume of Griess reagent (1% sulphanilamide, 0.1% *N*-(1-naphtyl)-ethylene diamine dihydrochloride, 5% phosphoric acid). The absorbance was measured at 540 nm using an automated microplate reader (EnSpire™, PerkinElmer, Waltham, MA, USA).

2.2.3. Measurement of Prostaglandin E_2 (PGE_2) Production

PGE_2 levels in the conditioned media were measured using a commercial competitive enzyme-linked immunosorbent assay (ELISA) kit (Prostaglandin E2 Parameter Assay Kit, R&D Systems, Minneapolis, MN, USA). Briefly, conditioned media and primary PGE_2 antibody were added to a goat anti-mouse-coated 96-well plate. After incubation on a microplate shaker, horseradish peroxidase (HRP)-labeled PGE_2 was added, and color development was initiated using a substrate solution. The reaction was stopped by adding 2 N H_2SO_4 solution. Absorbance was measured at a wavelength of 450 nm using an automated microplate reader (EnSpire™).

2.2.4. RNA Isolation and RT-qPCR

Total RNA was isolated using a TRI-Solution™ (Bioscience Technology, Daegu, Republic of Korea). An amount of 2 μg of RNA was reverse transcribed using oligo (dT) primer, Accupower® RT PreMix (Bioneer, Daejeon, Republic of Korea), and a SimpliAmp™ Thermal Cycler (Applied Biosystems, Waltham, MA, USA). Synthesized cDNA was amplified using SyBr green Ex-Taq master mix (Takara, Shiga, Japan), Quantstudio 5 Thermal cycler (Applied Biosystems), and specific primers for inducible *nitric oxide synthase (iNOS)*, *cyclooxygenase-2 (COX-2)*, *tumor necrosis factor-α (TNF-α)*, *interleukin (IL)-1β*, *monocyte chemoattractant protein-1 (MCP-1)*, and *glyceraldehyde-3-phosphate dehydrogenase (GAPDH)*. All sequences of the oligonucleotide primers used for PCR are listed in Table 1. After PCR amplification, a melting curve of each amplicon was determined to verify its accuracy. The expressions of each gene were normalized by *GAPDH* mRNA expression, using the comparative threshold cycle method [28].

Table 1. Information of oligonucleotides used in RT-qPCR.

Target Gene	Orientation	Sequence (5′–3′)	NCBI Accession No.
iNOS	Sense Antisense	GACAAGCTGCATGTGACATC, GCTGGTAGGTTCCTGTTGTT	NM_001313922.1
COX-2	Sense Antisense	TCCAGATCACATTTGATTGA, TCTTTGACTGTGGGAGGATA	NM_011198.5
TNF-α	Sense Antisense	ATGAGCACAGAAAGCATGAT, TACAGGCTTGTCACTCGAAT	NM_013693.3
IL-1β	Sense Antisense	ATGGCAACTGTTCCTGAACT, CAGGACAGGTATAGATTCTT	NM_008361.4
MCP-1	Sense Antisense	TGATCCCAATGAGTAGGCTGG, ATGTCTGGACCCATTCCTTCT	NM_011333.3
GAPDH	Sense Antisense	AACGACCCCTTCATTGAC, TCCACGACATACTCAGCAC	NM_001411843.1
RANKL	Sense Antisense	CTGATGAAAGGAGGGAGCAC, GAAGGGTTGGACACCTGAATGC	NM_057149.2
OPG	Sense Antisense	TCCTGGCACCTACCTAAAACAGCA, ACACTGGGCTGCAATACACA	U94331.1
β-actin	Sense Antisense	TCAGGTCATCACTATCGCCAAT, AAAGAAAGGGTGTAAAACGCA	NM_031144.3

RT-qPCR, reverse transcription quantitative polymerase chain reaction; *iNOS, inducible nitric oxide synthase; COX-2, cyclooxygenase-2; TNF-α, tumor necrosis factor-α; IL-1β, interleukin-1β; MCP-1, monocyte chemoattractant protein-1; GAPDH, glyceraldehyde-3-phosphate dehydrogenase; RANKL, receptor activator of nuclear factor-κB ligand; OPG, osteoprotegerin.*

2.2.5. Immunoblot Analysis

Cells were lysed with radioimmunoprecipitation assay buffer (RIPA) containing sodium fluoride, β-glycerophosphate, sodium orthovanadate, sodium pyrophosphate, and a protease inhibitor cocktail (GenDEPOT, Barker, TX, USA) and then incubated on ice for 1 h. After centrifugation at 15,000× *g* for 10 min, the supernatant was collected as whole cell lysates. Protein concentrations were determined using a bicinchoninic acid (BCA) assay (Thermo Fischer Scientific, Waltham, MA, USA). Equal amounts of proteins were resolved by sodium dodecyl sulfate-polyacrylamide gel electrophoresis (SDS-PAGE) and transferred to a nitrocellulose membrane (GE Healthcare Life Sciences, Buckinghamshire, UK). After blocking with 5% bovine serum albumin, the membrane was sequentially reacted with primary and secondary antibodies. Immunoreactive proteins were detected using West-Q Pico ECL solution (GenDEPOT) and visualized with a Fusion FX7 (Vilber Lourmat, Marne-la-Vallée, France). Densitometric analysis was performed using Image J software (ver. 1.53). Band intensity values were calculated by dividing the intensity of phosphorylated protein by the intensity of total target protein. Primary antibodies against phosphorylated c-Jun N-terminal kinase (JNK) 1/2, JNK 1/2, phosphorylated p38, p38, phosphorylated extracellular-signal regulated protein kinase (ERK) 1/2, ERK 1/2, phosphorylated p65, as well as HRP-conjugated secondary antibodies were obtained from Cell Signaling Technology (Danvers, MA, USA). The anti-p65 antibody was purchased from Santa Cruz Biotechnology (Santa Cruz, CA, USA).

2.2.6. Measurement of Radical Scavenging Activity

Radical scavenging activity was assessed using 2,2-diphenyl-1-picrylhydrazyl (DPPH). LRC was dissolved in water with concentrations ranging from 10 to 300 μg/mL. A 20 μL aliquot of each LRC sample was mixed with 180 μL of DPPH solution (150 μM, dissolved in ethanol) and incubated at room temperature for 30 min, protected from light. Absorbance was measured at a wavelength of 517 nm using an automated microplate reader (EnSpire™). The radical scavenging activity was calculated by comparing the absorbance of the DPPH solution with (S) and without the samples (S_0) to the absorbance of the solvent with (C) and without the sample (C_0). The percentage of radical scavenging activity is determined using the following equation: $[(S - S_0)/(C - C_0)] \times 100$ (%).

2.3. *Methods for In Vivo Experiments*

2.3.1. Preparation of Test Materials

The LRC powder was first dissolved in distilled water at a concentration of 40 mg/mL and further diluted in distilled water to concentrations of 20 and 10 mg/mL for the administration. Additionally, indomethacin (IND; Sigma-Aldrich) was prepared by suspending it in distilled water at a concentration of 1 mg/mL. Insadol^{TM} Plus (INP; Dongkook Pharmaceutical, Seoul, Republic of Korea), commonly used in the Korean market as a pharmaceutics agent for the relief of periodontal disease, was prepared by grinding tablets and suspending in distilled water at a concentration of 12.6 mg/mL.

2.3.2. Animal Husbandry and Grouping

A total of 70 six-week-old male Sprague-Dawley (Crl:CD) rats (body weight ranged from 150 to 210 g upon receipt) were obtained from OrientBio (Seungnam, Republic of Korea) and acclimated for 9 days. Animals were then divided into 7 groups (10 rats/group): intact vehicle control group, experimental periodontitis (EPD) control group, IND (EPD with IND oral administration) group, INP (EPD with INP oral administration) group, and three LRC groups (EPD with LRC administration at doses of 50, 100, and 200 mg/kg, respectively). All animal experiments were conducted according to the national regulation of the usage and welfare of laboratory animals and approved by the Institutional Animal Care and Use Committee in Daegu Haany University (Approval No. DHU2022-101).

2.3.3. Induction of EPD and Administration of Test Materials

EPD was induced by placing a sterile nylon suture (3–0) around the cervix of the upper left incisor in rats. The suture was tied on the buccal side of the tooth, based on the established method [29,30]. In the intact vehicle control groups, we only examined the cervix of the upper left incisor tooth, without applying any ligation. The procedures were conducted under inhalation anesthesia with 2–3% isoflurane (Hana Pharm. Co., Hwaseong, Republic of Korea) using a rodent inhalation anesthesia apparatus (Surgivet, Waukesha, WI, USA) and a rodent ventilator (Model 687, Harvard Apparatus, Cambridge, UK).

A total of 24 h after ligation placement, test materials including IND, INP, or three doses of LRC were administered orally by gastric gavage using a stainless steel zonde at a volume of 5 mL/kg, once daily for 10 consecutive days. To provide the same restraint stress for the intact vehicle control group and the EPD control group, an equal volume of distilled water was administered instead of IND, INP, or LRC.

2.3.4. Measurements of Alveolar Bone Loss Scores

The maxillary bone including the area with the ligation placement was photographed with a digital camera, and then excised. The distance from the cusp tip to the alveolar bone, indicative of horizontal alveolar bone loss, was measured using an electronic digital caliper along the axis of the root of the upper left incisor tooth as in previously described methods [29,31] and recorded in millimeters per rat.

2.3.5. Microbiological Analysis

The buccal gingival tissues from the ligated area were carefully excised and immediately homogenized with 0.3 mL of Brain Heart Infusion (BHI) broth. The homogenized tissue samples were then diluted 1:100 and 1:1000, and plated on BHI agar supplemented with defibrinated sheep blood to culture anaerobic bacteria. The plates were incubated at 37 °C for using anaerojars and anaerogen sachets. After 48 h, the numbers of the formed colonies were counted and represented as $\times 10^2$ CFU/g tissues [32].

2.3.6. Measurement of Myeloperoxidase (MPO) Activity

MPO activity was measured using a spectrophotometric assay [33]. The buccal gingival tissues around the left incisor were excised and suspended in 0.5% hexadecyltrimethylammonium bromide (Gibco, Carlsbad, CA, USA) in 50 mM potassium phosphate buffer (pH 6.0) to solubilize MPO, and then homogenized. After two freeze–thaw cycles, the homogenate was incubated with additional buffer (400 μL/15 mg of tissue) for 12 min, then centrifuged at 1000× *g* for 12 min at 4 °C. The supernatant (0.1 mL) was mixed with 2 mL of phosphate buffer (50 mM, pH 6.0), containing 0.167 mg/mL *o*-dianisidine dihydrochloride (Sigma-Aldrich), distilled water, and 0.0005% hydrogen peroxide. Absorbance was measured at 460 nm using a UV/Vis spectrometer (OPTIZEN POP, Mecasys, Daejeon, Republic of Korea). One unit of activity was defined as the amount needed to degrade 1 μM of hydrogen peroxide per min at 25 °C, and the results were expressed as units per mg of tissue.

2.3.7. Measurement of PGE_2, MMP-8, TNF-α, and IL-1β

To measure buccal gingival expressions of PGE_2, MMP-8, TNF-α, and IL-1β, tissues around the ligation placement were collected and homogenized. Each concentration of PGE_2 (R&D Systems), MMP-8 (MyBioSource, San Diego, CA, USA), TNF-α, and IL-1β (Abcam, Cambridge, UK) in the tissue homogenates was measured using a commercial ELISA kit according to the manufacturer's instructions.

2.3.8. RT-qPCR Analysis for *RANKL* and *Osteoprotegerin* (*OPG*) mRNA Expressions

The *RANKL* and *OPG* mRNA expressions in the maxillary gingival tissues were detected using RT-qPCR. Total RNA was extracted using Trizol reagent® (Invitrogen, Carlsbad, CA, USA). The extracted RNA was reverse transcribed using the High-Capacity cDNA Reverse Transcription Kit (Thermo Fisher Scientific). And the synthesized cDNA was amplified using the CFX96™ Real-Time System (Bio-Rad, Hercules, CA, USA). All sequences of the oligonucleotide primers used for PCR are listed in Table 1. The expressions of each gene were normalized by *β-actin* mRNA expression, using the comparative threshold cycle method [28].

2.3.9. Measurement of Lipid Peroxidation

Lipid peroxidation in the buccal gingival tissue was assessed by measuring the MDA levels. Buccal gingival tissues from the ligature placement site were collected and homogenized in a buffer composed of 50 mM Tris-HCl, 0.1 mM EGTA, and 1 mM phenylmethylsulfonylfluoride (pH 7.4). A 100 μL aliquot of the tissue homogenate was then added to a reaction mixture containing SDS, acetic acid, thiobarbituric acid, and distilled water. The samples were incubated at 95 °C for 1 h, and subsequently centrifuged at 3000× *g* for 10 min at 4 °C. The absorbance of the supernatant was measured at 650 nm by a UV/Vis spectrometer. Results are expressed as μM per mg of tissue.

2.3.10. Measurement of Inducible iNOS Activity

The gingival activity of iNOS was assessed by measuring the conversion of L-[^{3}H]-arginine to L-[^{3}H]-citrulline in tissue homogenates, following established protocols [29,30,34]. Briefly, tissue homogenates were incubated with a reaction mixture containing L-[^{3}H]-arginine (10 mM, 5 kBq/tube), NADPH (1 mM), calmodulin (30 nM), tetrahydrobiopterin (5 mM), and calcium (2 mM) for 30 min at 22 °C. The reaction was terminated by the addition of 0.5 mL of ice-cold HEPES buffer (pH 5.5) containing EGTA (2 mM) and EDTA (2 mM), which act as calcium chelators. To determine NOS-independent activity, the formation of L-[^{3}H]-citrulline was measured in the absence of NADPH. Additionally, to evaluate calcium-independent NOS activity, reactions were performed with NADH (1 mM) in the absence of calcium and in the presence of EGTA (5 mM). The reaction mixtures were then passed through Dowex 50W (Na^+ form) columns to separate L-[^{3}H]-citrulline, which was quantified using a liquid scintillation counter (Wallac, Annapolis, MD, USA). iNOS activity is expressed as fM/mg/min.

2.3.11. Histopathological Analysis

After the sacrifice, the maxilla region around the placed ligature including the left and right incisors was excised. After fixation in 10% neutral buffered formalin, tissues were incubated in a decalcifying solution containing 24.4% formic acid and 0.5 N sodium hydroxide for 5 days. Decalcified tissues were cross trimmed, embedded in paraffin, and sectioned at a thickness of 3–4 μm. Representative sections were then stained with hematoxylin and eosin for histological observation. The regions between the left and right incisor teeth, where the ligature was applied, were evaluated histologically using a previously established scoring system (0 to 3) [29,30,35], considering the infiltration of inflammatory cells and the integrity of the alveolar bone and cementum. In addition, the numbers of infiltrated inflammatory cells (numbers/mm^2) and collagen-occupied areas (%/mm^2) on the gingival tissues around the left incisor teeth were quantified. Furthermore, the volume of alveolar bone (%/mm^2), as well as the numbers (numbers/mm^2) and occupied percentages of osteoclasts (%/mm^2) on the alveolar bone surface between the upper incisor teeth were also measured. Histological analyses were conducted using a light microscope (Eclipse 80*i*, Nikon, Tokyo, Japan)-equipped histological camera system (ProgRes^{TM} C5, Jenoptik Optical Systems GmbH, Jena, Germany) and computer-assisted image analyzer (*i*Solution FL ver 9.1, IMT *i*-solution Inc., Bernaby, BC, Canada), with the histopathologist remaining blinded to the group assignments during evaluation.

2.4. Statistical Analysis

All numeric data were expressed as mean ± standard deviation. A one-way analysis of variance (ANOVA) test was performed to determine the statistical significance of differences among the experimental groups. According to the variance homogeneity result by Levene's test, Tukey's honestly significant difference (HSD) or Dunnett's T3 test were conducted as a *post hoc* analysis. p values under 0.05 were considered statistically significant. All statistical analyses were performed using SPSS 18.0 software (SPSS, Chicago, IL, USA).

3. Results

3.1. Anti-Inflammatory Effect of LRC on the LPS-Stimulated RAW 264.7 Cells

3.1.1. LRC Decreased the Expression of Proinflammatory Mediators

To confirm the anti-inflammatory effect of LRC, we conducted in vitro experiments using LPS-stimulated RAW 264.7 cells. Prior to evaluating the anti-inflammatory effect of LRC, we assessed the cytotoxicity of LRC on HaCaT (a human keratinocyte cell), HDFn (a human primary dermal fibroblast-neonatal cell), and RAW 264.7 (a murine macrophage-derived cell) cells. As a result, no significant changes on the survivability of HaCaT, HDFn, and RAW 264.7 cells by LRC treatment were observed (Supplementary Figure S2). To investigate the inhibitory effect of LRC on NO and PGE_2 production, RAW 264.7 cells were pretreated with LRC (0.3–3 mg/mL) or dexamethasone (1 μM, as a positive control) for 1 h. And cells were further stimulated with LPS (0.3 μg/mL, 18 h). After treatment, the levels of NO and PGE_2 in the conditioned media were measured. As expected, LPS stimulation significantly increased the production of NO and PGE_2 compared to the vehicle-treated group. However, LRC (1 and 3 mg/mL) pretreatment significantly reduced the NO and PGE_2 production (Figure 1a). Similarly, the LPS-induced increased mRNA levels of *iNOS* and *COX-2*, enzymes responsible for producing NO and PGE_2, respectively, were successfully decreased by LRC (3 mg/mL) treatment (Figure 1b). IL-1β, TNF-α, and MCP-1, produced from innate immune cells including macrophages, have been known as major inflammatory mediators for exaggerating acute inflammation [36]. To investigate the effects of LRC on the production of proinflammatory cytokines by LPS, RAW 264.7 cells were pretreated with LRC (0.3–3 mg/mL) for 1 h, and subsequently exposed to LPS (0.3 μg/mL, 6 h). As a result, the increased mRNA levels of proinflammatory cytokines including *IL-1β*,

TNF-α, and *MCP-1* by LPS stimulation were dose-dependently reduced by LRC treatment (Figure 1c).

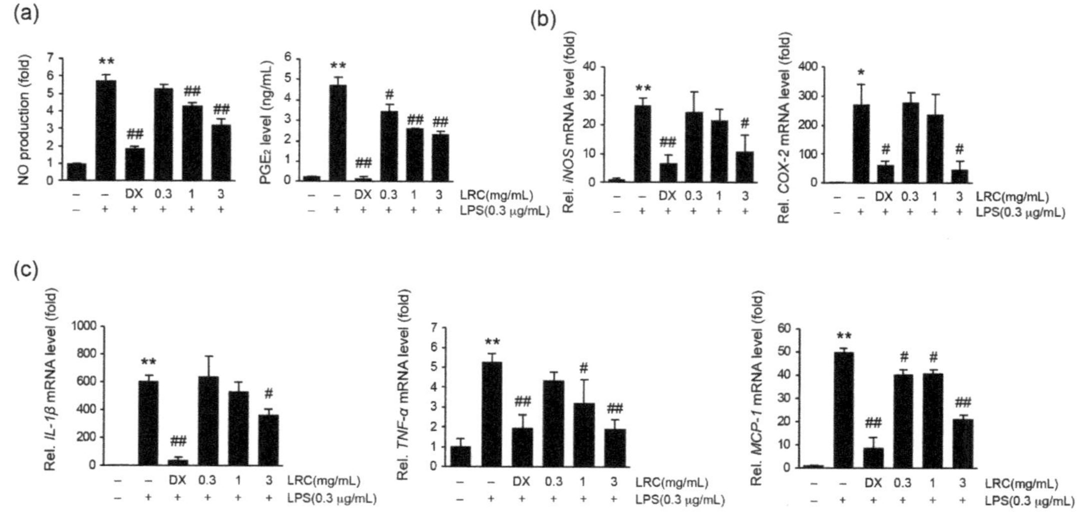

Figure 1. Effect of Lycii Radicis Cortex extract (LRC) on inflammatory mediators' expression in lipopolysaccharide (LPS)-stimulated RAW 264.7 cells. RAW 264.7 cells were pretreated with LRC (0.3–3 mg/mL) or dexamethasone (DX, 1 μM) for 1 h, and further stimulated with lipopolysaccharide (LPS, 0.3 μg/mL) for (**a**) 18 h or (**b**,**c**) 6 h. After treatment, (**a**) levels of nitric oxide (NO) and prostaglandin E_2 (PGE_2) were detected in collected conditioned media. (**b**) Relative mRNA levels of *iNOS*, *COX-2*, (**c**) *IL-1β*, *TNF-α*, and *MCP-1* were measured using RT-qPCR. The expressions of each gene were normalized by the expression of *GAPDH*. Significant versus vehicle group, * $p < 0.05$, ** $p < 0.01$; versus LPS-treated group, # $p < 0.05$, ## $p < 0.01$.

3.1.2. LRC Inhibited the Activation of Mitogen-Activated Protein Kinases (MAPKs) and Nuclear Factor-κB (NF-κB) Signaling Pathways

To elucidate the molecular mechanism underlying the anti-inflammatory effects of LRC, we investigated its effect on MAPK and NF-κB signaling pathways (Figure 2). For the immunoblotting, cells were pretreated with LRC (0.3–3 mg/mL) for 1 h, and further incubated with LPS (0.3 μg/mL, 0.5 h). In the MAPK signaling pathway, LRC treatment at 1 and 3 mg/mL significantly reduced the phosphorylation of p38. And the phosphorylation of JNK was significantly reduced by only 3 mg/mL of LRC treatment. But no significant effect of LRC (0.3–3 mg/mL) treatment on the phosphorylation of ERK was observed (Figure 2a). In the NF-κB pathway, the increase in p65 phosphorylation by LPS, which indicates transcriptional activation of NF-κB [37], was significantly and dose-dependently reduced by LRC (0.3–3 mg/mL) treatment (Figure 2b). These results suggest that LRC exhibits anti-inflammatory effects by inhibiting the production of inflammatory mediators through the suppression of MAPK and NF-κB signaling pathways.

3.2. The Effect of LRC on the EPD Ligation Rats

3.2.1. Experiment Procedure and Body Weight Gains

EPD was induced on day 0 by placing a sterilized nylon thread ligature around the upper left incisor of rats. At 24 h after the ligation placement, IND (5 mg/kg), INP (63 mg/kg), and three different doses of LRC (50, 100, and 200 mg/kg) were administered orally once a day for 10 consecutive days. On day 11, all animals were sacrificed, and tissues were harvested (Figure 3a). Over the 11 days of the experimental period, there were no significant body weight changes in the EPD control group compared to the intact vehicle control group in the present study. Similarly, rats administered with IND, INP, or LRC (50,

100, and 200 mg/kg) did not show any significant changes in body weight compared to the EPD group (Figure 3b).

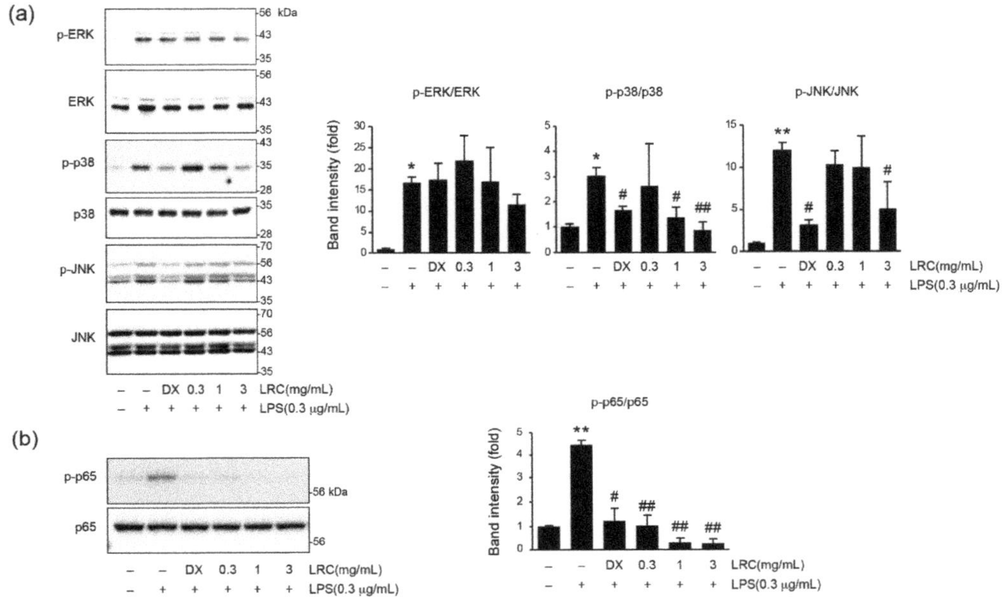

Figure 2. Effect of LRC on mitogen-activated protein kinases (MAPKs) and nuclear factor-κB (NF-κB) activation in LPS-stimulated RAW 264.7 cells. For immunoblot analysis, RAW 264.7 cells were pretreated with LRC (0.3–3 mg/mL) for 1 h and further incubated with LPS (0.3 μg/mL, 0.5 h). (**a**) MAPKs phosphorylation. (**b**) p65 phosphorylation. Significant versus vehicle group, * $p < 0.05$, ** $p < 0.01$; versus LPS-treated group, # $p < 0.05$, ## $p < 0.01$.

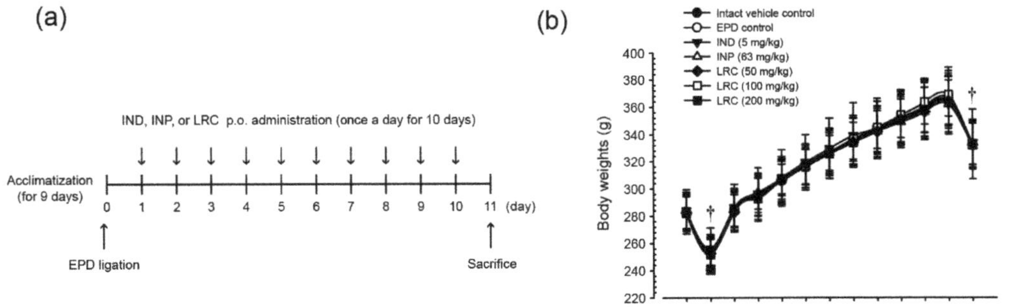

Figure 3. Procedure of experiment and body weights. (**a**) Scheme of animal experiment. At 24 h after experimental periodontitis (EPD) ligation, rats (10/group) were orally administered with indomethacin (IND, 5 mg/kg), Insadol Plus^{TM} (INP, 63 mg/kg), or three doses of LRC (50, 100, or 200 mg/kg) once a day for 10 days. Day 0 means the day of ligature placement. (**b**) Animal body weights. Body weight was measured from day −1 to day 11. All values are expressed as mean ± SD of ten rats. † All rats were fasted overnight before both the ligature placement and sacrifice.

3.2.2. LRC Ameliorated Alveolar Bone Loss Scores

Exposed teeth root areas were detected as the alveolar bone loss scores. In the EPD control group, alveolar bone loss scores were significantly increased compared to the intact vehicle control. As positive controls, both the IND (5 mg/kg) and the INP (63 mg/kg) groups exhibited a significant reduction in alveolar bone loss scores compared to the EPD group. All three dosages of LRC (50, 100, and 200 mg/kg) also showed a significant reduction in alveolar bone loss scores against ligation-induced EPD. In particular, the 200 mg/kg of LRC treatment showed a favorable inhibitory effect on ligation-induced EPD, comparable to those of IND or INP administration (Figure 4).

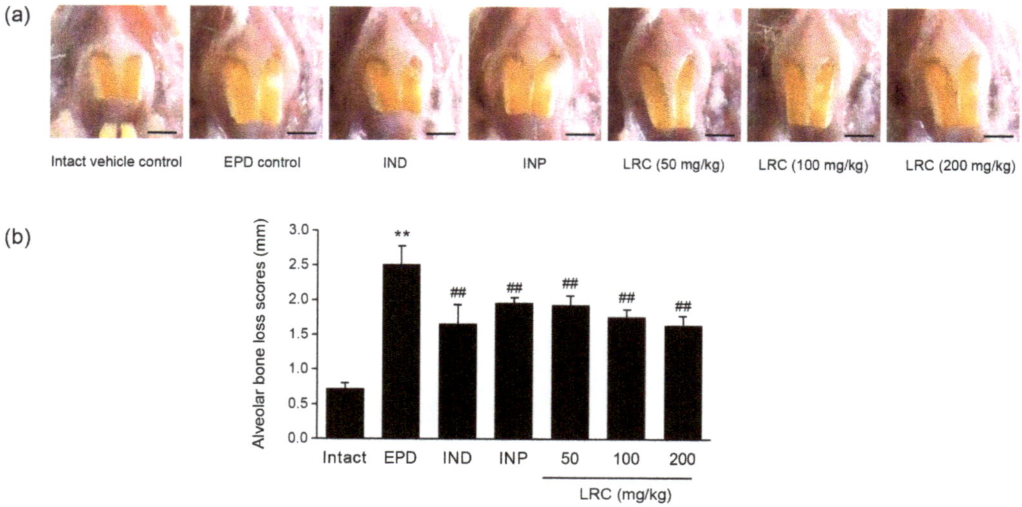

Figure 4. Effect of LRC on the alveolar bone loss. (**a**) Representative photos taken from upper left incisor teeth. Scale bars indicate 3.00 mm. (**b**) Alveolar bone loss scores. Horizontal alveolar bone loss, the distance from the cusp tip to the alveolar bone, was measured. All values are expressed as mean ± SD of ten rats. Significant versus intact vehicle control group, ** $p < 0.01$; versus EPD control group, ## $p < 0.01$.

3.2.3. LRC Decreased Gingival Anaerobic Bacterial Count and MPO Activity

In the EPD control group, the number of total anaerobic bacteria in the buccal gingiva was significantly higher compared to the intact vehicle control group. Administration of all three doses of LRC (50, 100, and 200 mg/kg) significantly reduced the viable bacterial count. However, IND (5 mg/kg) showed no significant changes in anaerobic bacterial count against ligation-induced EPD. Notably, LRC at 100 and 200 mg/kg showed a more favorable inhibitory effect on the total anaerobic bacteria in the buccal gingiva, comparable to that of INP (63 mg/kg) in the current experiment (Figure 5a). To measure neutrophil accumulation in gingival tissues, the MPO activity in tissue homogenates was assessed. Gingival MPO activity was significantly increased in the EPD control group compared to the intact vehicle control group. However, administration with all three doses of LRC (50, 100, and 200 mg/kg) significantly and dose-dependently decreased MPO activity. Specifically, LRC at 200 mg/kg showed the most potent inhibitory effect on ligation-induced EPD, comparable to that observed in the IND group (Figure 5b).

3.2.4. LRC Decreased the Gingival Expression of Proinflammatory Mediators

The buccal gingival expressions of proinflammatory mediators including PGE_2, IL-1β, and TNF-α were also detected using ELISA. Consistent with the findings from in vitro experiments, increased levels of PGE_2, IL-1β, and TNF-α in the EPD control group were significantly reduced by administration of all three doses of LRC (50, 100, and 200 mg/kg). In particular, LRC administration at 200 mg/kg showed favorable inhibitory effects on the gingival expression of PGE_2, IL-1β, and TNF-α as comparable to those of IND (5 mg/mL) group (Figure 6).

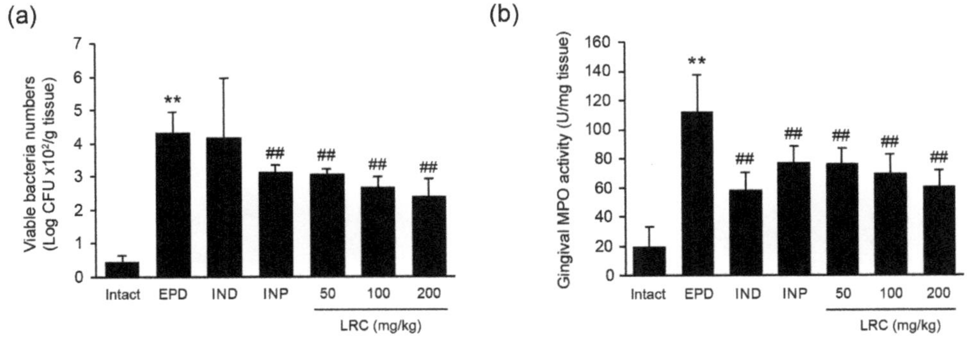

Figure 5. Effect of LRC on the buccal gingival total anaerobic bacterial counts and myeloperoxidase (MPO) activity. (**a**) Viable bacteria numbers. Excised buccal gingival tissues were homogenized and diluted for plating BHI agar to culture anaerobes. Formed colonies were counted as $\times 10^2$ CFU/g tissue. (**b**) MPO activity in gingival tissues. One unit of activity was defined as the amount required to degrade 1 μM of hydrogen peroxide per min at 25°C, with results expressed in units per milligram of tissue. All values are expressed as mean ± SD of ten rats. Significant versus intact vehicle control group, ** $p < 0.01$; versus EPD control group, ## $p < 0.01$.

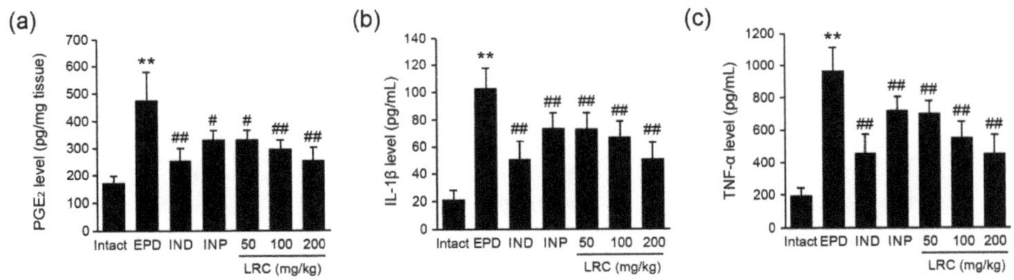

Figure 6. Effect of LRC on the gingival expression of inflammatory mediators. The expression levels of (**a**) PGE_2, (**b**) IL-1β, and (**c**) TNF-α in the gingival tissues were measured using a commercial ELISA kit. All values are expressed as mean ± SD of ten rats. Significant versus intact vehicle control group, ** $p < 0.01$; versus EPD control group, # $p < 0.05$, ## $p < 0.01$.

3.2.5. LRC Decreased MDA Levels and iNOS Activity

Oxidative stress is considered one of the key mechanisms involved in the progression of periodontal disease [16]. To investigate the antioxidant properties of LRC, we first evaluated its radical scavenging activity using the DPPH assay (Figure 7a). LRC exhibited a dose-dependent increase in radical scavenging activity, with statistically significant differences observed at concentrations ranging from 10 to 300 μg/mL. Additionally, we measured the MDA level in the gingival tissues as a maker of lipid peroxidation (Figure 7b). The ligation placement led to a significant increase in MDA level in the gingival tissue of the EPD control group compared to the intact vehicle control group. However, the administration of LRC (50, 100, and 200 mg/kg) significantly and dose-dependently reduced gingival MDA levels. Notably, LRC at 200 mg/kg showed the most potent inhibitory effect on ligation-induced MDA levels, comparable to that observed in the IND (5 mg/kg) group. Moreover, we assessed the gingival activity of iNOS, which is primarily responsible for NO formation in periodontitis (Figure 7c). NO not only indicates the intensity of the inflammatory response within tissues, but also reflects the level of oxidative stress. Excessive NO production by phagocytic cells can lead to the formation of NO derivatives such as peroxynitrite, contributing to tissue damage through oxidation and nitrosylation [38]. In the present experiments, ligation significantly increased gingival iNOS activity in the EPD control group compared to the intact vehicle group. Similarly to NO production and *iNOS* mRNA levels in LPS-stimulated RAW 264.7 cells (Figure 1), administration of LRC (50, 100, and 200 mg/kg) significantly reduced elevated iNOS activity. Specifically, LRC at 200 mg/kg exhibited the most potent inhibitory effect, comparable to that observed in the IND group.

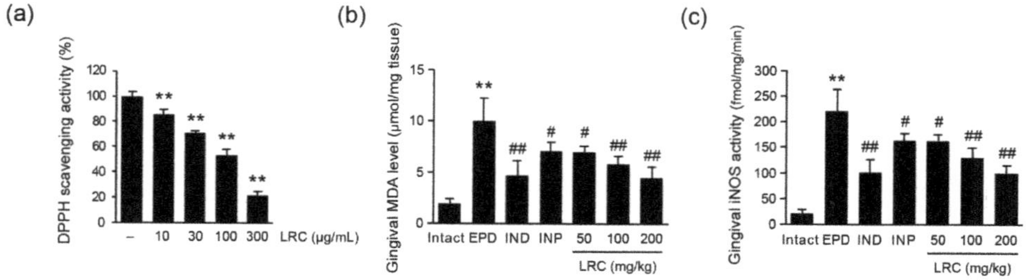

Figure 7. Effect of LRC on oxidative stress. (**a**) DPPH (2,2-diphenyl-1-picrylhydrazyl) radical scavenging activity. Significant versus vehicle group, ** $p < 0.01$. (**b**) Malondialdehyde (MDA) level and (**c**) iNOS activity were assessed in gingival tissue. All values are expressed as mean ± SD of ten rats. Significant versus intact vehicle control group, ** $p < 0.01$; versus EPD control group, # $p < 0.05$, ## $p < 0.01$.

3.2.6. LRC Decreased MMP-8 Levels in Gingival Tissue

MMPs, proteolytic enzymes, are responsible for the degradation of periodontal connective tissue. Since type I collagen is a major component of periodontal extracellular matrix, the role of MMP-8 in periodontal tissue destruction has received significant attention. Previous studies have shown that the concentration of MMP-8 is associated with the severity of tissue destruction, suggesting its potential as a biomarker for periodontal disease [11]. Therefore, we assessed the MMP-8 level in gingival tissue using ELISA (Figure 8). The ligation placement significantly increased the MMP-8 level in the gingival tissue of the EPD control group compared to the intact vehicle control group. However, the administration of LRC dose-dependently reduced gingival MMP-8 levels with statistically significant differences observed at 100 and 200 mg/mL. Notably, LRC at 200 mg/kg showed the most potent inhibitory effect on ligation-induced MMP-8 level, comparable to that observed in the IND (5 mg/kg) group.

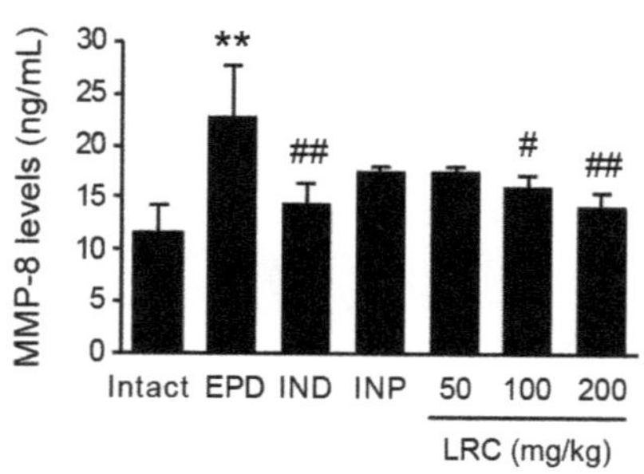

Figure 8. Effect of LRC on matrix metalloproteinase-8 (MMP-8) expression. MMP-8 level in gingival tissues was assessed using a commercial ELISA kit. All values are expressed as mean ± SD of ten rats. Significant versus intact vehicle control group, ** $p < 0.01$; versus EPD control group, # $p < 0.05$, ## $p < 0.01$.

3.2.7. LRC Decreased RANKL/OPG Ratio

The RANKL/RANK/OPG system plays a crucial role in the maturation of osteoclasts, as well as in bone remodeling. Therefore, we assessed the mRNA expression of *RANKL* and *OPG* in gingival tissue (Figure 9). Ligation placement significantly increased the *RANKL* mRNA level in the gingival tissue of the EPD control group compared to the intact vehicle control group. However, the administration of LRC (50, 100, and 200 mg/kg) significantly and dose-dependently reduced gingival *RANKL* mRNA levels (Figure 9a). In contrast, the administration of LRC (50, 100, and 200 mg/kg) significantly increased the mRNA level of *OPG* compared to the EPD group (Figure 9b). Next, since an increase in the RANKL/OPG ratio is known to reflect the occurrence of periodontitis [39], we represented the RANKL/OPG ratio by normalizing the *RANKL* mRNA level to the *OPG* mRNA level (Figure 9c). The ratio, which increased due to ligation, was significantly and dose-dependently decreased with LRC administration (50, 100, and 200 mg/kg).

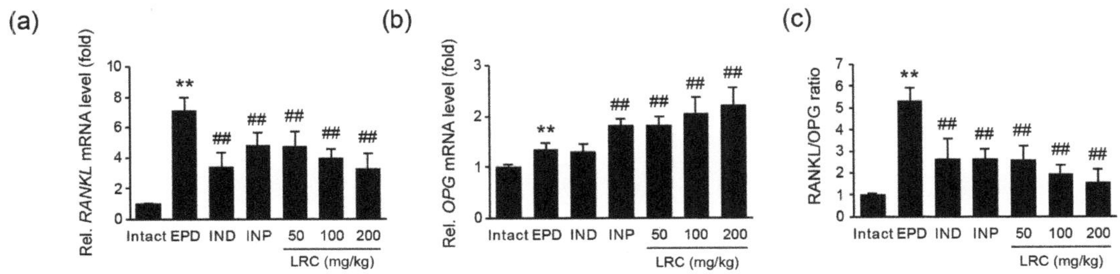

Figure 9. Effect of LRC on *receptor activator of nuclear factor-κB ligand* (*RANKL*) and *osteoprotegerin* (*OPG*) mRNA expressions. Relative mRNA levels of (**a**) *RANKL* and (**b**) *OPG* in gingival tissues were measured using RT-qPCR. The expressions of each gene were normalized by the expression of *β-actin*. All values are expressed as mean ± SD of ten rats. (**c**) RANKL/OPG ratio. The mRNA level of *RANKL* was divided by that of *OPG*. Significant versus intact vehicle control group, ** $p < 0.01$; versus EPD control group, ## $p < 0.01$.

3.2.8. Histopathological Changes of Maxillary Regions

Histopathological analyses were performed on both the gingival tissue and alveolar bone regions (Figure 10). The results showed a significant increase in inflammatory cell infiltrations, mostly PMNs, within the gingival tissues located between the upper left and right incisors in the EPD group. This was accompanied by severe edematous changes, including the loosening of collagen fibers and a loss of their compactness. In the alveolar bone regions of the EPD group, there was notable activation of osteoclasts, characterized by an increase in the numbers of osteoclast cells and the percentage of the alveolar bone surface occupied by these cells (OC/BS), along with a marked decrease in alveolar processes. These findings suggest that periodontitis and related bone loss were induced by the ligature placement, as previously reported [29,30].

However, oral administration of LRC at all doses (50, 100, and 200 mg/kg) significantly and dose-dependently reduced ligation-induced histopathological changes in gingival tissues, including histological scores, inflammatory cell infiltrations, and regions occupied by collagen fibers (Table 2). Moreover, all doses of LRC significantly restored alveolar bone volumes, and reduced the numbers and percentages of osteoclast-occupied regions on the alveolar bone surface (Table 3). Notably, LRC at 200 mg/kg exhibited a strong inhibitory effect on ligation-induced histological changes in the gingival tissue and alveolar bone regions, comparable to the effects observed with IND (5 mg/kg).

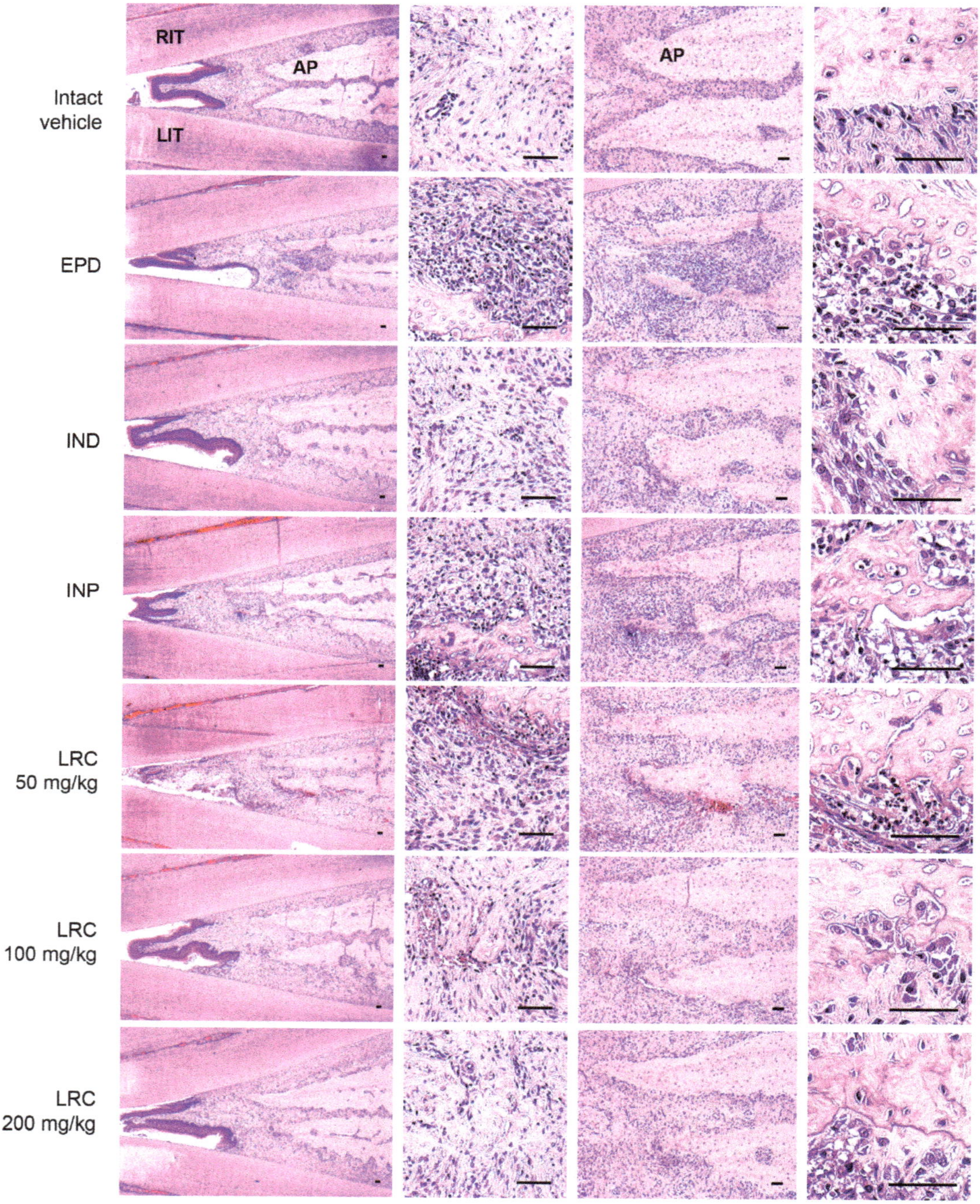

Figure 10. Representative histological images of gingival tissue and alveolar bone areas. Tissues were stained with hematoxylin and eosin. Images of gingival tissue and alveolar bone areas were captured between upper incisor teeth. Scale bars indicate 50 μm. RIT, right incisor tooth; LIT, left incisor tooth; AP, alveolar process.

Table 2. Histological scores and histomorphometrical analysis of maxillary regions around ligation placement—gingival tissues.

Group	In Gingival Tissues		
	Histological Scores (Max =3)	Inflammatory Cells (cells/mm^2)	Collagen Fibers (%/mm^2)
Intact vehicle	0.30 ± 0.48	58.60 ± 17.02	76.50 ± 10.58
EPD	2.90 ± 0.32 **	743.40 ± 67.48 **	14.16 ± 4.20 **
IND (5 mg/kg)	1.50 ± 0.53 ##	257.40 ± 94.73 ##	51.20 ± 10.11 ##
INP (63 mg/kg)	2.10 ± 0.57 #	532.80 ± 100.41 ##	31.48 ± 7.46 ##
LRC (50 mg/kg)	2.00 ± 0.67 ##	524.90 ± 101.41 ##	31.54 ± 10.01 ##
LRC (100 mg/kg)	1.70 ± 0.48 ##	387.80 ± 110.89 ##	41.86 ± 12.35 ##
LRC (200 mg/kg)	1.40 ± 0.52 ##	242.20 ± 84.66 ##	52.16 ± 10.29 ##

Values are expressed as mean ± SD of 10 rats. Significant versus intact vehicle control group, ** $p < 0.01$; versus EPD control group, # $p < 0.05$, ## $p < 0.01$.

Table 3. Histomorphometrical analysis of maxillary regions around ligation placement—alveolar bone areas.

Group	In Alveolar Bone Regions		
	Alveolar Bone Volume (%)	Osteoclast Cell (cells/mm^2)	OC/BS (%)
Intact vehicle	78.59 ± 6.50	7.80 ± 2.57	3.89 ± 1.47
EPD	20.65 ± 6.22 **	51.40 ± 7.72 **	68.42 ± 6.45 **
IND (5 mg/kg)	53.30 ± 6.41 ##	26.00 ± 5.89 ##	29.20 ± 8.95 ##
INP (63 mg/kg)	36.80 ± 4.61 ##	37.40 ± 2.84 ##	47.56 ± 6.20 ##
LRC (50 mg/kg)	36.76 ± 7.69 ##	37.20 ± 3.55 ##	47.45 ± 7.31 ##
LRC (100 mg/kg)	41.78 ± 6.92 ##	31.60 ± 5.48 ##	36.05 ± 6.83 ##
LRC (200 mg/kg)	53.89 ± 12.74 ##	25.80 ± 3.82 ##	29.10 ± 11.92 ##

Values are expressed as mean ± SD of 10 rats. OC/BS, the percentage of osteoclast cell-occupied regions on the alveolar bone surface. Significant versus intact vehicle control group, ** $p < 0.01$; versus EPD control group, ## $p < 0.01$.

4. Discussion

Along with conventional approaches, herbal medicine and natural phytochemicals have gained attention for their efficacy, safety, and cost effectiveness as adjuncts [7]. Despite the ethnopharmacological use of LRC, research validating its efficacy against periodontitis and related alveolar bone loss through animal studies remains limited. Therefore, we investigated the preventive and therapeutic efficacy of LRC on ligation-induced EPD rats. While molars are often preferred for EPD induction due to their clinical relevance including severe inflammation [40], we selected an incisor ligation model to align with our focus on developing preventive measures using medicinal herbs, aiming for a milder model. The incisor ligation model offers improved visibility and ease of procedure, and histopathological and biochemical analyses have confirmed its ability to induce periodontal inflammation and alveolar bone resorption [40,41]. Additionally, the incisor ligation model has been effectively used in prior studies on natural products and herbs, including our own previous research, to evaluate their efficacy [29,30,42,43]. In the present study, INP (containing Magnolia bark 75% ethanol soft extract) was used as a positive control representing a natural preventive agent, while IND (a non-selective cyclooxygenase inhibitor) served as a positive control representing anti-inflammatory agents.

Periodontal disease is initiated by localized gingival inflammation, which arises from the surge of pathogenic bacteria in the dental plaque and the subsequent host responses [10]. Because the outgrowth of keystone pathogens within the oral biofilm contributes to the pathogenesis of periodontitis, antimicrobial approaches using antibiotics are conventionally employed with scaling and root planning [44]. However, the systemic and direct use of antibiotics presents challenges such as disruption of the gut microbiota and development of antibiotic resistance [45]. Therefore, as one of the alternative options, natural products with antimicrobial effects have been investigated as potential candidates [8]. In our experiment, ligature placement significantly increased the anaerobic bacteria count as previously reported [29,30]. However, LRC administration showed a significant inhibitory effect on the viable anaerobic bacteria number (Figure 5a). Although antimicrobial effects have been reported for some isolated ingredients of LRC [46], further research on LRC and its constituents targeting specific keystone anaerobic bacteria related to periodontitis such as *Porphyromonas gingivalis*, *Treponema denticola*, and *Tannerella forsythia* is necessary to validate its antimicrobial effect [9].

In periodontal disease, persistent and chronic inflammation leads to tissue destruction. The excessive production of cytokines and eicosanoids by infiltrated immune cells accelerates the secretion of proteolytic enzymes and promotes bone resorption. In vitro studies have shown that IL-1β and TNF-α upregulate MMPs expression in gingival fibroblasts [10], while IL-1β, TNF-α, and PGE_2 increase osteoclast formation [47,48]. Therefore, pharmacological intervention to inhibit the production of inflammatory mediators is essential to prevent periodontal tissue destruction. In the present study, we assessed the levels of inflammatory mediators including TNF-α, IL-1β, and PGE_2 in gingival tissue. The elevated levels of TNF-α, IL-1β, and PGE_2 observed in the EPD control group were significantly reduced by LRC administration (Figure 6). Moreover, histopathological analyses revealed that the increased numbers of infiltrated immune cells from ligation placement were diminished in the LRC administration groups (Table 2). In LPS-stimulated RAW 264.7 cells, LRC treatment significantly reduced the production of inflammatory mediators including PGE_2, IL-1β, TNF-α, and MCP-1 (Figure 1). Furthermore, our results demonstrated that the anti-inflammatory effects of LRC were mediated by the inhibition of MAPK and NF-κB signaling pathways (Figure 2). These findings suggest that LRC has a potent anti-inflammatory role in periodontitis by inhibiting the expression of inflammatory mediators.

As discussed above, the loss of connective tissue and alveolar bone is a significant hall mark of periodontitis progression. During the inflammatory process, the degradation of periodontal extracellular matrix is mediated by proteolytic enzymes such as MMPs, elastase, and tryptase from host cells. In particular, MMP-8 has been reported to have elevated expression levels in patients with periodontal disease and has also been found to be associated with pocket depth [11,49]. In the present study, we found that the increased gingival level of MMP-8 by ligation placement was significantly reduced by LRC administration (Figure 8). Furthermore, the decreased collagen-occupied area observed in the EPD group was significantly restored by LRC administration (Table 2).

Alveolar bone destruction is mediated by osteoclasts, the principal bone resorptive cell. The binding of RANKL on osteoblasts to RANK, a receptor expressed in osteoclast precursor cells, induces osteoclast differentiation and activation, leading to bone loss. This interaction between RANKL and RANK can be abrogated by OPG, a decoy receptor of RANKL. In an experimental ligature-induced periodontitis model, subcutaneous injection of OPG blocked the alveolar bone loss [50]. As a key regulator of osteoclastogenesis and bone resorption, the RANKL/OPG ratio is considered as a biomarker of periodontitis [39]. In our experiment, the increase in RANKL/OPG mRNA expression ratio by ligature placement in the EPD group was significantly reduced by LRC administration (Figure 9). Moreover, the increased numbers and occupied regions of osteoclasts on the alveolar bone surface by ligation placement were also significantly reduced by LRC administration (Table 3). Accordingly, the recovery of alveolar bone loss scores by LRC administration was also observed (Figure 4). Consistent with these findings, LRC inhibited RANKL-induced osteoclastogenesis in RAW 264.7 cells in a previous report [25]. Taken together, these findings suggest that LRC has potential preventive effects against bone loss associated with osteoporosis and periodontitis by inhibiting osteoclastogenesis and regulating RANKL/OPG expression. Further studies that directly evaluate alveolar bone loss and recovery using radiographic imaging will enable a more accurate assessment of LRC's effects on periodontal destruction.

During periodontal pathogenesis, exposure to proinflammatory cytokines such as IL-1, IL-6, and TNF-α stimulates RANKL expression in osteoblasts [51]. In a rat ligature model, the exogenous application of cytokines including IL-1β and TNF-α not only intensified inflammation, but also accelerated alveolar bone destruction [52,53]. Moreover, exogenously applied PGE_2 increased alveolar bone resorption by increasing osteoclasts in the periodontal tissue of rats [54]. Conversely, the local injection of chemical inhibitors targeting IL-1, TNF-α, and PGE_2 decreased the resorption of alveolar bone in chronic periodontal disease models [55,56]. Therefore, these findings indicate that the inhibitory effects of LRC on the expression of inflammatory mediators can contribute to the prevention of alveolar bone loss.

Recently, oxidative stress has been increasingly recognized as a key factor in the pathogenesis of periodontitis. After initiation of the host response against pathogenic bacteria, PMNs are the most abundant inflammatory cells in the periodontal pocket and gingival cervices serving as a primary source of ROS [14]. ROS, generated by the response of neutrophils to periodontal pathogens, can be effective in killing microbes but also induce cytotoxicity to host cells through lipid peroxidation, protein damage, and DNA damage [14]. Studies have demonstrated that the peripheral neutrophils from periodontitis patients produce more ROS compared to those from healthy individuals [57]. This suggests that hyperactive neutrophils in periodontitis enhance oxidative burst and ROS release, exacerbating tissue inflammation and destruction. Consequently, levels of MDA, a well-established byproduct of lipid peroxidation, were found to be significantly higher in the gingival crevicular fluid, saliva, and serum of periodontitis patients compared to healthy individuals [14,58]. Therefore, studies validating the potential of MDA levels as a clinical biomarker of oxidative stress in periodontitis patients have been conducted [58,59]. In the present study, we showed that the activity of MPO, an enzyme produced by PMNs, was elevated in gingival tissue by ligation placement as previously reported [29,60]. The influx of PMNs in gingival tissues can be determined by assessing MPO activity. We found that LRC administration significantly reduced the elevated MPO activity observed in the EPD group (Figure 5b). Additionally, our results showed that LRC significantly inhibited the ligation-induced increase in MDA levels in gingival tissues (Figure 7b). Similarly, it has been reported that LRC application exerted an MDA-lowering effect against lipid peroxidation in mice skin [61]. Moreover, the antioxidant properties of LRC have been demonstrated in various experimental disease models including ethanol-induced gastric ulcer mice and dopaminergic neuronal cell death [24,62].

Kukoamines, phenolic alkaloids, are recognized as the predominant constituents of LRC and are used as markers for quality assessment of LRC [63]. Therefore, to ensure the quality of raw materials, we confirmed the kukoamine B content in the LRC extract using HPLC analysis (Supplementary Figure S1) before conducting efficacy tests. Additionally, as key bioactive constituents of LRC, kukoamines have been studied for their biological activities. Notably, kukoamines have been highlighted for their anti-inflammatory properties in various experimental models. For example, kukoamine A has demonstrated an anti-inflammatory effect in LPS-stimulated RAW 264.7 cells, osteoarthritis mice, and radiation-induced neuroinflammation in rats [64–66]. Kukoamine B has also shown an anti-inflammatory effect in LPS-induced septic mice and high fat and fructose fed mice [67,68]. In particular, kukoamine B has been considered as a potential candidate for treating sepsis and approved for clinical trials in China [69]. It neutralized LPS and CpG, which are major pathogenic molecules causing sepsis, by binding to them with high affinity, thereby preventing their interaction with toll-like receptors on macrophages [70,71]. In addition to anti-inflammatory effects, trans-*N*-caffeoyltyramine and kukoamine have shown protective effects against hydrogen peroxide-induced oxidative stress [72,73]. Moreover, kukoamine A and B exerted antioxidant and cytoprotective effects in Fenton-induced damage through radical scavenging effects [74]. Furthermore, scopolin and kukoamine A and B have shown the regulatory effects on the osteoblast or osteoclast differentiation in ovariectomized mice models [75–77]. Therefore, this evidence suggests that LRC can effectively improve periodontitis based on the biological activities of its bioactive constituents, particularly kukoamines.

5. Conclusions

In summary, our investigation revealed that LRC effectively inhibited ligation-induced anaerobic bacteria proliferation, immune cell infiltration, production of inflammatory mediators, and oxidative stress. Moreover, LRC administration ameliorated the degradation of connective tissue and the loss of alveolar bone associated with periodontitis. These findings suggest that LRC is a promising medicinal herb for alleviating periodontitis and related alveolar bone loss through its antimicrobial, anti-inflammatory, and antioxidant properties, and it could potentially be developed as a functional food for preventive measure in dental health.

Supplementary Materials: The following supporting information can be downloaded at https://www.mdpi.com/article/10.3390/antiox13111332/s1, Figure S1: Identification of kukoamine B in LRC using high performance liquid chromatography (HPLC) analysis; Figure S2: Effects of LRC on the viability of HaCaT, HDFn, and RAW 264.7 cells.

Author Contributions: Conceptualization, S.-K.K.; methodology, J.L., H.C. and S.-K.K.; validation, J.L., H.C. and J.-K.K.; formal analysis, K.-H.J.; investigation, J.Y., H.S., Y.-S.K., K.-H.J., J.-K.K. and S.-K.K.; writing—original draft preparation, J.Y., H.S. and J.-K.K.; writing—review and editing, J.-K.K.; supervision, S.-K.K. All authors have read and agreed to the published version of the manuscript.

Funding: This research received no external funding.

Institutional Review Board Statement: All animal experiments were conducted according to the national regulation of the usage and welfare of laboratory animals and approved by the Institutional Animal Care and Use Committee in Daegu Haany University (Approval No. DHU2022-101).

Informed Consent Statement: Not applicable.

Data Availability Statement: The raw data supporting the conclusions of this article will be made available by the authors on request.

Conflicts of Interest: J.L. and H.C. are employed by GAPI BIO Co., Ltd.; however, in this research, they were only involved in analysis of the raw materials to a limited extent. Remaining authors declare no conflicts of interest.

References

1. Meyle, J.; Chapple, I. Molecular aspects of the pathogenesis of periodontitis. *Periodontology 2000* **2015**, *69*, 7–17. [CrossRef] [PubMed]
2. Janakiram, C.; Dye, B.A. A public health approach for prevention of periodontal disease. *Periodontology 2000* **2020**, *84*, 202–214. [CrossRef] [PubMed]
3. Durham, J.; Fraser, H.M.; McCracken, G.I.; Stone, K.M.; John, M.T.; Preshaw, P.M. Impact of periodontitis on oral health-related quality of life. *J. Dent.* **2013**, *41*, 370–376. [CrossRef] [PubMed]
4. Buchwald, S.; Kocher, T.; Biffar, R.; Harb, A.; Holtfreter, B.; Meisel, P. Tooth loss and periodontitis by socio-economic status and inflammation in a longitudinal population-based study. *J. Clin. Periodontol.* **2013**, *40*, 203–211. [CrossRef] [PubMed]
5. Preshaw, P.M.; Bissett, S.M. Periodontitis and diabetes. *Br. Dent. J.* **2019**, *227*, 577–584. [CrossRef] [PubMed]
6. Sanz, M.; Marco Del Castillo, A.; Jepsen, S.; Gonzalez-Juanatey, J.R.; D'Aiuto, F.; Bouchard, P.; Chapple, I.; Dietrich, T.; Gotsman, I.; Graziani, F.; et al. Periodontitis and cardiovascular diseases: Consensus report. *J. Clin. Periodontol.* **2020**, *47*, 268–288. [CrossRef]
7. Eid Abdelmagyd, H.A.; Ram Shetty, D.S.; Musa Musleh Al-Ahmari, D.M. Herbal medicine as adjunct in periodontal therapies- A review of clinical trials in past decade. *J. Oral Biol. Craniofac. Res.* **2019**, *9*, 212–217. [CrossRef]
8. Lopez-Valverde, N.; Lopez-Valverde, A.; Montero, J.; Rodriguez, C.; Macedo de Sousa, B.; Aragoneses, J.M. Antioxidant, anti-inflammatory and antimicrobial activity of natural products in periodontal disease: A comprehensive review. *Front. Bioeng. Biotechnol.* **2023**, *11*, 1226907. [CrossRef]
9. Hajishengallis, G. Periodontitis: From microbial immune subversion to systemic inflammation. *Nat. Rev. Immunol.* **2015**, *15*, 30–44. [CrossRef]
10. Yucel-Lindberg, T.; Bage, T. Inflammatory mediators in the pathogenesis of periodontitis. *Expert Rev. Mol. Med.* **2013**, *15*, e7. [CrossRef]

11. de Morais, E.F.; Pinheiro, J.C.; Leite, R.B.; Santos, P.P.A.; Barboza, C.A.G.; Freitas, R.A. Matrix metalloproteinase-8 levels in periodontal disease patients: A systematic review. *J. Periodontal Res.* **2018**, *53*, 156–163. [CrossRef] [PubMed]
12. Graves, D. Cytokines that promote periodontal tissue destruction. *J. Periodontol.* **2008**, *79*, 1585–1591. [CrossRef] [PubMed]
13. Miyasaki, K.T. The neutrophil: Mechanisms of controlling periodontal bacteria. *J. Periodontol.* **1991**, *62*, 761–774. [CrossRef] [PubMed]
14. Wang, Y.; Andrukhov, O.; Rausch-Fan, X. Oxidative Stress and Antioxidant System in Periodontitis. *Front. Physiol.* **2017**, *8*, 910. [CrossRef]
15. Chapple, I.L.; Matthews, J.B. The role of reactive oxygen and antioxidant species in periodontal tissue destruction. *Periodontology 2000* **2007**, *43*, 160–232. [CrossRef]
16. Sczepanik, F.S.C.; Grossi, M.L.; Casati, M.; Goldberg, M.; Glogauer, M.; Fine, N.; Tenenbaum, H.C. Periodontitis is an inflammatory disease of oxidative stress: We should treat it that way. *Periodontology 2000* **2020**, *84*, 45–68. [CrossRef]
17. Jiang, T.T.; Li, J.C. Review on the systems biology research of Yin-deficiency-heat syndrome in traditional Chinese medicine. *Anat. Rec.* **2023**, *306*, 2939–2944. [CrossRef]
18. Wu, Y.; Liu, M.; He, X.; Zhou, H.; Wei, J.; Li, H.; Yuan, Q.; Zuo, Y.; Zhao, L.; Xie, Y. A breakthrough in periodontitis treatment: Revealing the pharmacodynamic substances and mechanisms of Kouqiangjie formula. *J. Ethnopharmacol.* **2024**, *323*, 117738. [CrossRef]
19. Jin, W.; Li, L.; Ai, H.; Jin, Z.; Zuo, Y. Efficacy and Safety of Traditional Chinese Medicine Based on the Method of "Nourishing Kidney and Clearing Heat" as Adjuvant in the Treatment of Diabetes Mellitus Patients with Periodontitis: A Systematic Review and Meta-Analysis. *Evid. Based Complement. Alternat. Med.* **2022**, *2022*, 3853303. [CrossRef]
20. Zee, K.Y.; Chan, P.S.; Ho, J.C.S.; Lai, S.M.L.; Corbet, E.F.; Leung, W.K. Adjunctive use of modified Yunu-Jian in the non-surgical treatment of male smokers with chronic periodontitis: A randomized double-blind, placebo-controlled clinical trial. *Chin. Med.* **2016**, *11*, 40. [CrossRef]
21. Cho, H.; Lee, D.H.; Jeong, D.H.; Jang, J.H.; Son, Y.; Lee, S.Y.; Kim, H.J. Study on Betaine and Growth Characteristics of *Lycium chinense* Mill. in Different Cultivation Environments in South Korea. *Plants* **2024**, *13*, 2316. [CrossRef] [PubMed]
22. Jeong, J.C.; Kim, S.J.; Kim, Y.K.; Kwon, C.H.; Kim, K.H. *Lycii cortex radicis* extract inhibits glioma tumor growth in vitro and in vivo through downregulation of the Akt/ERK pathway. *Oncol. Rep.* **2012**, *27*, 1467–1474. [CrossRef] [PubMed]
23. Son, R.H.; Kim, M.I.; Kim, H.M.; Guo, S.; Lee, D.H.; Lim, G.M.; Kim, S.M.; Kim, J.Y.; Kim, C.Y. Potential of Lycii Radicis Cortex as an Ameliorative Agent for Skeletal Muscle Atrophy. *Pharmaceuticals* **2024**, *17*, 462. [CrossRef] [PubMed]
24. Chen, H.; Olatunji, O.J.; Zhou, Y. Anti-oxidative, anti-secretory and anti-inflammatory activities of the extract from the root bark of *Lycium chinense* (Cortex Lycii) against gastric ulcer in mice. *J. Nat. Med.* **2016**, *70*, 610–619. [CrossRef] [PubMed]
25. Kim, J.H.; Kim, E.Y.; Lee, B.; Min, J.H.; Song, D.U.; Lim, J.M.; Eom, J.W.; Yeom, M.; Jung, H.S.; Sohn, Y. The effects of Lycii Radicis Cortex on RANKL-induced osteoclast differentiation and activation in RAW 264.7 cells. *Int. J. Mol. Med.* **2016**, *37*, 649–658. [CrossRef]
26. Park, E.; Jin, H.S.; Cho, D.Y.; Kim, J.; Kim, M.C.; Choi, C.W.; Jin, Y.; Lee, J.W.; Park, J.H.; Chung, Y.S.; et al. The effect of *Lycii Radicis Cortex* extract on bone formation in vitro and in vivo. *Molecules* **2014**, *19*, 19594–19609. [CrossRef]
27. Lee, B.; Hong, S.; Kim, M.; Kim, E.Y.; Park, H.J.; Jung, H.S.; Kim, J.H.; Sohn, Y. *Lycii radicis* cortex inhibits glucocorticoid-induced bone loss by downregulating Runx2 and BMP-2 expression. *Int. J. Mol. Med.* **2021**, *48*, 155. [CrossRef]
28. Livak, K.J.; Schmittgen, T.D. Analysis of relative gene expression data using real-time quantitative PCR and the $2^{-\Delta\Delta CT}$ Method. *Methods* **2001**, *25*, 402–408. [CrossRef]
29. Park, S.-I.; Kang, S.-J.; Han, C.-H.; Kim, J.-W.; Song, C.-H.; Lee, S.-N.; Ku, S.-K.; Lee, Y.-J. The Effects of Topical Application of Polycal (a 2:98 (g/g) Mixture of Polycan and Calcium Gluconate) on Experimental Periodontitis and Alveolar Bone Loss in Rats. *Molecules* **2016**, *21*, 527. [CrossRef]
30. Kim, T.G.; Park, M.-R.; Ku, S.-K.; Heo, S.-M.; Kim, J.-L. Effects of *Moringa oleifera* L. and *Eucommia ulmoides* Oliver Mixed Formula on Ligation-Induced Experimental Periodontitis and Alveolar Bone Loss in Rats. *J. Korean Soc. Food Sci. Nutr.* **2022**, *51*, 765–779. [CrossRef]
31. Samejima, Y.; Ebisu, S.; Okada, H. Effect of infection with Eikenella corrodens on the progression of ligature-induced periodontitis in rats. *J. Periodontal Res.* **1990**, *25*, 308–315. [CrossRef] [PubMed]
32. Botelho, M.A.; Rao, V.S.; Carvalho, C.B.; Bezerra-Filho, J.G.; Fonseca, S.G.; Vale, M.L.; Montenegro, D.; Cunha, F.; Ribeiro, R.A.; Brito, G.A. *Lippia sidoides* and *Myracrodruon urundeuva* gel prevents alveolar bone resorption in experimental periodontitis in rats. *J. Ethnopharmacol.* **2007**, *113*, 471–478. [CrossRef] [PubMed]
33. Bradley, P.P.; Christensen, R.D.; Rothstein, G. Cellular and extracellular myeloperoxidase in pyogenic inflammation. *Blood* **1982**, *60*, 618–622. [CrossRef] [PubMed]
34. Cuzzocrea, S.; Zingarelli, B.; Hake, P.; Salzman, A.L.; Szabo, C. Antiinflammatory effects of mercaptoethylguanidine, a combined inhibitor of nitric oxide synthase and peroxynitrite scavenger, in carrageenan-induced models of inflammation. *Free Radic. Biol. Med.* **1998**, *24*, 450–459. [CrossRef] [PubMed]
35. Menezes, A.M.; Rocha, F.A.; Chaves, H.V.; Carvalho, C.B.; Ribeiro, R.A.; Brito, G.A. Effect of sodium alendronate on alveolar bone resorption in experimental periodontitis in rats. *J. Periodontol.* **2005**, *76*, 1901–1909. [CrossRef]
36. Zhang, G.; Ghosh, S. Toll-like receptor-mediated NF-κB activation: A phylogenetically conserved paradigm in innate immunity. *J. Clin. Investig.* **2001**, *107*, 13–19. [CrossRef]

37. Yang, F.; Tang, E.; Guan, K.; Wang, C.Y. IKKβ plays an essential role in the phosphorylation of RelA/p65 on serine 536 induced by lipopolysaccharide. *J. Immunol.* **2003**, *170*, 5630–5635. [CrossRef]
38. Toczewska, J.; Konopka, T.; Zalewska, A.; Maciejczyk, M. Nitrosative Stress Biomarkers in the Non-Stimulated and Stimulated Saliva, as well as Gingival Crevicular Fluid of Patients with Periodontitis: Review and Clinical Study. *Antioxidants* **2020**, *9*, 259. [CrossRef]
39. Belibasakis, G.N.; Bostanci, N. The RANKL-OPG system in clinical periodontology. *J. Clin. Periodontol.* **2012**, *39*, 239–248. [CrossRef]
40. Tomina, D.C.; Petrutiu, S.A.; Dinu, C.M.; Crisan, B.; Cighi, V.S.; Ratiu, I.A. Comparative Testing of Two Ligature-Induced Periodontitis Models in Rats: A Clinical, Histological and Biochemical Study. *Biology* **2022**, *11*, 634. [CrossRef]
41. Ionel, A.; Lucaciu, O.; Moga, M.; Buhatel, D.; Ilea, A.; Tabaran, F.; Catoi, C.; Berce, C.; Toader, S.; Campian, R.S. Periodontal disease induced in Wistar rats— experimental study. *Hum. Vet. Med. Bioflux* **2015**, *7*, 90–95.
42. Azeez, S.H.; Gaphor, S.M.; Sha, A.M.; Garib, B.T. Effect of Pistacia atlantica subsp. kurdica Gum in Experimental Periodontitis Induced in Wistar Rats by Utilization of Osteoclastogenic Bone Markers. *Molecules* **2020**, *25*, 5819. [CrossRef] [PubMed]
43. Gimenez-Siurana, A.; Gomez Garcia, F.; Pagan Bernabeu, A.; Lozano-Perez, A.A.; Aznar-Cervantes, S.D.; Cenis, J.L.; Lopez-Jornet, P. Chemoprevention of Experimental Periodontitis in Diabetic Rats with Silk Fibroin Nanoparticles Loaded with Resveratrol. *Antioxidants* **2020**, *9*, 85. [CrossRef] [PubMed]
44. Feres, M.; Figueiredo, L.C.; Soares, G.M.; Faveri, M. Systemic antibiotics in the treatment of periodontitis. *Periodontology 2000* **2015**, *67*, 131–186. [CrossRef] [PubMed]
45. Elashiry, M.; Morandini, A.C.; Cornelius Timothius, C.J.; Ghaly, M.; Cutler, C.W. Selective Antimicrobial Therapies for Periodontitis: Win the "Battle and the War". *Int. J. Mol. Sci.* **2021**, *22*, 6459. [CrossRef]
46. Lee, D.G.; Jung, H.J.; Woo, E.R. Antimicrobial property of (+)-lyoniresinol-3α-O-β-D-glucopyranoside isolated from the root bark of *Lycium chinense* Miller against human pathogenic microorganisms. *Arch. Pharm. Res.* **2005**, *28*, 1031–1036. [CrossRef]
47. Pfeilschifter, J.; Chenu, C.; Bird, A.; Mundy, G.R.; Roodman, G.D. Interleukin-1 and Tumor Necrosis Factor Stimulate the Formation of Human Osteoclastlike Cells In Vitro. *J. Bone Miner. Res.* **1989**, *4*, 113–118. [CrossRef]
48. Lader, C.S.; Flanagan, A.M. Prostaglandin E_2, interleukin 1α, and tumor necrosis factor-α increase human osteoclast formation and bone resorption in vitro. *Endocrinology* **1998**, *139*, 3157–3164. [CrossRef]
49. Kraft-Neumarker, M.; Lorenz, K.; Koch, R.; Hoffmann, T.; Mantyla, P.; Sorsa, T.; Netuschil, L. Full-mouth profile of active MMP-8 in periodontitis patients. *J. Periodontal Res.* **2012**, *47*, 121–128. [CrossRef]
50. Jin, Q.; Cirelli, J.A.; Park, C.H.; Sugai, J.V.; Taba, M., Jr.; Kostenuik, P.J.; Giannobile, W.V. RANKL inhibition through osteoprotegerin blocks bone loss in experimental periodontitis. *J. Periodontol.* **2007**, *78*, 1300–1308. [CrossRef]
51. AlQranei, M.S.; Chellaiah, M.A. Osteoclastogenesis in periodontal diseases: Possible mediators and mechanisms. *J. Oral Biosci.* **2020**, *62*, 123–130. [CrossRef] [PubMed]
52. Koide, M.; Suda, S.; Saitoh, S.; Ofuji, Y.; Suzuki, T.; Yoshie, H.; Takai, M.; Ono, Y.; Taniguchi, Y.; Hara, K. *In vivo* administration of IL-1β accelerates silk ligature-induced alveolar bone resorption in rats. *J. Oral Pathol. Med.* **1995**, *24*, 420–434. [CrossRef] [PubMed]
53. Gaspersic, R.; Stiblar-Martincic, D.; Osredkar, J.; Skaleric, U. Influence of subcutaneous administration of recombinant TNF-α on ligature-induced periodontitis in rats. *J. Periodontal Res.* **2003**, *38*, 198–203. [CrossRef] [PubMed]
54. Miyauchi, M.; Ijuhin, N.; Nikai, H.; Takata, T.; Ito, H.; Ogawa, I. Effect of exogenously applied prostaglandin E_2 on alveolar bone losss—Histometric analysis. *J. Periodontol.* **1992**, *63*, 405–411. [CrossRef]
55. Williams, R.C.; Jeffcoat, M.K.; Kaplan, M.L.; Goldhaber, P.; Johnson, H.G.; Wechter, W.J. Flurbiprofen: A potent inhibitor of alveolar bone resorption in beagles. *Science* **1985**, *227*, 640–642. [CrossRef]
56. Assuma, R.; Oates, T.; Cochran, D.; Amar, S.; Graves, D.T. IL-1 and TNF antagonists inhibit the inflammatory response and bone loss in experimental periodontitis. *J. Immunol.* **1998**, *160*, 403–409. [CrossRef]
57. Matthews, J.B.; Wright, H.J.; Roberts, A.; Cooper, P.R.; Chapple, I.L. Hyperactivity and reactivity of peripheral blood neutrophils in chronic periodontitis. *Clin. Exp. Immunol.* **2007**, *147*, 255–264. [CrossRef]
58. Mohideen, K.; Chandrasekar, K.; Ramsridhar, S.; Rajkumar, C.; Ghosh, S.; Dhungel, S. Assessment of Oxidative Stress by the Estimation of Lipid Peroxidation Marker Malondialdehyde (MDA) in Patients with Chronic Periodontitis: A Systematic Review and Meta-Analysis. *Int. J. Dent.* **2023**, *2023*, 6014706. [CrossRef]
59. Cherian, D.A.; Peter, T.; Narayanan, A.; Madhavan, S.S.; Achammada, S.; Vynat, G.P. Malondialdehyde as a Marker of Oxidative Stress in Periodontitis Patients. *J. Pharm. Bioallied Sci.* **2019**, *11*, S297–S300. [CrossRef]
60. Kang, S.J.; Lee, E.K.; Han, C.H.; Lee, B.H.; Lee, Y.J.; Ku, S.K. Inhibitory effects of Persicariae Rhizoma aqueous extracts on experimental periodontitis and alveolar bone loss in Sprague-Dawley rats. *Exp. Ther. Med.* **2016**, *12*, 1563–1571. [CrossRef]
61. Ahn, B.Y.; Gwak, J.S.; Ryu, S.H.; Moon, G.S.; Choi, D.S.; Park, S.H.; Han, J.H. Protective effect of water extract of Lycii Cortex Radicis on lipid peroxidation of rat skin exposed to ultraviolet B radiation. *Appl. Biol. Chem.* **2002**, *45*, 218–222.
62. Kim, H.G.; Oh, M.S. Protective Effect of Lycii Radicis Cortex against 6-Hydroxydopamine-Induced Dopaminergic Neuronal Cell Death. *J. Food Biochem.* **2015**, *39*, 281–288. [CrossRef]
63. Li, Y.-Y.; Di, R.; Baibado, J.T.; Cheng, Y.-S.; Huang, Y.-Q.; Sun, H.; Cheung, H.-Y. Identification of kukoamines as the novel markers for quality assessment of Lycii Cortex. *Food Res. Int.* **2014**, *55*, 373–380. [CrossRef]

64. Wang, L.; Wang, P.; Wang, D.; Tao, M.; Xu, W.; Olatunji, O.J. Anti-Inflammatory Activities of Kukoamine A From the Root Bark of *Lycium chinense* Miller. *Nat. Prod. Commun.* **2020**, *15*, 1934578X20912088. [CrossRef]
65. Sun, J.; Zhang, Y.; Wang, C.; Ruan, Q. Kukoamine A protects mice against osteoarthritis by inhibiting chondrocyte inflammation and ferroptosis via SIRT1/GPX4 signaling pathway. *Life Sci.* **2023**, *332*, 122117. [CrossRef]
66. Zhang, Y.; Gao, L.; Cheng, Z.; Cai, J.; Niu, Y.; Meng, W.; Zhao, Q. Kukoamine A Prevents Radiation-Induced Neuroinflammation and Preserves Hippocampal Neurogenesis in Rats by Inhibiting Activation of NF-κB and AP-1. *Neurotox. Res.* **2017**, *31*, 259–268. [CrossRef]
67. Qin, W.T.; Wang, X.; Shen, W.C.; Sun, B.W. A novel role of kukoamine B: Inhibition of the inflammatory response in the livers of lipopolysaccharide-induced septic mice via its unique property of combining with lipopolysaccharide. *Exp. Ther. Med.* **2015**, *9*, 725–732. [CrossRef]
68. Zhao, Q.; Li, L.; Zhu, Y.; Hou, D.; Li, Y.; Guo, X.; Wang, Y.; Olatunji, O.J.; Wan, P.; Gong, K. Kukoamine B Ameliorate Insulin Resistance, Oxidative Stress, Inflammation and Other Metabolic Abnormalities in High-Fat/High-Fructose-Fed Rats. *Diabetes Metab. Syndr. Obes.* **2020**, *13*, 1843–1853. [CrossRef]
69. Wang, H.; Wang, T.; Hu, X.; Deng, C.; Jiang, J.; Qin, H.; Dong, K.; Chen, S.; Jin, C.; Zhao, Q.; et al. Fixed dosing of kukoamine B in sepsis patients: Results from population pharmacokinetic modelling and simulation. *Br. J. Clin. Pharmacol.* **2022**, *88*, 4111–4120. [CrossRef]
70. Liu, X.; Zheng, X.; Wang, N.; Cao, H.; Lu, Y.; Long, Y.; Zhao, K.; Zhou, H.; Zheng, J. Kukoamine B, a novel dual inhibitor of LPS and CpG DNA, is a potential candidate for sepsis treatment. *Br. J. Pharmacol.* **2011**, *162*, 1274–1290. [CrossRef]
71. Liu, X.; Zheng, X.; Long, Y.; Cao, H.; Wang, N.; Lu, Y.; Zhao, K.; Zhou, H.; Zheng, J. Dual targets guided screening and isolation of Kukoamine B as a novel natural anti-sepsis agent from traditional Chinese herb Cortex lycii. *Int. Immunopharmacol.* **2011**, *11*, 110–120. [CrossRef] [PubMed]
72. Olatunji, O.J.; Chen, H.; Zhou, Y. Neuroprotective effect of trans-*N*-caffeoyltyramine from *Lycium chinense* against H_2O_2 induced cytotoxicity in PC12 cells by attenuating oxidative stress. *Biomed. Pharmacother.* **2017**, *93*, 895–902. [CrossRef] [PubMed]
73. Li, Y.-Y.; Hu, S.; Huang, Y.-Q.; Han, Y.; Cheung, H.-Y. Preventing H_2O_2-induced toxicity in primary cerebellar granule neurons via activating the PI3-K/Akt/GSK3β pathway by kukoamine from Lycii Cortex. *J. Funct. Foods* **2015**, *17*, 709–721. [CrossRef]
74. Li, X.; Lin, J.; Chen, B.; Xie, H.; Chen, D. Antioxidant and Cytoprotective Effects of Kukoamines A and B: Comparison and Positional Isomeric Effect. *Molecules* **2018**, *23*, 973. [CrossRef] [PubMed]
75. Park, E.; Kim, J.; Kim, M.C.; Yeo, S.; Kim, J.; Park, S.; Jo, M.; Choi, C.W.; Jin, H.S.; Lee, S.W.; et al. Anti-Osteoporotic Effects of Kukoamine B Isolated from Lycii Radicis Cortex Extract on Osteoblast and Osteoclast Cells and Ovariectomized Osteoporosis Model Mice. *Int. J. Mol. Sci.* **2019**, *20*, 2784. [CrossRef]
76. Park, E.; Kim, J.; Jin, H.S.; Choi, C.W.; Choi, T.H.; Choi, S.; Huh, D.; Jeong, S.Y. Scopolin Attenuates Osteoporotic Bone Loss in Ovariectomized Mice. *Nutrients* **2020**, *12*, 3565. [CrossRef]
77. Luo, L.; Guan, Z.; Jin, X.; Guan, Z.; Jiang, Y. Identification of kukoamine a as an anti-osteoporosis drug target using network pharmacology and experiment verification. *Mol. Med.* **2023**, *29*, 36. [CrossRef]

antioxidants

MDPI

Article

Exploratory Role of Flavonoids on Metabolic Dysfunction-Associated Steatotic Liver Disease (MASLD) in a South Italian Cohort

Caterina Bonfiglio [1,*,†], Rossella Tatoli [1,*,†], Rossella Donghia [1], Davide Guido [1] and Gianluigi Giannelli [2]

[1] Unit of Data Science, National Institute of Gastroenterology—IRCCS "Saverio de Bellis", Castellana Grotte, 70013 Bari, Italy; rossella.donghia@irccsdebellis.it (R.D.); davide.guido@irccsdebellis.it (D.G.)

[2] Scientific Direction, National Institute of Gastroenterology—IRCCS "Saverio de Bellis", Castellana Grotte, 70013 Bari, Italy; gianluigi.giannelli@irccsdebellis.it

* Correspondence: catia.bonfiglio@irccsdebellis.it (C.B.); rossella.tatoli@irccsdebellis.it (R.T.)

† These authors contributed equally to this work.

Abstract: Background: Metabolic dysfunction-associated steatotic liver disease (MASLD) is the most recent definition for steatotic liver disease associated with metabolic syndrome. The results of recent metabolic and observational studies suggest a potential beneficial effect of food-derived flavonoids in some chronic diseases, including MASLD. The study aims to evaluate the protective role of diet flavonoids in subjects with and without MASLD belonging to a cohort living in the South of Italy. Methods: The study cohort comprised 1297 participants assessed in the NUTRIHEP cohort (2015–2018), divided into two groups, based on presence or absence of MASLD. Results: The results indicated statistically significant flavonoid consumption, showing a protective role against MASLD, at an optimal concentration of 165 mg/day, with an OR value of 0.63, (p = 0.001, 95% C.I.: 0.47; 0.83 t). The OR remained almost unchanged when the intake increased from 165 mg per day to 185 mg per day. Conclusions: In conclusion, our study results show a protective role of flavonoids against MASLD. Consuming only 165 mg of flavonoids daily can activate this protective function, reducing the risk of MASLD.

Keywords: MASLD; flavonoids; Mediterranean diet

1. Introduction

Metabolic dysfunction-associated steatotic liver disease (MASLD) is the most recent definition for steatotic liver disease associated with metabolic syndrome [1].

In 1980, Ludwing et al. first reported the existence of a liver condition similar to alcoholic hepatitis diagnosed by ultrasound in moderately obese subjects who did not consume significant amounts of alcohol [2]. In 2007, Farrel et al. proposed the working definition of NAFLD for this condition [3]. The close association between NAFLD and metabolic syndrome soon led to the need for a change in the nomenclature [4]. MASLD was introduced in June 2023 with a multi-society Delphi consensus statement on a new fatty liver disease nomenclature that led to the final withdrawal of the term NAFLD [5]. Recent studies have suggested an association between NAFLD and metabolic syndrome [6].

MASLD is characterized by fat accumulation in the liver detected by imaging or biopsy and is observed in individuals with little or no alcohol consumption, generally affected by obesity, type 2 diabetes mellitus (T2DM), dyslipidemia, and/or hypertension [7]. Currently, the pathogenesis of MASLD is not well defined. Probably, a role is played by insulin resistance and interactions of genetic and environmental factors, as well as metabolic dysfunction [4]. Metabolic dysfunction-associated steatohepatitis (MASH), the more severe form of MASLD, is histologically defined by the presence of lobular inflammation and hepatocyte ballooning, and is associated with a greater risk of fibrosis progression [8].

MASLD is the most common cause of chronic liver disease and the leading cause of liver-related morbidity and mortality [1]. Cardiovascular diseases are the leading cause of death for patients with MASLD [9]. NAFLD affects about 30% of the world's adult population, and the prevalence has risen from 22% to 37% in 20 years [10,11].

The prevalence of MSLD increases in parallel with the prevalence of obesity and related diseases, confirming the close connection between the two conditions [1]. The incidence of MASLD and MASH is strongly related to a sedentary lifestyle and excess dietary energy intake [12].

There is currently no Food and Drug Administration (FDA)-approved pharmacological therapy for MASLD and MASH, but a lifestyle intervention approach is suggested.

A randomized controlled trial showed the positive effect of a lifestyle modification program in patients with nonalcoholic fatty liver disease [13].

A healthy diet can significantly improve MASLD, based on the Mediterranean diet (MedDiet) model and associated with moderate physical activities [1,14].

The MedDiet owes its protective and preventive effects against various chronic diseases to the bioactive components in its foods [14]. These include vegetables and fruits, rich in fiber and polyphenols, fish, nuts, and extra virgin oil, rich in monounsaturated and polyunsaturated fatty acids [15]. Flavonoids are bioactive polyphenols ubiquitously present in plants [16,17].

The results of recent metabolic and observational studies suggest a potential beneficial effect of food-derived flavonoids in some chronic diseases, including NAFLD [16–18]. Many studies have been conducted in animal models to investigate the impact of flavonoids on fatty liver disease, but very few have been conducted in humans [18–20].

The scientific literature lacks studies evaluating the link between flavonoids consumption and MASLD.

The study aims to evaluate the protective role of diet flavonoids in subjects with and without MASLD belonging to a cohort living in the South of Italy.

2. Materials and Methods

2.1. Study Population

The NUTRIHEP study was a cohort created in 2005–2006 using the medical records of general practitioners in Putignano for persons aged 18 years and over. The study design involved selecting a sample of the general population aged 18 years and over through a systematic random sampling procedure from the lists of GP registers. Instead of using census data, we used the GP registers because no significant differences were found between the age and gender distribution of the general population of Putignano and that recorded in the GP registers. In Italy, since it is a legal requirement that everyone should have a general practitioner, the lists of the general population at GP surgeries correspond to the complete census list [21].

Trained physicians and/or nutritionists interviewed the participants at baseline in 2005–2006 to gather information on sociodemographic characteristics, health status, personal history, and lifestyle factors. This included a history of tobacco use, food intake, educational level [22], work profile [23], and marital status. Weight and height were measured with the participants wearing underclothing and no shoes. Weights were recorded to the nearest 0.1 kg using an electronic balance (SECA©), while height was recorded to the nearest 1 cm using a wall-mounted stadiometer (SECA©). Blood pressure (BP) was measured following international guidelines [24], and the average of three measurements was calculated.

The European Prospective Investigation into Cancer and Nutrition (EPIC) food frequency questionnaire (FFQ) was used to document the usual food intake of participants at baseline [25,26]. Nutritionists conducted an in-person structured interview asking participants to report on their frequency of usual intake of 260 food items over the past year; they reported intakes per day, per week, or per year. They were also asked to estimate their portion sizes from photographs; questions were referred to the usual intake in the last year.

From 2015 to 2018, all subjects participating in Nutrihep were recalled for the first follow-up of this cohort study. A total of 1426 subjects responded, and the respondents were subjected to the same protocol as the first enrolment.

All participants signed informed consent acknowledgements after receiving full information about their medical data to be studied. In this paper, we consider the data collected during the follow-up.

Eligible subjects were found to number 1297 (90.9%) (MASLD No, 668 subjects and MASLD Yes, 629 subjects) out of 1426 total. Eighty participants who had not completed the dietary questionnaire and 49 with hepatic steatosis and Hcv+ or Alcoholic Fatty Liver Disease were excluded (Figure 1).

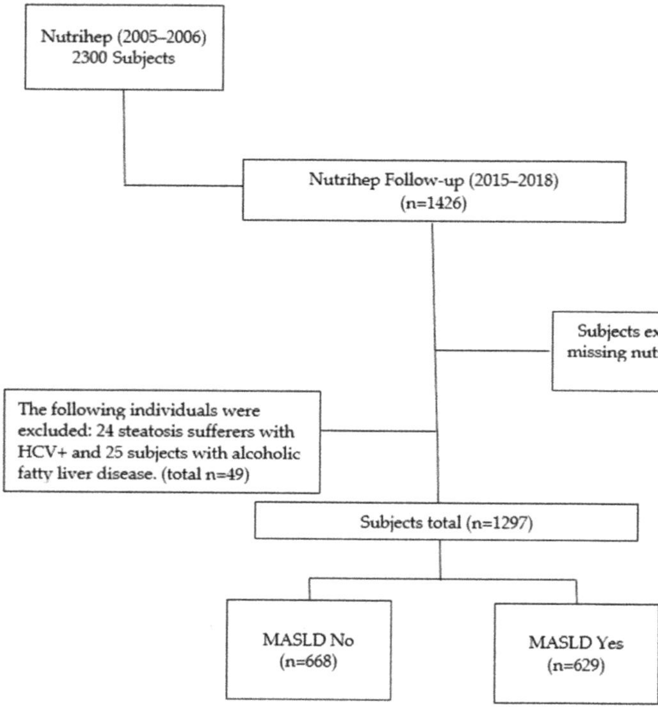

Figure 1. Flow chart.

The study was approved by the Ethical Committee of the Minister of Health (DDG-CE-792/2014, on 14 February 2014).

2.2. Outcome Assessment

As published in previous studies [1,27], the definition of MASLD was based on the presence of hepatic steatosis plus at least 1 of the following 5 conditions: (1) BMI > 25 kg/m^2 or waist circumference > 94 cm in men and >80 in women; (2) fasting serum glucose $\geq$ 100 mg/dL ($\geq$5.6 mmol/L), 2 h post-load glucose level $\geq$ 140 mg/dL ($\geq$7.8 mmol/L), HbA1c $\geq$ 5.7%, or in specific drug treatment; (3) blood pressure $\geq$ 130/85 mmHg or in specific drug treatment; (4) plasma triglycerides $\geq$ 150 mg/dL ($\geq$1.70 mmol/L) or in specific drug treatment; and (5) plasma HDL Cholesterol < 40 mg/dL (<1.0 mmol/L) for men and <50 mg/dL (<1.3 mmol/L) for women or in specific drug treatment.

Table S1 shows the distribution between MASLD No and MASLD Yes groups according to positive diagnostic criteria.

Furthermore, the definition of MASLD continued to limit alcohol intake (as previously for NAFLD) in the context of steatosis to an average daily intake of 20–50 g for women and 30–60 g for men [5].

Finally, other forms of liver disease coexisting with MASLD, such as HCV+, were ruled out to avoid altering the natural history of the disease [7].

Hepatic steatosis for individuals in the Nutrihep study was assessed.

2.3. Assessment of Flavonoid Intake

Flavonoid intake was calculated based on the daily consumption of the following healthy foods, typical of the Mediterranean diet: apples, grapes, oranges, orange juice, pears, peaches, strawberries, walnuts, broccoli, onion, cooked onion, aubergines, courgettes, spinach, celery, and tomato. The amount of flavonoids (mg/day) was obtained from the database Phenol-Explorer: an online comprehensive database on polyphenol contents in foods [28,29]. Table S2 displays the average amount of flavonoids (mg/day) and standard deviation (SD) in the health foods from the Nutrihep cohort.

2.4. Statistical Analysis

The individual characteristics are reported as means and standard deviations (M $\pm$ SD) or medians and interquartile ranges for continuous variables and as frequencies and percentages (%) for categorical variables.

We fitted a logistic regression model with MASLD as the outcome variable and flavonoid intake (both continuous and categorical) as predictors.

Flavonoid intake was categorized according to 10 mg/day intervals from 165 (lower) to 235 (upper) mg/day. Initially, confounding variables were selected from the existing literature. Then, the procedure of minimum absolute reduction and selection (LASSO) was adopted to reduce the number of candidate predictors and select those most useful for model construction (Table S3) [30]. In selecting variables to be added to the model as confounders, those already included in the definition of MASLD, such as BMI, waist, HDL, triglycerides, glucose, HbA1c, and blood pressure, were not considered. The models were adjusted for the following variables: gender, age (<65 vs. $\geq$65 years), daily calories, weight (kg), γGT, ALT, HOMA (<2.5 vs. $\geq$2.5), job, and marital status. Estimated coefficients were transformed into odds ratios (OR).

In addition, the variance inflation factor (VIF) was also evaluated to check multicollinearity, and confounders with VIF > 5 were discarded (Table S4) [31].

A forest plot was drawn to compare the OR values obtained by the model fitting, considering flavonoid consumption as a categorical variable.

In addition, dose–response modelling for dichotomous outcomes was fitted to determine the shape and magnitude of the relationship between exposure and MASLD [32] firstly, by considering the flavonoids intake categorized in deciles and secondly, by using a smoothing cubic spline function, with 9 knots on the percentiles 0% (or 1%), 12.5%, 25%, 37.5%, 50%, 62.5%, 75%, 87.5%, and 100% (or 99%) [33,34]. The reference value of the flavonoid intake regressor was the median. In this case, the odds ratios (ORs) of the dose–response relationships (with 95% confidence bounds, i.e., confidence intervals) were plotted in continuous shape around flavonoid intake (mg/day), and the evaluation in relation to OR = 1 was considered to denote significance or not. The adjustment covariates were selected by LASSO regression and fixed to the median values, mode, and reference category in relation to their continuous, categorical and dichotomous nature [35]. The two-tailed probability level was set at 0.05 to test the null hypothesis of non-association.

The analyses were conducted with StataCorp 2023 Stata Statistical Software: Release 18 (College Station, TX, USA: StataCorp LLC), while the forest plots [36] and dose–response modelling were created using RStudio and its packages forest plot, rms [35].

3. Results

Table 1 shows the main characteristics of the 1297 participants, classified according to MASLD (absence or presence); 48.50% of the sample had MASLD, most of whom (54.6%) were male.

Table 1. Characteristics of participants by MASLD. Nutrihep study. Putignano (BA). Italy. 2015–2018.

Parameters [a]		MASLD	
	Whole Sample [b]	No	Yes
N (%)	1297	668 (51.50)	629 (48.50)
Age (years)	54.33 (14.34)	49.24 (13.80)	59.74 (12.86)
Age categories (years) (%)			
<65	955 (73.6)	565 (59.2)	390 (40.8%)
≥65	342 (26.4)	103 (30.1)	239 (69.9%)
Gender (%)			
Female	744 (57.4)	417 (56.0)	327 (44.0%)
Male	553 (42.6)	251 (45.4)	302 (54.6%)
Flavonoids (mg/day)	203.89 (126.36)	192.30 (123.03)	216.19 (128.75)
rMED (median (IQR))	8.00 (6.00; 10.00)	8.00 (6.00; 10.00)	8.00 (6.00; 10.00)
BMI (kg/m^2)	27.58 (5.05)	25.04 (3.59)	30.28 (4.97)
Weight (kg)	72.93 (14.87)	66.66 (12.02)	79.58 (14.73)
Waist (cm)	90.45 (13.46)	83.04 (10.38)	98.32 (11.79)
SBP (mmHg)	120.93 (15.81)	115.64 (15.35)	126.52 (14.30)
DBP (mmHg)	77.68 (8.00)	75.69 (7.88)	79.78 (7.58)
HbA1c (mmol/mol)	38.07 (6.87)	36.59 (5.05)	39.64 (8.09)
HOMA	1.89 (1.88)	1.33 (0.90)	2.43 (2.38)
ALT (U/L)	22.20 (16.21)	19.70 (8.27)	24.86 (21.37)
γGT (U/L)	17.58 (13.46)	14.80 (7.67)	20.54 (17.16)
AST (U/L)	21.74 (10.87)	20.70 (5.94)	22.85 (14.29)
TG (mg/dL)	98.41 (69.23)	80.73 (58.55)	117.22 (74.60)
C-reactive protein (mg/dL)	0.26 (0.55)	0.21 (0.52)	0.31 (0.58)
TC (mg/dL)	191.35 (35.36)	188.90 (33.06)	193.96 (37.50)
HDL (mg/dL)	50.79 (12.59)	53.18 (12.80)	48.24 (11.85)
Glucose (mg/dL)	95.34 (17.34)	90.13 (10.54)	100.89 (21.06)
ALP (U/L)	52.98 (16.10)	50.10 (15.56)	56.04 (16.11)
Alcohol intake (g/day)	10.58 (12.72)	10.74 (13.41)	10.42 (11.96)
Kcal (day)	2056.26 (750.22)	2100.33 (724.88)	2009.46 (774.05)
Smoker (%)			
Never/Former	1137 (87.7)	587 (51.6)	550 (48.4)
Current	159 (12.3)	81 (50.9)	78 (49.1)
Hypertension (%)			
No	847 (68.8)	517 (61.0)	330 (39.0)
Yes	385 (31.2)	115 (29.9)	270 (70.1)
Dyslipidemia (%)			
No	1047 (85.1)	561 (53.6)	486 (46.4)
Yes	184 (14.9)	71 (38.6)	113 (61.4)
Diabetes (%)			
No	1148 (93.2)	620 (54.0)	528 (46.0)
Yes	84 (6.8)	12 (14.3)	72 (85.7)
Marital Status (%)			
Single	181 (14.0)	115 (63.5)	66 (36.5)
Married or living together	1034 (79.7)	519 (50.2)	515 (49.8)
Separated or divorced	28 (2.2)	20 (71.4)	8 (28.6)
Widow/er	54 (4.2)	14 (25.9)	40 (74.1)
Education (%)			
Primary school	282 (21.8)	71 (25.2)	211 (74.8)
Secondary school	383 (29.5)	171 (44.6)	212 (55.5)
High school	460 (35.5)	307 (66.7)	153 (33.3)

Table 1. *Cont.*

Parameters [a]	MASLD		
	Whole Sample [b]	No	Yes
Graduate	172 (13.3)	119 (69.2)	53 (30.8)
Job (%)			
Managers and professionals	102 (7.9)	57 (55.9)	45 (44.1)
Craft, agricultural, and sales Workers	469 (36.2)	285 (60.8)	184 (39.2)
Elementary occupations	185 (14.1)	93 (50.3)	92 (49.7)
Housewife	141 (10.9)	74 (52.5)	67 (47.5)
Pensioner	325 (25.1)	110 (33.8)	215 (66.2)
Unemployed	75 (5.8)	49 (65.3)	26 (34.7)
Family income assessment (%)			
Insufficient	27 (2.1)	10 (37.0)	17 (63.0)
Just sufficient	167 (12.9)	81 (48.5)	86 (51.5)
Sufficient	1019 (78.6)	521 (51.1)	498 (48.9)
More than sufficient	64 (4.9)	44 (68.8)	20 (31.2)
Good	20 (1.5)	12 (60.0)	8 (40.0)

[a] As means and standard deviations. MASLD: metabolic dysfunction-associated steatotic liver disease; rMED: relative Mediterranean diet; BMI: body mass index; SBP: systolic blood pressure; DBP: diastolic blood pressure; HbA1c: glycosylated hemoglobin; HOMA: homeostasis model assessment; ALT: alanine amino transferase; γGT: gamma glutamyl transferase; AST: aspartate amino transferase; TG: triglycerides; TC: total cholesterol; HDL: high-density lipoprotein cholesterol; ALP: alkaline phosphatase level. [b] Percentages calculated for the column. Otherwise, percentages are calculated for the row.

Table S5 presents the main characteristics of the 1297 participants, categorized based on the median value of flavonoid intake (<185 vs. ≥185 mg/day).

The logistic regression models are reported in Table 2.

Table 2. Logistic regression analysis of MASLD on flavonoids intake as continuous and categorical variables inserted in the models.

Flavonoids (mg/day)	MASLD			
	OR	SE (OR)	*p-Value*	95% C.I.
Categories [b]:				
≥165 vs. <165	0.63	0.09	0.001	0.47; 0.83
≥175 vs. ≤175	0.64	0.09	0.002	0.48; 0.84
≥185 vs. <185 [a]	0.64	0.09	0.002	0.48; 0.85
≥195 vs. <195	0.66	0.09	0.004	0.50; 0.87
≥205 vs. <205	0.70	0.10	0.014	0.53; 0.93
≥215 vs. <215	0.72	0.10	0.023	0.54; 0.95
≥225 vs. <225	0.74	0.11	0.045	0.56; 0.99
≥235 vs. <235	0.81	0.12	0.170	0.60; 1.09
continuous	1.001	0.001	0.016	1.000; 1.003

[a] Median value. [b] ≥referent categories. Models adjusted for job, daily calories, weight (kg), gamma glutamyl transferase, alanine aminotransferase, gender (female vs. male), age (<65 vs. ≥65 years), marital status, and HOMA (<2.5 vs. ≥2.5). SE: standard error; MASLD: metabolic dysfunction-associated steatotic liver disease; OR: odds ratio.

We found a statistically significant protective effect up to a maximum daily consumption of 225 mg, with an odds ratio (OR) of 0.74 (at 95% C.I. 0.56; 0.99).

The OR remained almost unchanged when the intake increased from 165 mg per day to 185 mg per day [OR: 0.63 (95% C.I. 0.47; 0.83) and 0.64 (95% C.I. 0.48; 0.85), respectively]. An increase by 20 mg daily did not change the protective effect against the risk of developing MASLD.

Table S7 shows the results of the regression model with the exposure variable divided into deciles and adjusted for the same variables as the models shown in Table 2, from which the statistically significant no longer protective effect is shown from decile 6 onward.

The distribution of deciles by the presence and absence of MASLD is shown in Table S6.

In Figure 2, the protective role of flavonoids has been shown across various consumption categories.

	MASLD		
Flavonoids Categories (mg/day)	**Yes**	**No**	**OR**
≥ 165 vs <165	318	240	0.63
≥ 175 vs <175	345	257	0.64
≥ 185 vs <185	367	281	0.64
≥ 195 vs <195	395	305	0.66
≥ 205 vs <205	417	334	0.70
≥ 215 vs <215	434	351	0.72
≥ 225 vs <225	454	378	0.74
≥ 235 vs <235	470	403	0.81

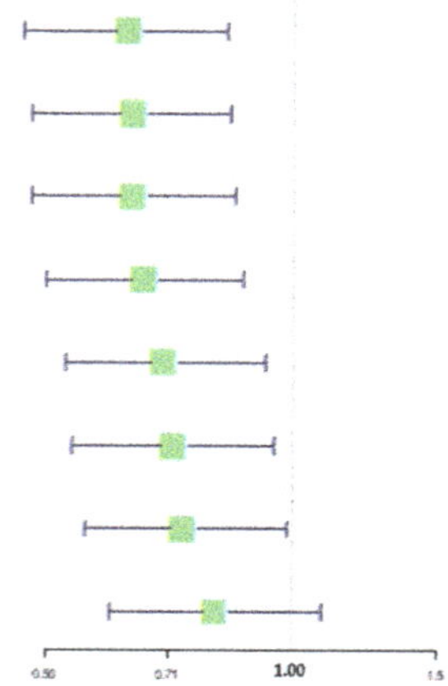

Figure 2. Forest plot of OR values in the MASLD by flavonoid categories (mg/day).

Figure 2 was useful to represent, in a graphical manner, the estimation as odds ratios (ORs). It provides a visual way to compare the size and direction of the effects. On the right side of the plot, the numerical values of the ORs are present along with their confidence intervals at 95%, corresponding to each flavonoids category. Furthermore, the absolute frequency of subjects divided of MASLD categories was present. In this case, all OR values were <1.0, indicating the protective role of flavonoids intake, and the value of 1.00 was the null effect on MASLD outcome.

Figure 3 illustrates that the smoothing dose–response relationship of the daily flavonoid intake on MASLD is increasing up to about 240–250 mg, by indicating a protective role of flavonoid intake up 185 mg. However, the OR is significant (or quasi-) up to 80–90 mg. However, beyond 185 mg/day, the dose–response relationship is centered around OR = 1, indicating no effect.

Concerning that, the shaded area on the graph shows the confidence bands, representing the confidence intervals for each OR value. The OR is considered statistically significant for statistical significance if 95% C.I. does not include the value '1'. Notably, a large 95% confidence interval indicates a small sample size for the corresponding flavonoid intake values.

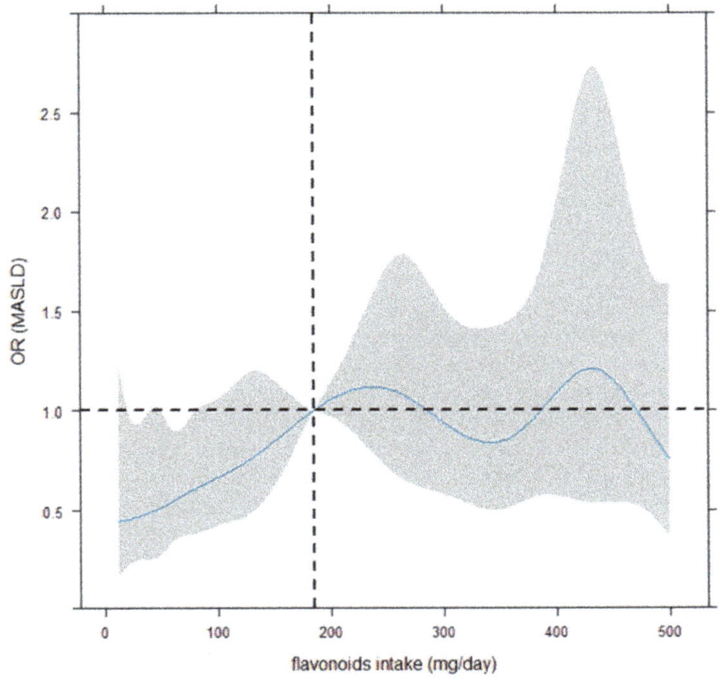

Figure 3. Dose–response curves for flavonoid intake-MASLD by restricted cubic spline. The reference value of the flavonoid intake regressor was the median (185 mg/day, vertical dotted line). The odds ratios (ORs) of the dose–response relationship (with 95% confidence bounds shown as shaded zones) were plotted in continuous shape around flavonoid intake (mg/day), and the evaluation in relation to OR = 1 (i.e., horizontal dotted line) was considered to denote significance. The logistic model was adjusted for job, daily calories, weight (kg), gamma glutamyl transferase, alanine amino transferase, gender (female vs. male), age (<65 vs. ≥65 years), marital status, and HOMA (<2.5 vs. ≥2.5).

4. Discussion

This study, carried out in a population of 1297 Italian middle-aged participants from Putignano (Puglia, Italy), describes the effect of dietary flavonoid intake on the risk of MASLD.

Our results show that consumption of only 165 mg per day is already protective against MASLD. Since, until now, lifestyle intervention has been the cornerstone in the management of patients with MASLD, this finding may make an important contribution [1].

Given the recent emergence of MASLD, the scientific literature is lacking in studies evaluating the effect of bioactive components such as polyphenols on the risk of this liver condition. Studies evaluating the relationship with NAFLD are also very scarce and often report controversial results [37–39].

Given the recent new fatty liver disease nomenclature, from NAFLD to MASLD, we will refer to the studies in the scientific literature on NAFLD and flavonoids to discuss our findings.

Flavonoids are a large group of natural substances with variable phenolic structures [40]. They can be subdivided into flavonoids, flavonols, orange ketones, isoflavones, anthocyanins, chalcones, and dihydrogen derivatives [41]. These classes are amply present in fruits, vegetables, grains, bark, roots, stems, flowers, tea, and wine [40]. Flavonoids have various positive effects on human health, including antitumor, antioxidant, antibacterial, antiviral, anti-inflammatory, and analgesic effects [42,43]. These properties make flavonoids potentially beneficial substances for liver diseases as well [44].

The excessive intake of flavonoids may cause harmful effects that are not yet well known. At higher doses, flavonoids may act as mutagens, free radical-generating pro-oxidants, and inhibitors of key enzymes involved in several metabolic processes. Therefore, at high doses, the adverse effects of flavonoids may outweigh the beneficial ones. Although further studies are needed to precisely define a toxic dose, it is hypothesized that it is difficult to incur these harmful effects with the intake of dietary flavonoids alone [45,46].

Initially, flavonoids' health benefits were attributed to their potent antioxidant capacity. However, other studies also suggest the importance of their anti-inflammatory and metabolic effects [18]. Flavonoids positively intervene in various forms of liver steatosis, such as by regulating lipid metabolism, insulin resistance, inflammation, and oxidative stress [45].

Wang et al. observed a reduction in hepatic fat content in mice treated with a high-fat diet after flavonoids from Broussonetia papyrifera treatment [47].

This flavonoid treatment also inhibited the production of ROS, reduced the content of myeloperoxidase, and improved the activity of superoxide dismutase (SOD). These results demonstrate the ability of flavonoids to reduce fat accumulation and oxidative stress.

Another study evaluated the effects of licorice chalcone [48]. This inhibits adipogenesis and increases lipid decomposition and fatty acid β-oxidation in fatty-liver mice by promoting the Sirtuin1/AMP-activated protein kinase pathway. Zhu et al. identified the same effect on lipid metabolism mediated by luteolin, lycopene, and their combination. They indirectly activate the SIRT1/AMPK pathway, thus inhibiting lipogenesis and increasing β-oxidation [49].

Yin et al. studied the effects of Cyanidin-3-O-glucoside, the most abundant anthocyanin in the flavonoid family, in fatty liver disease. They found that this kind of flavonoid is able to eliminate damaged mitochondria to maintain mitochondrial homeostasis and alleviate oxidative stress [50].

Positive effects on improving oxidative stress at the liver level are induced by quercetin, which has been demonstrated to restore the levels of superoxide dismutase, catalase, and glutathione in the liver of NAFLD mice [51]. Flavonoids may inhibit oxidative stress by regulating malondialdehyde (MDA), superoxide dismutase (SOD), and catalase (CAT) [52].

In liver steatosis, oxidative stress promotes inflammatory responses promoting liver injury. When the level of oxidative stress increases, it can promote the expression of IL-6, IL-1β, and TNF-α [53].

Wang et al. found that the levels of IL-1β, IL-6, and TNF-α, higher in the liver tissue of rats in the NAFLD model group, could be reduced by the total flavonoids of *Scutellaria baicalensis* [52]. Subsequent studies have found the same results [54]. These results support the idea that the anti-inflammatory effect of flavonoids occurs mainly through the inhibition of the NF-κβ pathway [55]. Luteolin has also been shown to reduce a variety of inflammatory factors in rats with liver steatosis, proving to have not only an antioxidant effect but also a good anti-inflammatory effect [56]. Thus, the main pathways through which flavonoids exert their protective effect toward liver steatosis are anti-inflammatory and antioxidant [57].

They can also decrease oxidative damage through a free radical scavenging activity because they have hydroxyl groups [58].

The main food source of flavonoids for the human body is fruits and vegetables [40]. Our results are in line with other studies that have reported a link between vegetable and fruits intake and a lower risk of liver steatosis [44,59].

They are also in agreement with the results that suggested that the Mediterranean diet's beneficial effect is due to its high content of bioactive phytochemicals [60]. Fruits and vegetables are at the base of the Mediterranean diet food pyramid, which includes their consumption at every main meal [61]. Our study population is strongly tied to the Mediterranean food culture and habitually eats foods typical of this dietary pattern.

Consuming one apple per day is sufficient to achieve the flavonoid intake we have identified as protective toward MASLD [27]. The apple is now readily available all year round and it is affordable, allowing the entire population to consume it regularly.

Strengths and Limitations

The present study evaluated the effect of dietary flavonoid intake and risk of MASLD in a middle-aged cohort from Putignano.

The strengths of the present study include the generalizability of the results to southern Mediterranean populations. The main strength of our study is that no other study in the scientific literature has analyzed this aspect of MASLD, nor has it done so in a similar population. However, some limitations must be considered. One of the main limitations is the absence of data on individual flavonoid classes. This did not allow us to investigate whether there is a class of flavonoids that has a greater protective effect toward MASLD than others. Another potential limitation is related to the self-reporting of the diet, although to remedy canvas, each FFQ was checked by our dietitian upon delivery of the questionnaire. We were also unable to examine the effect of parsley consumption on the risk of MASLD. Parsley is rich in flavonoids [62] and frequently used as an aromatic seasoning herb in the South of Italy.

5. Conclusions

In conclusion, our study results show a protective role of flavonoids against MASLD. Consuming only 165 mg of flavonoids daily can activate this protective function, reducing the risk of MASLD.

However, further studies are needed to support the validity of our results also in different populations, identify the classes of flavonoids with potentially greater protective roles, and define the proportions of single classes to be eaten that may enhance the protective effect.

Supplementary Materials: The following supporting information can be downloaded at: https://www.mdpi.com/article/10.3390/antiox13111286/s1, Table S1: Amount (mean ± SD) of flavonoids (mg/day) contained in the following health foods by MASLD in the Nutrihep cohort; Table S2: Collinearity diagnostics by VIF for the independent variables; Table S3: Characteristics of participants by median value of flavonoids in the Nutrihep cohort. Table S4: Collinearity Diagnostics by VIF for the independent variables. Table S5: Characteristics of Participants by Median Value of Flavonoids in the Nutrihep cohort. Table S6: Absolute frequencies on deciles of daily flavonoid. Table S7: Logistic regression analysis of MASLD on flavonoid intake divided into deciles.

Author Contributions: Conceptualization, C.B. and R.T.; methodology, R.D. and C.B.; software, R.D., D.G. and C.B.; validation, G.G.; formal analysis, C.B. and D.G.; resources, G.G.; data curation, R.D. and C.B.; writing—original draft preparation, R.T.; writing—review and editing, C.B. and R.T.; supervision, G.G.; project administration, R.D. and G.G.; funding acquisition, G.G. All authors have read and agreed to the published version of the manuscript.

Funding: This research was funded by the Italian Ministry of Health with Ricerca Corrente 2024.

Institutional Review Board Statement: The study was conducted according to the Declaration of Helsinki and approved by the Ethics Committee of the National Institute of Gastroenterology and Research Hospital (DDG-CE-792/2014, 14 February 2014).

Informed Consent Statement: Informed consent was obtained from all subjects involved in the study.

Data Availability Statement: The original contributions presented in this study are included in the article. Further inquiries can be directed to the corresponding author.

Acknowledgments: The authors are grateful to the NUTRIHEP Group.

Conflicts of Interest: The authors declare no conflicts of interest.

References

1. Chan, W.-K.; Chuah, K.-H.; Rajaram, R.B.; Lim, L.-L.; Ratnasingam, J.; Vethakkan, S.R. Metabolic Dysfunction-Associated Steatotic Liver Disease (MASLD): A State-of-the-Art Review. *J. Obes. Metab. Syndr.* **2023**, *32*, 197–213. [CrossRef] [PubMed]
2. Ludwig, J.; Viggiano, T.; McGill, D.; Oh, B. Nonalcoholic Steatohepatitis: Mayo Clinic Experiences with a Hitherto Unnamed Disease. *Mayo Clin. Proc.* **1980**, *55*, 434–438. [PubMed]
3. Farrell, G.C.; Chitturi, S.; Lau, G.K.; Sollano, J.D.; Asia–Pacific Working Party on NAFLD. Guidelines for the Assessment and Management of Non-alcoholic Fatty Liver Disease in the Asia–Pacific Region: Executive Summary. *J. Gastro. Hepatol.* **2007**, *22*, 775–777. [CrossRef] [PubMed]
4. Ratziu, V.; Bellentani, S.; Cortez-Pinto, H.; Day, C.; Marchesini, G. A Position Statement on NAFLD/NASH Based on the EASL 2009 Special Conference. *J. Hepatol.* **2010**, *53*, 372–384. [CrossRef]
5. Rinella, M.E.; Lazarus, J.V.; Ratziu, V.; Francque, S.M.; Sanyal, A.J.; Kanwal, F.; Romero, D.; Abdelmalek, M.F.; Anstee, Q.M.; Arab, J.P.; et al. A Multisociety Delphi Consensus Statement on New Fatty Liver Disease Nomenclature. *Hepatology* **2023**, *78*, 1966–1986. [CrossRef]
6. Powell, E.E.; Wong, V.W.-S.; Rinella, M. Non-Alcoholic Fatty Liver Disease. *Lancet* **2021**, *397*, 2212–2224. [CrossRef]
7. Eslam, M.; Newsome, P.N.; Sarin, S.K.; Anstee, Q.M.; Targher, G.; Romero-Gomez, M.; Zelber-Sagi, S.; Wai-Sun Wong, V.; Dufour, J.-F.; Schattenberg, J.M.; et al. A New Definition for Metabolic Dysfunction-Associated Fatty Liver Disease: An International Expert Consensus Statement. *J. Hepatol.* **2020**, *73*, 202–209. [CrossRef]
8. Singh, S.; Allen, A.M.; Wang, Z.; Prokop, L.J.; Murad, M.H.; Loomba, R. Fibrosis Progression in Nonalcoholic Fatty Liver vs Nonalcoholic Steatohepatitis: A Systematic Review and Meta-Analysis of Paired-Biopsy Studies. *Clin. Gastroenterol. Hepatol.* **2015**, *13*, 643–654.e9. [CrossRef]
9. Dulai, P.S.; Singh, S.; Patel, J.; Soni, M.; Prokop, L.J.; Younossi, Z.; Sebastiani, G.; Ekstedt, M.; Hagstrom, H.; Nasr, P.; et al. Increased Risk of Mortality by Fibrosis Stage in Nonalcoholic Fatty Liver Disease: Systematic Review and Meta-analysis. *Hepatology* **2017**, *65*, 1557–1565. [CrossRef]
10. Le, M.H.; Yeo, Y.H.; Li, X.; Li, J.; Zou, B.; Wu, Y.; Ye, Q.; Huang, D.Q.; Zhao, C.; Zhang, J.; et al. 2019 Global NAFLD Prevalence: A Systematic Review and Meta-Analysis. *Clin. Gastroenterol. Hepatol.* **2022**, *20*, 2809–2817.e28. [CrossRef]
11. Huh, Y.; Cho, Y.J.; Nam, G.E. Recent Epidemiology and Risk Factors of Nonalcoholic Fatty Liver Disease. *J. Obes. Metab. Syndr.* **2022**, *31*, 17–27. [CrossRef] [PubMed]
12. Farrell, G.C.; Wong, V.W.-S.; Chitturi, S. NAFLD in Asia—As Common and Important as in the West. *Nat. Rev. Gastroenterol. Hepatol.* **2013**, *10*, 307–318. [CrossRef] [PubMed]
13. Wong, V.W.-S.; Chan, R.S.-M.; Wong, G.L.-H.; Cheung, B.H.-K.; Chu, W.C.-W.; Yeung, D.K.-W.; Chim, A.M.-L.; Lai, J.W.-Y.; Li, L.S.; Sea, M.M.-M.; et al. Community-Based Lifestyle Modification Programme for Non-Alcoholic Fatty Liver Disease: A Randomized Controlled Trial. *J. Hepatol.* **2013**, *59*, 536–542. [CrossRef] [PubMed]
14. Lassailly, G.; Caiazzo, R.; Buob, D.; Pigeyre, M.; Verkindt, H.; Labreuche, J.; Raverdy, V.; Leteurtre, E.; Dharancy, S.; Louvet, A.; et al. Bariatric Surgery Reduces Features of Nonalcoholic Steatohepatitis in Morbidly Obese Patients. *Gastroenterology* **2015**, *149*, 379–388. [CrossRef]
15. Schwingshackl, L.; Morze, J.; Hoffmann, G. Mediterranean Diet and Health Status: Active Ingredients and Pharmacological Mechanisms. *Br. J. Pharmacol.* **2020**, *177*, 1241–1257. [CrossRef]
16. Corradini, E.; Foglia, P.; Giansanti, P.; Gubbiotti, R.; Samperi, R.; Laganà, A. Flavonoids: Chemical Properties and Analytical Methodologies of Identification and Quantitation in Foods and Plants. *Nat. Prod. Res.* **2011**, *25*, 469–495. [CrossRef]
17. Crozier, A.; Jaganath, I.B.; Clifford, M.N. Dietary Phenolics: Chemistry, Bioavailability and Effects on Health. *Nat. Prod. Rep.* **2009**, *26*, 1001. [CrossRef]
18. AboZaid, O.; AbdEl-hamid, O.; Atwa, S. Biochemical Alterations of Resveratrol on Oxidative Stress in Experimental Induced Non—Alcoholic Fatty Liver Disease in Rats. *Benha Vet. Med. J.* **2015**, *28*, 166–177. [CrossRef]
19. Schmatz, R.; Perreira, L.B.; Stefanello, N.; Mazzanti, C.; Spanevello, R.; Gutierres, J.; Bagatini, M.; Martins, C.C.; Abdalla, F.H.; Daci Da Silva Serres, J.; et al. Effects of Resveratrol on Biomarkers of Oxidative Stress and on the Activity of Delta Aminolevulinic Acid Dehydratase in Liver and Kidney of Streptozotocin-Induced Diabetic Rats. *Biochimie* **2012**, *94*, 374–383. [CrossRef]
20. Ying, H.-Z.; Liu, Y.-H.; Yu, B.; Wang, Z.-Y.; Zang, J.-N.; Yu, C.-H. Dietary Quercetin Ameliorates Nonalcoholic Steatohepatitis Induced by a High-Fat Diet in Gerbils. *Food Chem. Toxicol.* **2013**, *52*, 53–60. [CrossRef]
21. Cozzolongo, R.; Osella, A.R.; Elba, S.; Petruzzi, J.; Buongiorno, G.; Giannuzzi, V.; Leone, G.; Bonfiglio, C.; Lanzilotta, E.; Manghisi, O.G.; et al. Epidemiology of HCV Infection in the General Population: A Survey in a Southern Italian Town. *Am. J. Gastroenterol.* **2009**, *104*, 2740–2746. [CrossRef] [PubMed]
22. United Nations Educational, Scientific and Cultural Organization. International Standard Classification of Education, ISCED 1997. In *Advances in Cross-National Comparison: A European Working Book for Demographic and Socio-Economic Variables*; Springer: Boston, MA, USA, 2003. Available online: https://uis.unesco.org/sites/default/files/documents/international-standard-classification-of-education-1997-en_0.pdf (accessed on 20 October 2024).
23. International Standard Classification of Occupations, International Labour Office. Available online: https://www.ilo.org/sites/default/files/wcmsp5/groups/public/@dgreports/@dcomm/@publ/documents/publication/wcms_172572.pdf (accessed on 20 October 2024).

24. Sever, P. New Hypertension Guidelines from the National Institute for Health and Clinical Excellence and the British Hypertension Society. *J. Renin Angiotensin Aldosterone Syst.* **2006**, *7*, 61–63. [CrossRef] [PubMed]
25. Riboli, E. The EPIC Project: Rationale and Study Design. European Prospective Investigation into Cancer and Nutrition. *Int. J. Epidemiol.* **1997**, *26*, S6–S14. [CrossRef] [PubMed]
26. Riboli, E.; Hunt, K.; Slimani, N.; Ferrari, P.; Norat, T.; Fahey, M.; Charrondière, U.; Hémon, B.; Casagrande, C.; Vignat, J.; et al. European Prospective Investigation into Cancer and Nutrition (EPIC): Study Populations and Data Collection. *Public Health Nutr.* **2002**, *5*, 1113–1124. [CrossRef]
27. Kalligeros, M.; Vassilopoulos, A.; Vassilopoulos, S.; Victor, D.W.; Mylonakis, E.; Noureddin, M. Prevalence of Steatotic Liver Disease (MASLD, MetALD, and ALD) in the United States: NHANES 2017–2020. *Clin. Gastroenterol. Hepatol.* **2024**, *22*, 1330–1332.e4. [CrossRef]
28. Neveu, V.; Perez-Jimenez, J.; Vos, F.; Crespy, V.; Du Chaffaut, L.; Mennen, L.; Knox, C.; Eisner, R.; Cruz, J.; Wishart, D.; et al. Phenol-Explorer: An Online Comprehensive Database on Polyphenol Contents in Foods. *Database* **2010**, *2010*, bap024. [CrossRef]
29. Peterson, J.J.; Dwyer, J.T.; Jacques, P.F.; McCullough, M.L. Improving the Estimation of Flavonoid Intake for Study of Health Outcomes. *Nutr. Rev.* **2015**, *73*, 553–576. [CrossRef]
30. Tibshirani, R. Regression Shrinkage and Selection via the Lasso. *J. R. Stat. Soc.* **1996**, *58*, 267–288. [CrossRef]
31. Belsley, D.A.; Kuh, E.; Welsch, R.E. *Regression Diagnostics: Identifying Influential Data and Sources of Collinearity*; Wiley: New York, NY, USA, 1980.
32. Ritz, C.; Jensen, S.M.; Gerhard, D.; Streibig, J.C. *Dose-Response Analysis Using R*, 1st ed.; Chapman and Hall/CRC: Boca Raton, FL, USA, 2019; ISBN 978-1-315-27009-8.
33. Desquilbet, L.; Mariotti, F. Dose-response Analyses Using Restricted Cubic Spline Functions in Public Health Research. *Stat. Med.* **2010**, *29*, 1037–1057. [CrossRef]
34. Ruppert, D.; Wand, M.P.; Frontmatter, R.J.C. *Semiparametric Regression. Cambridge Series in Statistical and Probabilistic Mathematics*; Cambridge University Press: Cambridge, UK, 2003.
35. Harrell, F. Regression Modeling Strategies. R Package, Version 6.8-0; 2024. Available online: https://cran.r-project.org/web/packages/rms/rms.pdf (accessed on 20 October 2024).
36. R Core Team. *A Language and Environment for Statistical Computing*; R Foundation for Statistical Computing: Vienna, Austria, 2024.
37. Faghihzadeh, F.; Adibi, P.; Rafiei, R.; Hekmatdoost, A. Resveratrol Supplementation Improves Inflammatory Biomarkers in Patients with Nonalcoholic Fatty Liver Disease. *Nutr. Res.* **2014**, *34*, 837–843. [CrossRef]
38. Chen, S.; Zhao, X.; Ran, L.; Wan, J.; Wang, X.; Qin, Y.; Shu, F.; Gao, Y.; Yuan, L.; Zhang, Q.; et al. Resveratrol Improves Insulin Resistance, Glucose and Lipid Metabolism in Patients with Non-Alcoholic Fatty Liver Disease: A Randomized Controlled Trial. *Dig. Liver Dis.* **2015**, *47*, 226–232. [CrossRef] [PubMed]
39. Heebøll, S.; Kreuzfeldt, M.; Hamilton-Dutoit, S.; Kjær Poulsen, M.; Stødkilde-Jørgensen, H.; Møller, H.J.; Jessen, N.; Thorsen, K.; Kristina Hellberg, Y.; Bønløkke Pedersen, S.; et al. Placebo-Controlled, Randomised Clinical Trial: High-Dose Resveratrol Treatment for Non-Alcoholic Fatty Liver Disease. *Scand. J. Gastroenterol.* **2016**, *51*, 456–464. [CrossRef] [PubMed]
40. Panche, A.N.; Diwan, A.D.; Chandra, S.R. Flavonoids: An Overview. *J. Nutr. Sci.* **2016**, *5*, e47. [CrossRef] [PubMed]
41. Tsuchiya, H. Structure-Dependent Membrane Interaction of Flavonoids Associated with Their Bioactivity. *Food Chem.* **2010**, *120*, 1089–1096. [CrossRef]
42. Maleki, S.J.; Crespo, J.F.; Cabanillas, B. Anti-Inflammatory Effects of Flavonoids. *Food Chem.* **2019**, *299*, 125124. [CrossRef]
43. Kaigongi, M.M.; Lukhoba, C.W.; Ochieng', P.J.; Taylor, M.; Yenesew, A.; Makunga, N.P. LC-MS-Based Metabolomics for the Chemosystematics of Kenyan *Dodonaea Viscosa* Jacq (Sapindaceae) Populations. *Molecules* **2020**, *25*, 4130. [CrossRef]
44. Mazidi, M.; Katsiki, N.; Banach, M. A Higher Flavonoid Intake Is Associated with Less Likelihood of Nonalcoholic Fatty Liver Disease: Results from a Multiethnic Study. *J. Nutr. Biochem.* **2019**, *65*, 66–71. [CrossRef]
45. Van De Wier, B.; Koek, G.H.; Bast, A.; Haenen, G.R.M.M. The Potential of Flavonoids in the Treatment of Non-Alcoholic Fatty Liver Disease. *Crit. Rev. Food Sci. Nutr.* **2017**, *57*, 834–855. [CrossRef]
46. Skibola, C.F.; Smith, M.T. Potential Health Impacts of Excessive Flavonoid Intake. *Free Radic. Biol. Med.* **2000**, *29*, 375–383. [CrossRef]
47. Wang, S.; Sheng, F.; Zou, L.; Xiao, J.; Li, P. Hyperoside Attenuates Non-Alcoholic Fatty Liver Disease in Rats via Cholesterol Metabolism and Bile Acid Metabolism. *J. Adv. Res.* **2021**, *34*, 109–122. [CrossRef]
48. Liu, B.; Zhang, J.; Sun, P.; Yi, R.; Han, X.; Zhao, X. Raw Bowl Tea (Tuocha) Polyphenol Prevention of Nonalcoholic Fatty Liver Disease by Regulating Intestinal Function in Mice. *Biomolecules* **2019**, *9*, 435. [CrossRef] [PubMed]
49. Zhu, Y.; Liu, R.; Shen, Z.; Cai, G. Combination of Luteolin and Lycopene Effectively Protect against the "Two-Hit" in NAFLD through Sirt1/AMPK Signal Pathway. *Life Sci.* **2020**, *256*, 117990. [CrossRef] [PubMed]
50. Yin, Y.; Gao, L.; Lin, H.; Wu, Y.; Han, X.; Zhu, Y.; Li, J. Luteolin Improves Non-Alcoholic Fatty Liver Disease in Db/Db Mice by Inhibition of Liver X Receptor Activation to down-Regulate Expression of Sterol Regulatory Element Binding Protein 1c. *Biochem. Biophys. Res. Commun.* **2017**, *482*, 720–726. [CrossRef]
51. Yang, H.; Yang, T.; Heng, C.; Zhou, Y.; Jiang, Z.; Qian, X.; Du, L.; Mao, S.; Yin, X.; Lu, Q. Quercetin Improves Nonalcoholic Fatty Liver by Ameliorating Inflammation, Oxidative Stress, and Lipid Metabolism in *Db/Db* Mice. *Phytother. Res.* **2019**, *33*, 3140–3152. [CrossRef]

52. Tan, P.; Jin, L.; Qin, X.; He, B. Natural Flavonoids: Potential Therapeutic Strategies for Non-Alcoholic Fatty Liver Disease. *Front. Pharmacol.* **2022**, *13*, 1005312. [CrossRef]
53. Xiao, M.; Chen, G.; Zeng, F.; Qiu, R.; Shi, W.; Lin, J.; Cao, Y.; Li, H.; Ling, W.; Chen, Y. Higher Serum Carotenoids Associated with Improvement of Non-Alcoholic Fatty Liver Disease in Adults: A Prospective Study. *Eur. J. Nutr.* **2019**, *58*, 721–730. [CrossRef]
54. Wang, M.; Qiao, X.; Fang, Q.; Fu, S.; Li, X.; Huang, F.; Lin, Y. Effect of Different Polar Extracts of *Scutellaria baicalensis* Total Flavonoids on Non-Alcoholic Fatty Liver Model Rats. *Chin. Pharm.* **2022**, *33*, 1338–1342.
55. González-Gallego, J.; García-Mediavilla, M.V.; Sánchez-Campos, S.; Tuñón, M.J. Fruit Polyphenols, Immunity and Inflammation. *Br. J. Nutr.* **2010**, *104*, S15–S27. [CrossRef]
56. Abu-Elsaad, N.; El-Karef, A. Protection against Nonalcoholic Steatohepatitis through Targeting IL-18 and IL-1alpha by Luteolin. *Pharmacol. Rep.* **2019**, *71*, 688–694. [CrossRef]
57. Visioli, F.; Davalos, A. Polyphenols and Cardiovascular Disease: A Critical Summary of the Evidence. *Mini Rev. Med. Chem.* **2011**, *11*, 1186–1190. [CrossRef]
58. Middleton, E., Jr.; Kandaswami, C.; Theoharides, T.C. The Effects of Plant Flavonoids on Mammalian Cells: Implications for Inflammation, Heart Disease, and Cancer. *Pharmacol. Rev.* **2000**, *52*, 673–751. [PubMed]
59. Rietman, A.; Sluik, D.; Feskens, E.J.M.; Kok, F.J.; Mensink, M. Associations between Dietary Factors and Markers of NAFLD in a General Dutch Adult Population. *Eur. J. Clin. Nutr.* **2018**, *72*, 117–123. [CrossRef] [PubMed]
60. Papamiltiadous, E.S.; Roberts, S.K.; Nicoll, A.J.; Ryan, M.C.; Itsiopoulos, C.; Salim, A.; Tierney, A.C. A Randomised Controlled Trial of a Mediterranean Dietary Intervention for Adults with Non Alcoholic Fatty Liver Disease (MEDINA): Study Protocol. *BMC Gastroenterol.* **2016**, *16*, 14. [CrossRef] [PubMed]
61. D'Alessandro, A.; De Pergola, G. Mediterranean Diet Pyramid: A Proposal for Italian People. *Nutrients* **2014**, *6*, 4302–4316. [CrossRef] [PubMed]
62. Ferreira, F.S.; De Oliveira, V.S.; Chávez, D.W.H.; Chaves, D.S.; Riger, C.J.; Sawaya, A.C.H.F.; Guizellini, G.M.; Sampaio, G.R.; Torres, E.A.F.D.S.; Saldanha, T. Bioactive Compounds of Parsley (*Petroselinum crispum*), Chives (*Allium schoenoprasum* L.) and Their Mixture (*Brazilian cheiro-verde*) as Promising Antioxidant and Anti-Cholesterol Oxidation Agents in a Food System. *Food Res. Int.* **2022**, *151*, 110864. [CrossRef]

antioxidants

MDPI

Article

Protective Effect of Polyphenols Purified from *Mallotus oblongfolius* on Ethanol-Induced Gastric Mucosal Injury by Regulating Nrf2 and MAPKs Pathways

Shasha Yu [1,2,†], Zhouwei Duan [1,3,4,†], Peng Li [5], Shiping Wang [1,3], Lijun Guo [3], Guanghua Xia [2,*] and Hui Xie [1,3,*]

1 Institute of Agro-Products Processing and Design, Hainan Academy of Agricultural Science, Haikou 571100, China
2 College of Food Science and Technology, Hainan University, Haikou 570228, China
3 Sanya Institute, Hainan Academy of Agricultural Sciences, Sanya 572000, China
4 College of Food Science and Technology, Huazhong Agricultural University, Wuhan 430070, China
5 College of Feng Xiang, Hainan University, Haikou 570228, China
* Correspondence: xiaguanghua2011@126.com (G.X.); xiehuinky@163.com (H.X.)
† These authors contributed equally to this work.

Abstract: *Mallotus oblongifolius* (MO), which is rich in polyphenols, is a characteristic tea resource with medicinal value. In this study, a total of 45 polyphenolic components of MO, including narirutin, isoquercitrin, rutin and digallic acid, were identified by UPLC-Q-TOF/MS analysis. In addition, the gastroprotective effect of *Mallotus oblongifolius* polyphenols (MOP) on ethanol-induced gastric mucosal injury in rats was investigated. The rats received anhydrous ethanol after continuous gavage of MOP or lansoprazole for one week. In addition, the macro- and micro-damage induced by ethanol in the gastric tissue was significantly reduced after MOP pretreatment for one week. Further analysis showed that MOP prevented ethanol-induced acute gastric mucosal injury by increasing the expression of antioxidant enzymes (SOD, CAT, GSH-Px) and decreasing the expression of reactive oxygen species (ROS), lipid oxidation product (MDA) and myeloperoxidase (MPO). Meanwhile, MOP inhibited the phosphorylation of p38/ERK/JNK and promoted the activation of the Nrf2 pathway. These results suggested that MOP may be a promising therapeutic target for the prevention of ethanol-induced gastric mucosal injury by improving oxidative stress, inhibiting the p38/ERK/JNK signaling pathways and activating Nrf2 expression.

Keywords: *Mallotus oblongifolius* polyphenols (MOP); purification; gastric mucosal injury; oxidative stress; ethanol

1. Introduction

The gastric mucosa, which is the mucosa of the inner surface of the stomach, is irregularly wrinkled and has a special gastric protective effect. When an imbalance between attack factors (such as gastric acid, alcohol abuse, nonsteroidal anti-inflammatory drugs) and defense factors (such as prostaglandin, bicarbonate, mucin, mucosal blood flow) occurs in the stomach, the gastric mucosal barrier will be destroyed, resulting in gastric mucosal damage, specifically manifested as gastric mucosal swelling, bleeding, and even gastric ulcer [1]. However, with the accelerated pace of life and the increase in stress in life, alcohol consumption has become a common phenomenon, which is closely related to health problems [2]. Alcohol is a highly corrosive, fat-soluble substance that destroys the normal physiological environment and break the barrier of the gastric mucosa, and is an important factor in causing haemorrhagic gastric mucosal damage [3,4]. Previous studies have shown that lipid peroxidation and oxidative stress induced by alcohol are the cause of gastric mucosa injury [5]. As an invasion factor of the gastric mucosa, oxygen free radicals occupy a vital place in the pathophysiological changes of gastric mucosa injury by causing oxidative damage [6].

Several drugs, including antacids (sodium bicarbonate, aluminium hydroxide), proton pump inhibitors (omeprazole, lansoprazole), antibiotics (Helicobacter pylori inhibition) and H_2 receptor antagonists (cimetidine, ranitidine), have been used to treat and ameliorate gastric mucosal injury. However, there is growing evidence that these drugs have limited effects and may cause side effects in many aspects when taken over a long period of time [7–9]. Therefore, it is vital to find efficient, safe and economical alternative drugs to avoid the suffering of gastric mucosa injury. Natural products have become a hot topic of scientific research as possible alternatives to stomach protection. Among these, polyphenols have attracted particular attention and have been shown to exert a protective effect in the stomach by inhibiting oxidative stress.

Mallotus oblongifolius (MO) is a characteristic substitute for tea beverages in Hainan, China [10]. It is locally known as "glossy ganoderma grass" with important medicinal and economic value [11]. It is reported to have a variety of beneficial effects, including antioxidant, antibiosis, reducing blood pressure, nourishing the stomach and so on. Most of these effects are related to the active substances contained in MO, especially polyphenols. Recent studies have found that gallic acid has a gastric protective effect on ethanol-induced gastric mucosal injury in rats [12]. However, the composition analysis of *Mallotus oblongifolius* polyphenols (MOP) and its protective mechanism against gastric injury have not been investigated. Notably, MO is a traditional tea with medicinal value, but its value has not been fully exploited due to limited research and development. In this study, the purification method of MOP was optimized, the main components of MOP were identified by UPLC-Q-TOF/MS analysis and the potential mechanism of preventing ethanol-induced gastric mucosal injury in rats was explored by biomolecular detection, which provides reference values for the in-depth development of MO resources.

2. Materials and Methods

2.1. Materials and Chemicals

The MO was obtained from Hainan Academy of Agricultural Sciences (Hainan, China). The absolute alcohol was purchased from Xilong Science Co., Ltd. (Shantou, China). The enzyme-linked immunoassay assay (ELISA) kits for reactive oxygen species (ROS) and myeloperoxidase (MPO) were purchased from Shanghai X-Y Biotechnology Co., Ltd. (Shanghai, China). The assay kits for malondialdehyde (MDA), superoxide dismutase (SOD), glutathione peroxidase (GSH-Px) and catalase (CAT) were purchased from the Nanjing Jiancheng Bioengineering Institute (Nanjing, Jiangsu, China). The primers (Nrf2, Keap1, HO-1, p38, JNK and ERK) were obtained from Sangon Biotech Co., Ltd. (Shanghai, China). The antibodies (Nrf2: ab92946, Keap1: #8047S, HO-1: ab189491, p38: ab4822, p-p38: ab4822, JNK: ab179461, p-JNK: ab76572, ERK: ab184699 and p-ERK: ab201015) were purchased from Abcam (Cambridge, UK) and Cell Signaling Technology (Boston, MA, USA).

2.2. Extraction of Crude MOP

The methods used to prepare the polyphenol extracts were slightly improved, according to a previous study [13]. The MO powder (100 g) was extracted with 50% ethanol (1500 mL) at 55 °C for 2 h. Next, the solid-liquid mixture was filtered to obtain the MO extract. The filtrates were then evaporated at 55 °C by a rotary evaporator (RE52AA). Finally, the crude MOP was obtained by vacuum freeze-drying.

2.3. Purification of MOP

2.3.1. Static Adsorption and Desorption Kinetics

Following pretreatment, 5 g of NKA−9 macroporous resin was placed into a 250 mL triangular flask, and 100 mL of crude polyphenol extraction solution (5 mg/mL) was added. Adsorption equilibrium was achieved by shaking for 12 h on a shaker (30 °C, 180 r/min). The Folin-Ciocalteu method was used to determine the mass concentration of the polyphenols after adsorption, each at 0.5 h. Once adsorption equilibrium was achieved, the solution was desorbed with 100 mL 60% (*V*/*V*) ethanol solution on a shaker (30 °C, 180 r/min). The mass concentration of the polyphenols in the solution was measured, each at 0.2 h.

2.3.2. Effects of Different Factors on Adsorption and Desorption

As described in Section 2.3.1, the macroporous resin was mixed with the crude extract, and the polyphenol concentration (C_e, mg/mL) in the filtrate was determined after adsorption. Then, the macroporous resin was desorbed with ethanol solution on a shaker with the same rotational speed, and the concentration of polyphenols (C_d, mg/mL) in the filtrate was determined after desorption. The type of resin, concentration of crude extract and concentration of ethanol in the purification process were determined according to the adsorption and desorption effect. The following formulas were used to calculate the adsorption capacity, adsorption rate, resolution rate.

$$\text{The adsorption rate: } (\%) = (C_0 - C_e)/C_0 \times 100$$

$$\text{The desorption rate: } (\%) = (C_d \times V_d)/[(C_0 - C_e) \times V_0] \times 100$$

$$\text{The adsorption quantity: } (\text{mg/g}) = (C_0 - C_e) \times V_0/M$$

where C_0 and C_e are the mass concentration of the polyphenol at the beginning and adsorption equilibrium, respectively (mg/mL); C_d is the mass concentration of the polyphenol in the desorption solution (mg/mL); V_0 is the initial volume of the crude extract; V_d is the volume of the ethanol desorption solution (mL); and M is the weight of macroporous resin (g).

2.3.3. Dynamic Adsorption and Desorption

The glass column (16 mm × 500 mm) was filled with pretreated NKA−9 resin for dynamic adsorption and desorption experiments. The sample solution flowed through the glass column at different flow rates (1.0, 1.5, 2.0, 2.5, 3.0, 3.5 mL/min). The adsorption rate was calculated according to the concentration of polyphenol in the effluent sample solution. After the adsorption was complete, the glass column was eluted with 60% (*V*/*V*) alcohol solution at different flow velocities (1.0, 1.5, 2.0, 2.5, 3.0 mL/min). After desorption, the mass concentration of the polyphenols in the effluent solution was ascertained, and the desorption rate of the resin was calculated. The optimum flow rates of the sample solution and elution were determined according to the adsorption rate and desorption rate, respectively.

The sample solution was permitted to flow through the glass column at a rate of 1.5 mL/min. The effluent solution was collected. One tube was 9 mL, and 10 tubes were a bed volume (BV). The dynamic adsorption curve was drawn with BV as the abscissa and polyphenol concentration in the effluent as the ordinate. After the dynamic adsorption was complete, 60% (*V*/*V*) ethanol was used as the eluent at a flow rate of 2.0 mg/mL. The mass concentration of the polyphenols in the effluent solution was determined every 0.2 BV, and the dynamic elution curve was drawn with the above method.

2.4. Method for Composition Analysis of MOP

2.4.1. UPLC-Q-TOF/MS Analysis

The sample was analyzed by using an Acquity UPLC I-Class System high-performance fluid chromatograph equipped with a Xevo G2-XS QTOF mass spectrometer, coupled with an ACQUITY UPLC® HSS T3 column (100 mm × 2.1 mm, 1.8 μm) maintained at 35 °C. The purification of the polyphenols was redissolved in 70% (*v*/*v*) ethanol and then filtered with a 0.22 μm nylon filter. The ionization conditions were set as follows: collision energy ramp 30–60 eV; cone voltage of 40 V, trap collision energy of 6.0 V; ramp collision energy from 20 to 30 V; capillary voltage of 2.5 kV; sample temperature 20 °C; source temperature of 120 °C; desolvation temperature of 450 °C; cone gas flow of 50 L/h; and desolvation gas flow of 900 L/h. The scanning time of the whole detection process is 1.5 s. Furthermore, leucine-enkephalin (m/z 556.2771, [M + H]+) solution at a concentration of 200 ng/μL, with a flow rate of 10 μL/min, was used as the external reference to ensure accuracy during the analysis. Elution was performed with a mobile phase of A (0.1% formic acid) and B (acetonitrile) under a gradient program: 0–1 min, 5% B; 1–2 min, 5–10% B; 2–4 min, 10–15% B; 4–16 min, 15% B; 16–21 min, 15–17% B; 21–30 min, 17% B; 30–32 min, 17–22% B; 32–34 min, 22–30% B; 34–36 min, 30–32% B; 36–38 min, 32–35% B; 38–40 min, 35–40% B; 40–42 min, 40–45% B; 42–44 min, 45–50% B; 44–46 min, 50–60% B; 46–56 min, 60–70% B; 56–62 min, 70–80% B; 62–64 min, 80–95% B; 64–67 min, 95% B; 67–67.1 min, 95–5% B; 67.1–70 min, 5% B. The flow velocity was 0.3 mL/min, and the sample volume was 3 μL. In positive ion mode, the mass range was set to m/z 100–1500 Da. The data were recorded by the TOF MSE (MS at Elevated Fragmentation Energy) model.

2.4.2. Chemical Composition Analysis

The UPLC-Q-TOF/MS data acquisition from MasslClx software (version 4.1, Waters Corporation, Milford, MA, USA) was imported to Progenesis QI software (version 2.0, Waters Corporation) for data handling, involving peak detection, alignment and normalization. The compound information, scanned by UPLC-Q-TOF/MS, was compared with the database by UNIFI software. The analysis conditions were set as follows: mass error $\leq$ 3 mDa, chromatographic peak extraction time 0–70 min, positive ion mode adduct ions +H, +Na.

2.5. Animals and Experimental Design

2.5.1. Arrangement of Animals and Induction of the Model

Seven-week-old male Sprague-Dawley rats (200 $\pm$ 20 g) were purchased from Changsha Tianqin Biotechnology Co., Ltd. (Changsha, China, Certificate number: SCXK (Xiang) 2019-0013). All rats were fed a standard diet and were given free access to water. They acclimated for a week before the experiment; they were placed in a room with conditions of 23 °C and 12 h day/night cycle [14,15]. All animal procedures were executed on the basis of the National Guidelines for Experimental Animal Care and Use and were permitted by the ethical committee of experimental animal care of Hainan University (HNDX2020072).

After the seven-day adaptation period, all of the rats were distributed into 6 groups at random (n = 10): the normal control group (NC) and ulcer control group (UC) were treated with vehicle (5% Tween-80); the lansoprazole group (LAN) was treated with 30 mg/kg bw of lansoprazole; the low-dose group of MOP (MOP-L), median-dose group of MOP (MOP-M) and high-dose group of MOP (MOP-H). MOP-L, MOP-M and MOP-H were treated with 50 mg/kg bw, 100 mg/kg bw and 200 mg/kg bw MOP, respectively, which were dissolved in 5% Tween-80. After continuous gavage for a week, all of the rats were fasted but kept in water for 24 h. On the eighth day, with the exception of the NC group, all groups were given 5 mL/kg anhydrous ethanol by gavage. Four hours later, the rats were anesthetized with 10% chloral hydras and sacrificed by cervical dislocation [16,17]. The required samples were collected and preserved.

2.5.2. Sample Collection and Preservation

The collected blood were centrifuged at 4000$\times$ g for 10 min to obtain serum. The serum was divided into several portions and stored at -80 °C for next experiment. At the same time, the stomach was cut along a larger bend with scissors and rinsed with cold normal saline. Then, the stomach tissue was photographed after blotting the rest of the saline with filter paper. Thereafter, the gastric tissue was randomly divided into two parts. One part was immersed in 4% paraformaldehyde, and the other part was collected in a freezer tube and saved in a -80 °C refrigerator.

2.5.3. Macroscopic and Microscopic Injury Assessment of Gastric Tissue

Image J (1.47v, National Institutes of Health, Bethesda, MD, USA) was used to analyze the photos of the stomach tissue. The ulcer area (UA, mm^2) was measured, and the ulcer index was evaluated on account of the formula below:

$$UI = 10/X$$

where X is the ratio of the total stomach mucosa area to the total ulcer area

In addition, the stomach tissues, soaked in 4% paraformaldehyde, were embedded in paraffin and sliced. These sections were used for hematoxylin-eosin staining (H&E) to observe and analyze the microscopic lesions of gastric tissue [18,19].

2.5.4. Determination of Oxidative Stress Cytokines

The gastric tissues were mixed with cold phosphate buffer saline (PBS) to prepare 10% homogenate, which was centrifuged at $4000\times g$ at 4 °C for 10 min. The ROS (XYR9212641) and MPO (XY9R0142) levels in the stomach tissues were detected by ELISA kits. The expression of MDA (A003-1), SOD (A001-3), GSH-Px (A005-1) and CAT (A007-1-1) in the stomach tissues was determined by biochemical kits. All kits are operated in accordance with the manufacturer's instructions.

2.5.5. Real-Time Quantitative PCR

Total RNA extraction was performed from the gastric tissues using the RNA prep Pure Tissue Kit (Tiangen, Beijing, China). In brief, 10–20 mg of stomach tissues was mixed with 300 μL RL lysate and fully ground. Proteinase K was added, and the mixture was treated at 56 °C for 10–20 min. After centrifugation, they were treated with anhydrous ethanol, Buffer RW1, DNase I working solution and Buffer RW, respectively. Finally, 30–100 μL of RNase-Free ddH_2O was added to obtain total RNA samples of gastric tissues. The total RNA extracted was mixed with 5× FastKing-RT SuperMix and RNase-Free ddH_2O to establish a 20 μL reaction system by FastKing gDNA Dispelling RT SuperMix Kit (Tiangen, Beijing, China). Another 20 μL reaction system (10 μL 2× SuperReal PreMix Plus, 0.6 μL forward primer, 0.6 μL reverse primer, 1.5 μL cDNA, 7.3 μL RNase-Free ddH2O) was established to complete RT-PCR on the CFX Connect RT-PCR platform (Bio-Rad, Hercules, CA, USA). PCR amplification was performed using the following cycling conditions: activation of HotStar TaqDNA polymerase and initial denaturation at 95 °C for 15 min, and 40 cycles of PCR reactions at 95 °C, 60 °C and 72 °C. All kits were operated according to the manufacturer's instructions. Relative gene expression was standardized to GAPDH by the $2^{-\Delta\Delta CT}$ method [20,21]. The relevant primers information is supplemented in the additional document (Supplementary Table S1).

2.5.6. Western Blot Analysis

RIPA lysate buffer (Servicebio, Wuhan, China) was used to separate protein from gastric tissue, and the protein concentration was measured by a BCA kit (Servicebio, Wuhan, China). Prepared protein samples were separated by gel electrophoresis and transferred to solid phase carrier (PVDF) membranes, which were incubated with special antibodies and washed with TBST. After that, the membranes were soaked in ECL chemiluminescence solution to detect the target protein blots which were visualized by ImageQuant LAS4000mini system (GE, Boston, MA, USA). During the whole process, β-actin was treated as a control protein [22,23].

2.6. Statistical Analysis

The experimental data are expressed as the mean values ± standard deviation (SD) and were analyzed with SPSS 25.0 software (IBM Inc., Armonk, New York, NY, USA) or GraphPad Prism 8 (San Diego, CA, USA). One-way analysis of variance (ANOVA) was utilized to evaluate differences among groups. When the p value was less than 0.05, the difference was considered statistically significant.

3. Results

3.1. Static Adsorption and Desorption

3.1.1. Static Adsorption and Desorption Kinetics

The static adsorption and desorption curves of NKA−9 resin on the polyphenols of MO leaves are shown in Figure 1A. NKA−9 macroporous resin was used for the rapid adsorption of the polyphenols from the MO leaves. The adsorption capacity increased sharply within 6 h. After 8.5 h, the concentration of polyphenols did not change significantly, showing that the adsorption of the polyphenols had reached equilibrium. It could be seen from the desorption curve that the 60% (v/v) ethanol solution had a good desorption effect on polyphenols adsorbed by resin, and the desorption rate increased rapidly in the initial stage (within 0.6 h) and reached equilibrium after 2 h.

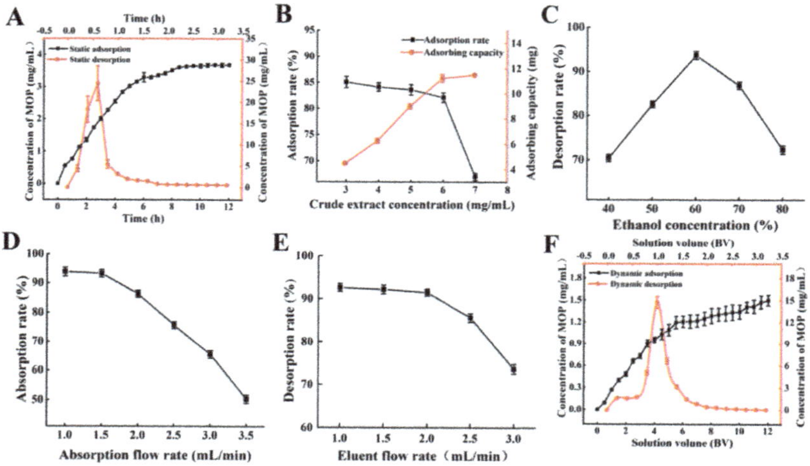

Figure 1. Adsorption and desorption capacities of NKA−9 resin for polyphenols from MO. (**A**) Kinetics for static adsorption and desorption on NKA−9 resin for polyphenols from MO. (**B**) Effect of concentration of initial MOP on adsorption rate on NKA−9 resin for polyphenols from MO. (**C**) Effect of eluent concentration on desorption rare on NKA−9 resin for polyphenols from MO. (**D**) Effect of adsorption flow rate on adsorption rate on NKA−9 resin for polyphenols from MO. (**E**) Effect of eluent flow rate on desorption rate on NKA−9 resin for polyphenols from MO. (**F**) Kinetics for dynamic adsorption and desorption on NKA−9 resin for polyphenols from MO.

3.1.2. Influence of Resin Type on Adsorption and Desorption Effect

The adsorption and dissolution of polyphenols are closely related to the spatial structure, polarity and water content of macroporous resin. As shown in Supplementary Table S2, among the five kinds of macroporous resins, the highest adsorption rate on MOP was D201 with 56.07%, followed by NKA−9 with 54.65%. However, the highest desorption rate on MOP was NKA−9, which was up to 92.35%. Therefore, NKA−9 was used to adsorb and desorb MOP. This was because the polarity, structure and physicochemical properties of the NKA−9 resin were more conducive to the adsorption of MOP. Van der Waals forces and hydrogen bonds were the main forces between the two and were weaker and more easily broken than ionic bonds, favoring elution.

3.1.3. Influence of Concentration of Initial MOP on Adsorption Effect

The influence of the initial MOP concentration on the adsorption effect is demonstrated in Figure 1B. When the initial MOP concentration was 6 mg/mL, a high adsorption rate and adsorption capacity were observed. The adsorption rate decreased significantly, while the adsorption amount tended to saturate if the initial MOP concentration surpassed 6 mg/mL. When the concentration of the crude extract was lower than 6 mg/mL, the NKA−9 resin did not reach saturation. With the increase in the crude extract concentration, the contact amount between the resin and polyphenols, per unit surface area, was augmented, and the adsorption amount increased. Nevertheless, the adsorption rate decreased slightly because of the reduction in free groups on the resin surface. When the concentration of the crude extract exceeded 6 mg/mL, the mass transfer rate slowed down with the gradual saturation of groups on the resin surface, and some polyphenols flowed out without being adsorbed, resulting in a decrease in the adsorption rate. The resin gradually reached saturation adsorption, and the adsorption amount changed very little.

3.1.4. Influence of Eluent Concentration on Desorption Effect

The effect of ethanol concentration on NKA−9 resin desorption can be seen in Figure 1C. The desorption rate of the NKA−9 resin increased with the eluent concentration, from 40% to 60%. The optimal desorption rate of 93.69% was observed at an eluent concentration of 60%. As the ethanol increased, the desorption rate began to decrease. The ethanol concentration was related to its polarity, which could change the molecular forces between the polyphenols and macroporous resins. In addition, the solubility of the polyphenols was altered in different concentrations of ethanol. The polyphenols were not easily dissolved in low concentrations of ethanol. However, the ethanol concentration was too high, and some impurities were also removed. The above two factors led to a great influence of ethanol concentration on the resin desorption.

3.2. Dynamic Adsorption and Desorption

3.2.1. Influence of Adsorption Flow Rate on Adsorption and Desorption Effect

The flow rate of the sample solution affected the reaction between the solute and NKA−9 resin and further affected the adsorption and desorption effect of the resin. Figure 1D demonstrated that the adsorption rate of the resin decreased with the increasing flow rate of the sample solution. As the sample flow rate increased, the time that the crude extract remained in the column decreased, and part of the MOP was not adsorbed by the resin in time and then flowed out of the column [24]. It is worth noting that the adsorption rate of the NKA−9 resin on the polyphenols decreased by 0.56% at a 1.5 mL/min flow rate compared with that at a 1.0 mL/min flow rate, and the difference was not significant; however, the sample loading amount of crude extract per unit time increased by 0.5 times. Therefore, the optimal sample flow rate was 1.5 mL/min. Moreover, the desorption rate gradually decreased with the increasing flow rate of the eluent (Figure 1E). When the elution flow rate was higher than 1 mL/min, the target substance in the resin was not desorbed by the eluent in time, and the elution was incomplete. A flow rate of 1.0 mL/min was conducive to the desorption of the polyphenols from the resin. This may be due to ethanol entering the resin micropores at low flow rates, allowing the polyphenols to dissolve and elute more thoroughly [25]. However, a slow flow rate would prolong the purification cycle. In comparison to the elution flow rate of 1.0 mL/min, the desorption rate of the resin on the polyphenols only decreased by 1.12% at 2.0 mL/min. It is evident tht there was no significant difference; however, the time taken to desorb the resin with the same volume of ethanol was halved. Hence, 2.0 mL/min was chosen as the most appropriate elution velocity.

3.2.2. Dynamic Adsorption and Desorption Kinetics

The sample solution was pressed into a glass column filled with NKA−9 resin through a peristaltic pump. The dynamic adsorption curve is shown in Figure 1F. When the inflow volume was 3.0 BV, the polyphenol concentration of the effluent solution was 0.729 mg/mL, which was approximately 46% of the initial sample, indicating that most of the polyphenols in the sample were adsorbed by NKA−9 resin. The injection volume was 11.5 BV, the concentration of polyphenols in the effluent solution was 1.475 mg/mL, which was close to the concentration of polyphenols in the initial sample. As the injection volume continued to increase, the polyphenol concentration of the effluent solution tended to be stable. The mass concentration of polyphenols in the effluent solution was close to that of the initial sample, indicating that the resin was saturated by adsorption. Thus, the saturated adsorption of the NKA−9 resin was achieved at 11.5 BV. Furthermore, Figure 1F also shows the results of the dynamic desorption kinetics. The 0–2.0 BV effluent solution contained most of the desorbed polyphenols. Subsequently, the concentration of polyphenols in the effluent solution decreased rapidly with the increasing ethanol dosage. When the consumption of ethanol was 2.8 BV, the concentration of polyphenols in the effluent solution was only 0.099 μg/mL, which was approximately 1/16 of the initial sample solution. At this point, the polyphenol concentration in the effluent solution tended to be stable, indicating that the adsorbed polyphenols were basically desorbed completely.

3.3. *MOP Composition Analysis*

Polyphenol Compositions of MO

With the verified UPLC-QTOF/MS means and UNIFI software, a total of 50 polyphenol components were identified in MO. All of the compounds were numbered according to their order of elution (Figure 2). The fragmentation details of 50 compounds are shown in Supplementary Table S3. According to the proposed tactics and comparison of collected data with existing publications, as well as the fragmentation pattern of different components generalized by former studies, the main compositions of the polyphenols in MO were narirutin, isoquercitrin, rutin, digallic acid, vitexin, D-catechin, quercetin, luteolin, kaempferol, myricetin (Supplementary Table S3). However, the components of Nos. 43, 45, 47, 48 and 50 were undefined compared to the database and required further study.

3.4. *Macroscopic and Microscopic Analysis of Gastric Tissue*

The macroscopic structure of the gastric tissue showed no tissue lesions in the NC group, while haemorrhagic lesions and elongated strip injuries appeared in the model group (Figure 3A). Lansoprazole or MOP pretreatment attenuated the gastric lesions caused by ethanol, and the effect of MOP at a high dose was improved. Similar results were obtained for the ulcer index (Figure 3B). Compared with the NC group, the ulcer index was significantly increased in the ethanol treatment group, while MOP or LAN pretreatment significantly improved this phenomenon. Meanwhile, the HE results showed that the gastric tissue had areas of necrosis, owing to intragastric alcohol administration, with shedding of many gastric glandular cells, cell nuclei fragmentation or dissolution and a small amount of bleeding, accompanied by lymphocyte infiltration, severe oedema in the submucosa and loose connective tissue arrangement (Figure 3C,D). After MOP or lansoprazole intervention in advance, the microscopic pathological injury of the gastric tissue was significantly improved.

3.5. *Oxidative Stress Factor Levels*

Compared with the NC group, the levels of SOD, CAT and GSH-Px in the UC group were remarkably diminished ($p < 0.05$, Figure 4A–C). Conversely, the SOD level, CAT level and GSH-Px level were increased to different degrees in the positive control LAN group, MOP-L group, MOP-M group and MOP-H group. The ROS and MDA activities showed opposite trends to antioxidant enzyme activities (Figure 4D–E). Compared with the NC group, the activity of ROS, MPO and MDA significantly increased in the gastric tissue of the UC group, and the LAN group and MOP groups at three doses were able to reduce their activity to different degrees ($p < 0.05$).

3.6. *Effect of MOP on the MAPK and Nrf2 Signaling Pathways*

To confirm whether the Nrf2 pathway protects the gastric mucosa, the gene and protein expression levels of Keap1, Nrf2 and HO-1 were investigated. The level of Keap1 in the gastric stomach was significantly increased only in the alcohol treatment group, while the expression of Keap1 was observably decreased in a dose-dependent manner in the polyphenol pretreatment groups (Figures 5A and 6A). Moreover, compared with the normal group, ethanol gavage resulted in an obvious decrease in Nrf2 but had little effect on HO-1 expression. In contrast, the gene and protein levels of Nrf2 and HO-1 were significantly upregulated by MOP pretreatment.

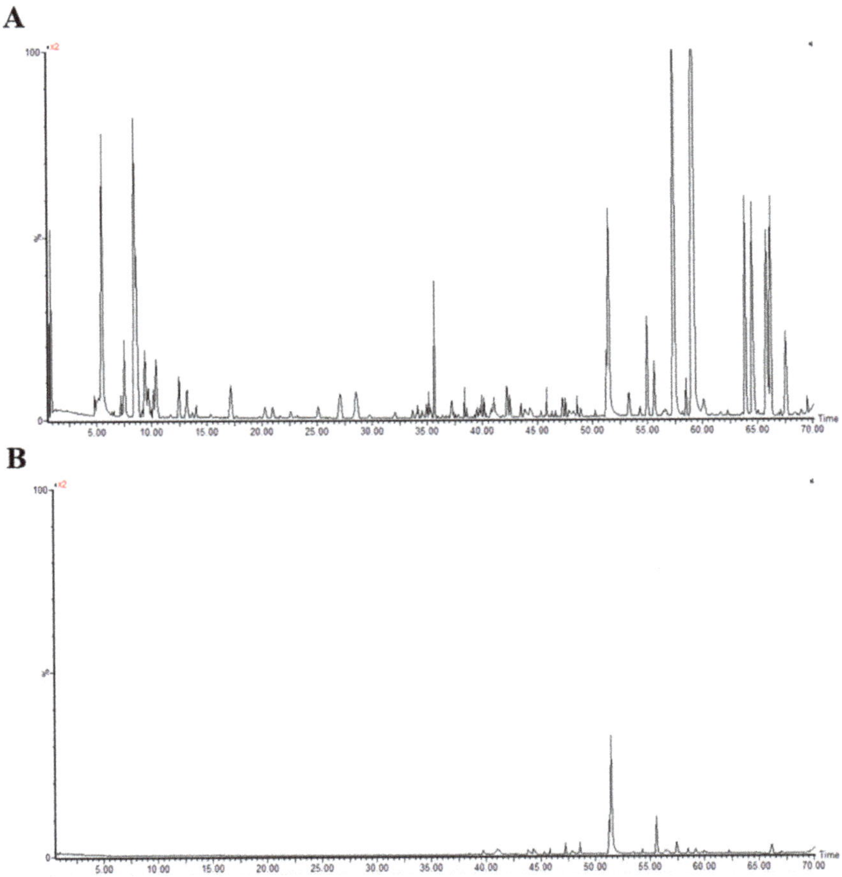

Figure 2. The base peak chromatograms of purified polyphenols from Mallotus oblongifolius (**A**) and the control of acetonitrile (**B**) by UPLC-Q-TOF/MS analysis under positive ion mode.

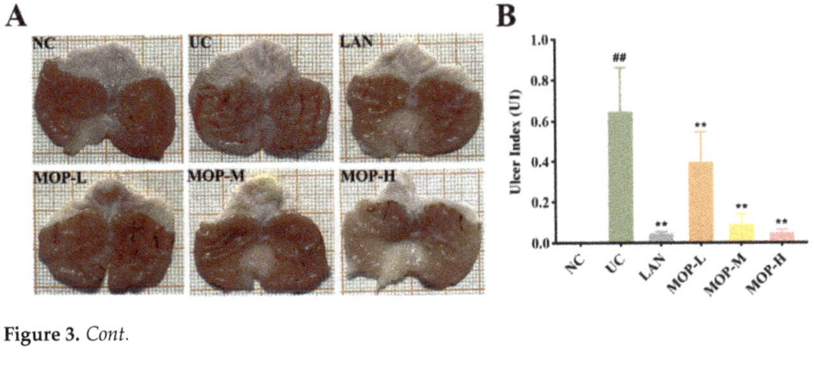

Figure 3. *Cont.*

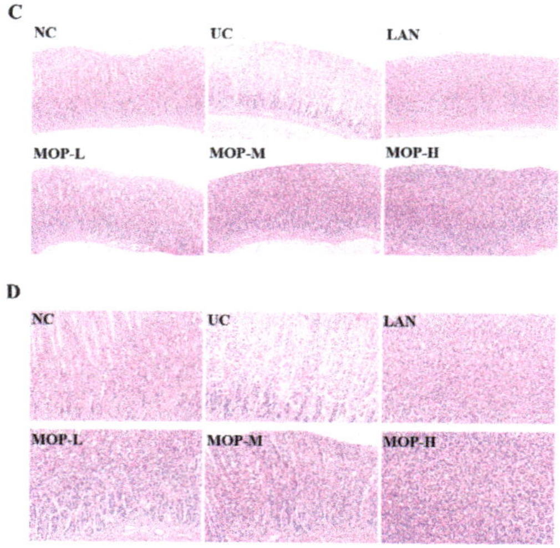

Figure 3. Macroscopic and microscopic injury of gastric mucosal of rats in all experimental groups (n = 10). (**A**) Representative macroscopic images of stomach tissue. (**B**) Effects of MOP on the ulcer index of ethanol-induced gastric mucosal injury. (**C**) Representative histological morphology of stomach tissues (magnification 200×). (**D**) Representative histological morphology of stomach tissues (magnification 400×). # and * denotes significant differences (n = 6). ## $p < 0.01$ vs. NC group; ** $p < 0.01$ vs. UC group.

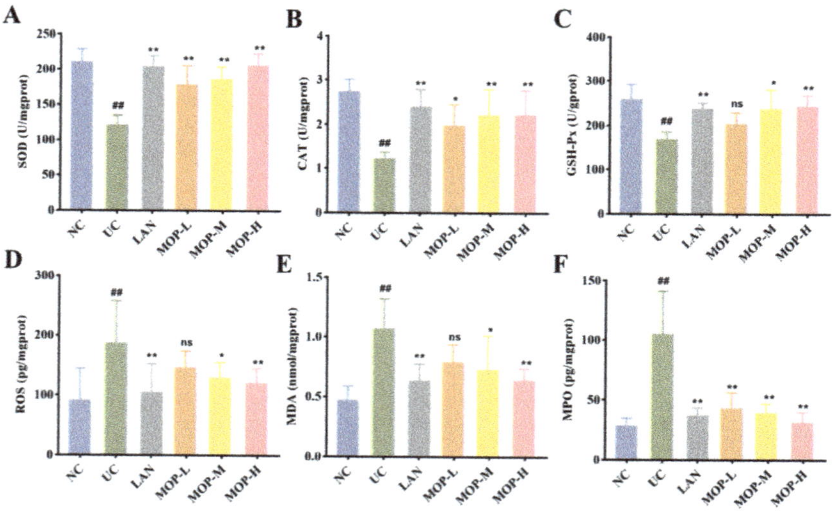

Figure 4. Effects of MOP on levels of SOD (**A**), CAT (**B**), GSH-Px (**C**), ROS (**D**), MDA (**E**), MPO (**F**) in stomach tissue of rats in vitro. # and * denotes significant differences (n = 6). ## $p < 0.01$ vs. NC group; * $p < 0.05$, ** $p < 0.01$ vs. UC group. 'ns' denotes no significant difference.

Furthermore, the MAPK signaling cascade is a major stress signaling pathway related to oxidative stress [26]. Next, in order to further verify the therapeutic target of MOP on ethanol-induced gastric mucosal injury, the effects of MOP on ethanol-induced gastric mucosal injury and the expression of JNK, ERK, and p38 were evaluated. The results showed that ethanol treatment alone significantly stimulated the expression of JNK, ERK and p38 (Figures 5B and 6B). However, pregavage of MOP apparently reversed this change. Among them, the MOP high-dose group had the best effect.

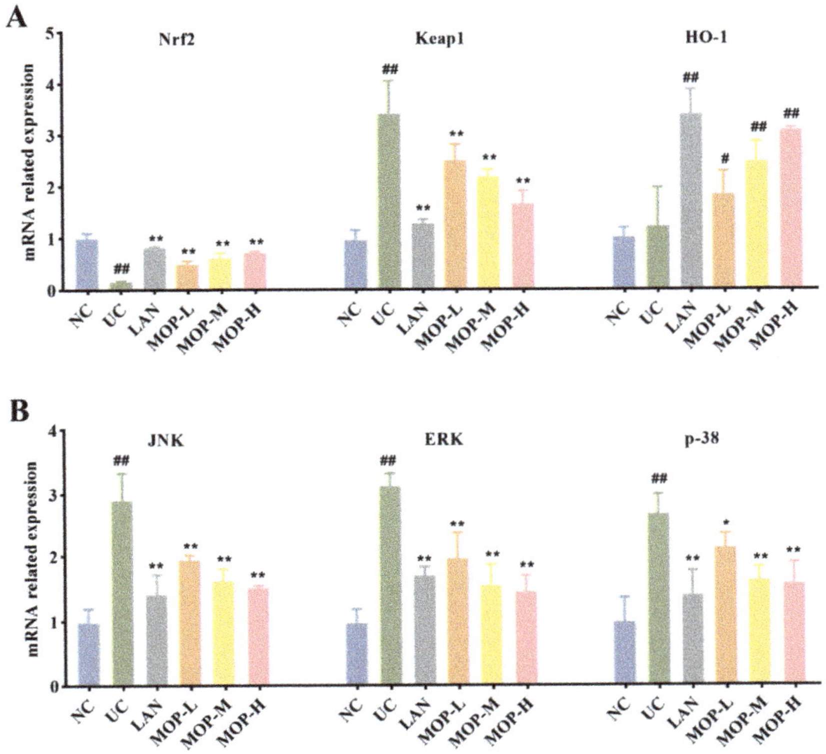

Figure 5. mRNA expression levels of Nrf2, HO-1 and Keap1 (**A**), p38, JNK and ERK (**B**) in stomach tissue of MOP pretreated ethanol-induced gastric mucosal injury. All experiments were performed in triplicate (n = 3) to determine the repeatability. # and * denotes significant differences. # $p < 0.05$, ## $p < 0.01$ vs. NC group; * $p < 0.05$, ** $p < 0.01$ vs. UC group.

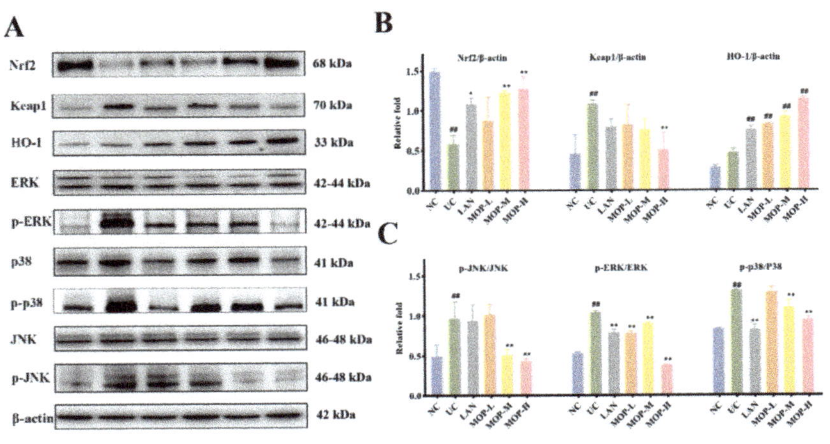

Figure 6. MOP attenuated ethanol-induced gastric mucosal injury by activating Nrf2/HO-1 and inhibiting MAPKs pathway. The western blot bands (**A**) and relative intensities of (**B**) Nrf2, Keap1 and HO-1, (**C**) p-JNK, p-ERK and p-p38. All experiments were performed in triplicate (n = 3) to determine the repeatability. # and * denotes significant differences. ## $p < 0.01$ vs. NC group; * $p < 0.05$, ** $p < 0.01$ vs. UC group.

4. Discussion

Polyphenols are natural active ingredients in plants and have potential antioxidant properties [27]. In this study, the purification method of polyphenols from MO was investigated to determine the influencing factors, such as resin type, crude extract concentration and flow rate, as well as eluent concentration and flow rate during the purification process. UPLC-Q-TOF/MS analysis demonstrated that the main composition of MOP was rutin, gallogen, vitexin and catechin. In addition, we investigated the effect of MOP pretreatment on ethanol-induced gastric mucosal damage and found that it exhibited good intervention.

Previous studies have shown that ethanol is a key factor in causing gastric mucosal injury and upper gastrointestinal bleeding [28,29]. Usually, an ethanol induced gastric mucosa injury model is used to study the pathogenesis of gastric ulcers and the intervention effect of active substances or drugs [30]. This caused infiltration of inflammatory cell and production of large amounts of ROS, which further triggered oxidative stress [31,32]. In this study, it was found that oxidative stress damage was particularly severe in the ethanol treatment group (Figures 3–6). These phenomena are consistent with previous studies and demonstrates that ethanol as an oxidative ulcer inducer is persuasive in evaluating the antioxidant role of MOP in gastric protection. Ethanol treatment caused a series of complex morphological and functional changes in the gastric tissue [33]. From a macro and micro perspective, MOP pretreatment effectively alleviated ethanol-induced gastric mucosa injury. Compared with the ethanol treatment group, the lesion area and ulcer index of gastric mucosa in the MOP treatment groups were significantly reduced in a dose-dependent manner. At the same time, this result was supported by histopathological changes, with the alleviation of symptoms such as bleeding, submucosal edema and inflammatory cell infiltration. These results were consistent with previous studies evaluating the gastric protective effect of different synthetic compounds and confirmed the gastric protective effect of MOP on ethanol-induced gastric mucosal injury.

The pathological changes to gastric mucosa injuries are closely related to oxidative stress. Ethanol is first metabolized in the stomach by alcohol dehydrogenase to produce a large amount of acetaldehyde, which generates oxygen free radicals under the action of xanthine oxidase [34]. Excessive oxygen free radicals notably enhance oxidation and lead to an oxidative stress state [35]. It has been reported that oxidative stress plays a key role in alcohol-induced gastric mucosal injury [36]. Alcohol causes the superabundant production of ROS in the gastric mucosa and leads to oxidative stress [37]. Despite the destructive aftermath of the overproduction of ROS, the defense systems of tissues and cells respond to oxidation-related cell and tissue damage. An excess of ROS consumes more antioxidant enzymes and nonenzymatic substances (such as SOD, CAT, GSH-Px) and destroys the antioxidant system of the gastric tissue, resulting in excessive lipid oxidation products (such as MDA) [38]. MDA cross-linked with biological macromolecules in vivo and further destroyed the structure and function of proteins. To some extent, the MDA reflected the damage degree of tissue oxidation. Studies have also shown that MOP catalyzed hydrogen peroxide to produce ROS [39]. Thus, these indices reflected the relationship between oxidative stress and gastric tissue injury to a certain extent. In this study, it was found that alcohol treatment significantly increased the levels of ROS, MPO and MDA, while markedly decreasing the levels of SOD, CAT and GSH-Px ($p < 0.05$, Figure 4A–E). Importantly, MOP intervention inhibited the above change, manifesting its antioxidant potential. These results confirmed that alcohol induced severe oxidative stress and gastric injury, and that MOP pretreatment effectively regulated the state of oxidative stress in gastric tissue.

In addition to causing pathological processes, ROS also plays an important role in cell signal transduction and regulation as second messengers, indirectly participating in immune responses or other cellular protective processes. The Keap1/Nrf2/HO-1 signaling pathway was activated during antioxidative damage processes in cells. Nrf2, belonging to the NF-E2 family, has been reported to play a crucial role in various kinds of cell protection mechanisms [40]. Under normal physiological conditions, Nrf2 is inactivated in the cytoplasm due to coupling with Keap1. With a large increase in ROS, the conformation of Keap1 is changed, and Nrf2 is transferred to the nucleus, where it combines with ARE (antioxidant reaction element) and activates the downstream factors [41]. In particular, HO-1, which is mainly activated by Nrf2, degrades heme and is an important enzyme for reversing oxidative stress damage [42]. HO-1 degrades heme to CO, biliverdin and Fe^{2+} and is an essential mediator of the antioxidant effect of Nrf2 [43]. It was reported that the enhanced expression of HO-1 had a protective effect on gastric mucosa in alcohol-induced gastric mucosa injury [6]. Our results showed that pretreatment with MOP enhanced the expression of Nrf2 and HO-1 and decreased the expression of Keap1. Thus, Keap1/Nrf2/HO-1 is a potential mechanism for antioxidative damage to gastric mucosa. This result suggested that MOP activated the Nrf2/HO-1 signaling pathway to protect the gastric mucosa.

Notably, MAPK carries signals from the cell surface to the nucleus, which are activated by oxidative stress and various proinflammatory cytokines [44]. It is widely involved in various kinds of endocellular signaling processes and plays a vital role in the activation and expression regulation of the Nrf2/HO-1 pathway [45]. JNK, ERK and p38 are the most studied among MAPK. The inhibition of p38 phosphorylation prevented the Keap1 degradation and nuclear translocation of Nrf2 and subsequently upregulated the induction of HO-1 expression [46]. At present, the regulatory mechanism of ERK and JNK on the Nrf2/HO-1 signaling pathway is still not completely clear. However, phosphorylation of ERK1/2 was reported to significantly reduce the quercetin-induced expression of Nrf2 and HO-1 [47]. Studies have also shown that the downregulation of JNK phosphorylation increased HO-1 expression and contributed to positive intervention in oxidative stress-related diseases [48,49]. Figures 5 and 6 showed that MOP pretreatment significantly inhibited the phosphorylation of JNK, ERK and p38 induced by ethanol treatment alone. These results suggest that MOP may mediate HO-1 expression by inhibiting JNK, ERK and p38 activation. In addition, many studies have shown that MAPKs are also involved in inflammation. The activation of MAPK regulated the expression of inflammatory factors and indirectly activated the downstream NF-κB pathway, which plays an important role in the process of inflammation and immune response. Therefore, the specific mechanism of MAPK-mediated HO-1 expression and its regulation of the inflammatory response in ethanol-induced gastric mucosal injury still need to be further study.

5. Conclusions

In this study, the extraction and purification process parameters of MOP were optimized and its preventive effect on ethanol-induced gastric mucosal damage was investigated. A total of 50 polyphenolic components were identified by UPLC-Q-TOF/MS analysis. In addition, the main components of MOP included narirutin, isoquercitrin, rutin, digallic acid, vitexin, D-catechin, quercetin, luteolin, kaempferol, myricetin and so on. The results of the animal experiments showed that MOP down-regulated the expression of MDA and increased the expression of SOD, CAT and GSH-Px. Meanwhile, MOP played a key role by inhibiting the MAPK signaling pathway and activating the Nrf2 signaling pathway during gastric defense and protection. These results provide a substantial contribution to MOP intervention on gastric injury induced by ethanol. Therefore, MOP is expected to provide reference value for the clinical prevention and treatment of gastric injury.

Supplementary Materials: The following supporting information can be downloaded at: https://www.mdpi.com/article/10.3390/antiox11122452/s1, Table S1: Genes and primers for PCR; Table S2: Effects of different resin types on adsorption and desorption of MOP; Table S3 Identification of chemical compounds by UHPLC/Q-TOF-MS/MS.

Author Contributions: S.Y.: Investigation, data curation, writing—original draft. Z.D.: Investigation, formal analysis, software. P.L.: Formal analysis. S.W.: Investigation. L.G.: Software. G.X.: Conceptualization, visualization, supervision. H.X.: Methodology, supervision. All authors have read and agreed to the published version of the manuscript.

Funding: Finance science and technology project of Hainan Province, China (ZDYF2020098, SQKY2022-0008, SQKY2022-0020, SQKY2022-004).

Institutional Review Board Statement: The study was approved by the National Guidelines for Experimental Animal Care and Use and were permitted by the ethical committee of experimental animal care of Hainan University (HNDX2020072).

Informed Consent Statement: Not applicable.

Data Availability Statement: Data is contained within the article and Supplementary Materials.

Acknowledgments: This work was supported by Finance science and technology project of Hainan Province, China (ZDYF2020098, SQKY2022-0008, SQKY2022-0020, SQKY2022-004).

Conflicts of Interest: The authors declare that they have no known competing financial interest or personal relationship that could have appeared to influence the work reported in this paper.

References

1. Amandeep, K.; Robin, S.; Ramica, S.; Sunil, K. Peptic Ulcer: A Review on Etiology and Pathogenesis. *Int. Res. J. Pharm.* **2012**, *3*, 34–38.
2. Chang, B.Y.; Kim, H.J.; Kim, T.Y.; Kim, S.Y. Enzyme-Treated Zizania latifolia Extract Protects against Alcohol-Induced Liver Injury by Regulating the NRF2 Pathway. *Antioxidants* **2021**, *10*, 960. [CrossRef] [PubMed]
3. Salim, A.S. Removing oxygen-derived free radicals stimulates healing of ethanol-induced erosive gastritis in the rat. *Digestion* **1990**, *47*, 24–28. [CrossRef] [PubMed]
4. Stermer, E. Alcohol consumption and the gastrointestinal tract. *Isr. Med. Assoc. J.* **2002**, *4*, 200–202.
5. Hajrezaie, M.; Salehen, N.; Karimian, H.; Zahedifard, M.; Shams, K.; Batran, R.A.; Majid, N.A.; Khalifa, S.A.M.; Ali, H.M.; El-Seedi, H.; et al. Biochanin A Gastroprotective Effects in Ethanol-Induced Gastric Mucosal Ulceration in Rats. *PLoS ONE* **2015**, *10*, e121529. [CrossRef]
6. Alarcon, D.L.L.C.; Nieto, A.; Martin, M.J.; Cabre, F.; Herrerias, J.M.; Motilva, V. Gastric toxicity of racemic ketoprofen and its enantiomers in rat: Oxygen radical generation and COX-expression. *Inflamm. Res.* **2002**, *51*, 51–57.
7. Cekin, A.H.; Sahinturk, Y.; Akbay, H.F.; Uyar, S.; Yolcular, B.O.; Cekin, Y. Use of probiotics as an adjuvant to sequential H. pylori eradication therapy: Impact on eradication rates, treatment resistance, treatment-related side effects, and patient compliance. *Turk. J. Gastroenterol.* **2017**, *28*, 3–11. [CrossRef]
8. Tran-Duy, A.; Spaetgens, B.; Hoes, A.W.; de Wit, N.J.; Stehouwer, C.D.A. Use of Proton Pump Inhibitors and Risks of Fundic Gland Polyps and Gastric Cancer: Systematic Review and Meta-analysis. *Clin. Gastroenterol. Hepatol.* **2016**, *14*, 1706–1719. [CrossRef]
9. Waldum, H.L.; Fossmark, R. Proton pump inhibitors and gastric cancer: A long expected side effect finally reported also in man. *Gut* **2018**, *67*, 199–200. [CrossRef]
10. Liu, T.; Chen, D.; Liu, Z.; Hou, J.M. First Report of Colletotrichum siamense Causing Anthracnose on Partridge tea (*Mallotus oblongifolius*) in China. *Plant Dis.* **2018**, *102*, 1669. [CrossRef]
11. Li, S.R.; Song, Y.J.; Deng, R.; Li, X.W.; Cheng, Y.; Zhang, Z.Q.; Sun, F.Y.; Liu, Q.S. Mallotus oblongifolius extracts ameliorate ischemic nerve damage by increasing endogenous neural stem cell proliferation through the Wnt/beta-catenin signaling pathway. *Food Funct.* **2020**, *11*, 1027–1036. [CrossRef] [PubMed]
12. Zhou, D.; Yang, Q.; Tian, T.; Chang, Y.; Li, Y.; Duan, L.R.; Li, H.; Wang, S.W. Gastroprotective effect of gallic acid against ethanol-induced gastric ulcer in rats: Involvement of the Nrf2/HO-1 signaling and anti-apoptosis role. *Biomed. Pharmacother.* **2020**, *126*, 110075. [CrossRef] [PubMed]
13. Xi, L.; Mu, T.; Sun, H. Preparative purification of polyphenols from sweet potato (*Ipomoea batatas* L.) leaves by AB-8 macroporous resins. *Food Chem.* **2015**, *172*, 166–174. [CrossRef] [PubMed]
14. Hei, Z.; Zhao, M.; Tian, Y.; Chang, H.; Shen, X.; Xia, G.; Wang, J. Isolation and Characterization of a Novel Sialoglycopeptide Promoting Osteogenesis from *Gadus morhua* Eggs. *Molecules* **2019**, *25*, 156. [CrossRef] [PubMed]
15. Xia, G.; Yu, Z.; Zhao, Y.; Wang, Y.; Wang, S.; He, M.; Wang, J.; Xue, C. Sialoglycoproteins isolated from the eggs of Carassius auratus prevents osteoporosis by suppressing the activation of osteoclastogenesis related NF-κB and MAPK pathways. *J. Funct. Foods* **2015**, *17*, 491–503. [CrossRef]
16. Lu, Y.; Zhao, M.; Peng, Y.; He, S.; Zhu, X.; Hu, C.; Xia, G.; Zuo, T.; Zhang, X.; Yun, Y.; et al. A physicochemical double-cross-linked gelatin hydrogel with enhanced antibacterial and anti-inflammatory capabilities for improving wound healing. *J. Nanobiotechnol.* **2022**, *20*, 426. [CrossRef] [PubMed]
17. Meng, K.; Mei, F.; Zhu, L.; Xiang, Q.; Quan, Z.; Pan, F.; Xia, G.; Shen, X.; Yun, Y.; Zhang, C.; et al. Arecanut (*Areca catechu* L.) seed polyphenol improves osteoporosis via gut-serotonin mediated Wnt/β-catenin pathway in ovariectomized rats. *J. Funct. Foods* **2021**, *84*, 104598. [CrossRef]
18. Gu, Z.; Zhu, Y.; Mei, F.; Dong, X.; Xia, G.; Shen, X. Tilapia head glycolipids protect mice against dextran sulfate sodium-induced colitis by ameliorating the gut barrier and suppressing NF-kappa B signaling pathway. *Int. Immunopharmacol.* **2021**, *96*, 107802. [CrossRef]

19. Mei, F.; Duan, Z.; Chen, M.; Lu, J.; Zhao, M.; Li, L.; Shen, X.; Xia, G.; Chen, S. Effect of a high-collagen peptide diet on the gut microbiota and short-chain fatty acid metabolism. *J. Funct. Foods* **2020**, *75*, 104278. [CrossRef]
20. Wei, J.; Zhao, M.; Meng, K.; Xia, G.; Pan, Y.; Li, C.; Zhang, W. The Diuretic Effects of Coconut Water by Suppressing Aquaporin and Renin-Angiotensin-Aldosterone System in Saline-Loaded Rats. *Front. Nutr.* **2022**, *9*, 930506. [CrossRef]
21. Zhao, M.; Mei, F.; Lu, J.; Xiang, Q.; Xia, G.; Zhang, X.; Liu, Z.; Zhang, C.; Shen, X.; Zhong, Q. Gadus morhua Eggs Sialoglycoprotein Prevent Estrogen Deficiency-Induced High Bone Turnover by Controlling OPG/RANKL/TRAF6 Pathway and Serum Metabolism. *Front. Nutr.* **2022**, *9*, 871521. [CrossRef] [PubMed]
22. Mei, F.; Meng, K.; Gu, Z.; Yun, Y.; Zhang, W.; Zhang, C.; Zhong, Q.; Pan, F.; Shen, X.; Xia, G.; et al. Arecanut (*Areca catechu* L.) Seed Polyphenol-Ameliorated Osteoporosis by Altering Gut Microbiome via LYZ and the Immune System in Estrogen-Deficient Rats. *J. Agric. Food Chem.* **2021**, *69*, 246–258. [CrossRef] [PubMed]
23. Zhu, Y.; Liu, S.; Mei, F.; Zhao, M.; Xia, G.; Shen, X. Tilapia nilotica Head Lipids Improved Bone Loss by Regulating Inflammation and Serum Metabolism through Gut Microbiota in Ovariectomized Rats. *Front. Nutr.* **2021**, *8*, 792793. [CrossRef]
24. Fu, Y.; Zu, Y.; Liu, W.; Efferth, T.; Zhang, N.; Liu, X.; Kong, Y. Optimization of luteolin separation from pigeonpea [*Cajanus cajan* (L.) Millsp.] leaves by macroporous resins. *J. Chromatogr. A* **2006**, *1137*, 145–152. [CrossRef] [PubMed]
25. Jia, G.; Lu, X. Enrichment and purification of madecassoside and asiaticoside from Centella asiatica extracts with macroporous resins. *J. Chromatogr. A* **2008**, *1193*, 136–141. [CrossRef]
26. Nair, V.D.; Yuen, T.; Olanow, C.W.; Sealfon, S.C. Early single cell bifurcation of pro- and antiapoptotic states during oxidative stress. *J. Biol. Chem.* **2004**, *279*, 27494–27501. [CrossRef]
27. Boby, N.; Abbas, M.A.; Lee, E.B.; Im, Z.E.; Hsu, W.H.; Park, S.C. Protective Effect of Pyrus ussuriensis Maxim. Extract against Ethanol-Induced Gastritis in Rats. *Antioxidants* **2021**, *10*, 439. [CrossRef]
28. Franke, A.; Teyssen, S.; Singer, M.V. Alcohol-related diseases of the esophagus and stomach. *Dig. Dis.* **2005**, *23*, 204–213. [CrossRef]
29. Simoes, S.; Lopes, R.; Campos, M.; Marruz, M.J.; Da, C.M.; Corvo, L. Animal models of acute gastric mucosal injury: Macroscopic and microscopic evaluation. *Anim. Models Exp. Med.* **2019**, *2*, 121–126. [CrossRef]
30. Aziz, R.S.; Siddiqua, A.; Shahzad, M.; Shabbir, A.; Naseem, N. Oxyresveratrol ameliorates ethanol-induced gastric ulcer via downregulation of IL-6, TNF-alpha, NF-kB, and COX-2 levels, and upregulation of TFF-2 levels. *Biomed. Pharmacother.* **2019**, *110*, 554–560. [CrossRef]
31. Wu, X.; Huang, Q.; Xu, N.; Cai, J.; Luo, D.; Zhang, Q.; Su, Z.; Gao, C.; Liu, Y. Antioxidative and Anti-Inflammatory Effects of Water Extract of *Acrostichum aureum* Linn. against Ethanol-Induced Gastric Ulcer in Rats. *Evid. Based Complement. Altern. Med.* **2018**, *2018*, 3585394. [CrossRef] [PubMed]
32. Byeon, S.; Oh, J.; Lim, J.S.; Lee, J.S.; Kim, J.S. Protective Effects of Dioscorea batatas Flesh and Peel Extracts against Ethanol-Induced Gastric Ulcer in Mice. *Nutrients* **2018**, *10*, 1680. [CrossRef] [PubMed]
33. Glavin, G.B.; Szabo, S. Experimental gastric mucosal injury: Laboratory models reveal mechanisms of pathogenesis and new therapeutic strategies. *FASEB J.* **1992**, *6*, 825–831. [CrossRef] [PubMed]
34. Teschke, R. Alcoholic Liver Disease: Alcohol Metabolism, Cascade of Molecular Mechanisms, Cellular Targets, and Clinical Aspects. *Biomedicines* **2018**, *6*, 106. [CrossRef]
35. Suzuki, H.; Nishizawa, T.; Tsugawa, H.; Mogami, S.; Hibi, T. Roles of oxidative stress in stomach disorders. *J. Clin. Biochem. Nutr.* **2012**, *50*, 35–39. [CrossRef]
36. Zhang, Y.; Wang, B.Y.; Jia, Z.; Scarlett, C.J.; Sheng, Z.L. Adsorption/desorption characteristics and enrichment of quercetin, luteolin and apigenin from Flos populi using macroporous resin. *Rev. Bras. Farmacogn.* **2019**, *29*, 69–76. [CrossRef]
37. Hui, S.; Fangyu, W. Protective effects of bilobalide against ethanol-induced gastric ulcer in vivo/vitro. *Biomed. Pharmacother.* **2017**, *85*, 592–600. [CrossRef]
38. Barboza, K.; Coco, L.Z.; Alves, G.M.; Peters, B.; Vasquez, E.C.; Pereira, T.; Meyrelles, S.S.; Campagnaro, B.P. Gastroprotective effect of oral kefir on indomethacin-induced acute gastric lesions in mice: Impact on oxidative stress. *Life Sci.* **2018**, *209*, 370–376. [CrossRef]
39. Witko-Sarsat, V.; Rieu, P.; Descamps-Latscha, B.; Lesavre, P.; Halbwachs-Mecarelli, L. Neutrophils: Molecules, functions and pathophysiological aspects. *Lab. Investig.* **2000**, *80*, 617–653. [CrossRef]
40. Singh, A.; Rangasamy, T.; Thimmulappa, R.K.; Lee, H.; Osburn, W.O.; Brigelius-Flohe, R.; Kensler, T.W.; Yamamoto, M.; Biswal, S. Glutathione peroxidase 2, the major cigarette smoke-inducible isoform of GPX in lungs, is regulated by Nrf2. *Am. J. Respir. Cell Mol. Biol.* **2006**, *35*, 639–650. [CrossRef]
41. Kansanen, E.; Kuosmanen, S.M.; Leinonen, H.; Levonen, A.L. The Keap1-Nrf2 pathway: Mechanisms of activation and dysregulation in cancer. *Redox Biol.* **2013**, *1*, 45–49. [CrossRef] [PubMed]
42. Guo, X.; Shin, V.Y.; Cho, C.H. Modulation of heme oxygenase in tissue injury and its implication in protection against gastrointestinal diseases. *Life Sci.* **2001**, *69*, 3113–3119. [CrossRef] [PubMed]
43. Vijayan, V.; Wagener, F.; Immenschuh, S. The macrophage heme-heme oxygenase-1 system and its role in inflammation. *Biochem. Pharmacol.* **2018**, *153*, 159–167. [CrossRef] [PubMed]
44. Chang, X.; Luo, F.; Jiang, W.; Zhu, L.; Gao, J.; He, H.; Wei, T.; Gong, S.; Yan, T. Protective activity of salidroside against ethanol-induced gastric ulcer via the MAPK/NF-κB pathway in vivo and in vitro. *Int. Immunopharmacol.* **2015**, *28*, 604–615. [CrossRef] [PubMed]

45. Kim, E.K.; Choi, E.J. Compromised MAPK signaling in human diseases: An update. *Arch. Toxicol.* **2015**, *89*, 867–882. [CrossRef]
46. Wang, Z.; Ka, S.O.; Lee, Y.; Park, B.H.; Bae, E.J. Butein induction of HO-1 by p38 MAPK/Nrf2 pathway in adipocytes attenuates high-fat diet induced adipose hypertrophy in mice. *Eur. J. Pharmacol.* **2017**, *799*, 201–210. [CrossRef]
47. Sun, G.Y.; Chen, Z.H.; Jasmer, K.J.; Chuang, D.Y.; Gu, Z.Z.; Hannink, M.; Simonyi, A. Quercetin Attenuates Inflammatory Responses in BV-2 Microglial Cells: Role of MAPKs on the Nrf2 Pathway and Induction of Heme Oxygenase-1. *PLoS ONE* **2015**, *10*, e0141509. [CrossRef]
48. Wang, J.; Huang, X.; Zhang, K.; Mao, X.; Ding, X.; Zeng, Q.; Bai, S.; Xuan, Y.; Peng, H. Vanadate oxidative and apoptotic effects are mediated by the MAPK-Nrf2 pathway in layer oviduct magnum epithelial cells. *Metallomics* **2017**, *9*, 1562–1575. [CrossRef]
49. Khan, N.M.; Haseeb, A.; Ansari, M.Y.; Devarapalli, P.; Haynie, S.; Haqqi, T.M. Wogonin, a plant derived small molecule, exerts potent anti-inflammatory and chondroprotective effects through the activation of ROS/ERK/Nrf2 signaling pathways in human Osteoarthritis chondrocytes. *Free Radic. Biol. Med.* **2017**, *106*, 288–301. [CrossRef]

antioxidants

MDPI

Article

Quercus acuta Thunb. Suppresses LPS-Induced Neuroinflammation in BV2 Microglial Cells via Regulating MAPK/NF-κB and Nrf2/HO-1 Pathway

Jae Kwang Kim, Hye Jin Yang and Younghoon Go *

Korean Medicine (KM)-Application Center, Korea Institute of Oriental Medicine (KIOM), Daegu 41062, Korea
* Correspondence: gotra827@kiom.re.kr

Abstract: Microglial activation-mediated neuroinflammation is associated with the pathogenesis of neurodegenerative disorders. Therefore, the management of microglial cell activation and their inflammatory response is an important therapeutic approach for preventing neurodegenerative diseases. *Quercus acuta* Thunb. (QA) (*Fagaceae*) is a tree found in Korea, China, and Japan. The current study investigated the anti-neuroinflammatory effects of QA and its mechanism of action in lipopolysaccharide (LPS)-stimulated BV2 microglial cells. Pretreatment with a methanol extract of dried QA stems (QAE) inhibited the production of nitric oxide and proinflammatory cytokines and decreased the expression of inducible nitric oxide synthase, cyclooxygenase-2 in LPS-stimulated BV2 microglial cells. Furthermore, it inhibited the phosphorylation and degradation of inhibitory κBα and decreased the nuclear translocation and phosphorylation of nuclear factor-κB (NF-κB). Moreover, QAE inhibited the phosphorylation of extracellular signal-regulated kinase, p38 and c-Jun N-terminal kinase, which is known as mitogen-activated protein kinase (MAPK). Additionally, QAE treatment increased heme oxygenase-1 (HO-1) expression by activating the nuclear factor erythroid 2-related factor 2 (Nrf2) signaling, thereby ameliorating LPS-induced intracellular hydrogen peroxide production. Finally, it was found that catechin and taxifolin, two phytochemicals of QAE, also reduced the expression of inflammatory mediators. These findings suggest that QA is beneficial for preventing microglia-mediated neuroinflammatory response through the inhibition of NF-κB, MAPK and the activation of Nrf2/HO-1 signaling pathways.

Keywords: *Quercus acuta* Thunb.; neuroinflammation; microglia; NF-κB; MAPK; Nrf2; HO-1

1. Introduction

Microglia, innate immune effector cells that reside in the central nerve system (CNS), perform pivotal roles in brain homeostasis. Microglia exert neuroprotective functions in the healthy brain by providing innate immunity, removing apoptotic cells and modifying synaptic connectivity [1]. On the other hand, stimuli such as pathogenic insults, traumatic brain injury and amyloid, tau protein activate the microglia, prompting them to release cytokines, chemokines and reactive oxygen species (ROS) [2]. Microglial activation-driven neuroinflammation eventually causes neuronal cell death and synaptic dysfunction since neurons are vulnerable to these inflammatory mediators [1,3]. Given that increased microglial activation is positively correlated with neuronal dysfunction and abnormal protein aggregation in patients with Alzheimer's and Parkinson's disease, regulating the inflammatory response of microglia is regarded as an important prophylactic target to prevent the development and progress of neurodegenerative disease [4,5].

Microglial cells mediate an inflammatory response against pathogenic stimuli by recognizing pathogen-associated molecular patterns or damage-associated molecular patterns via toll-like receptor (TLR) [6]. Endotoxin binding to TLR on the surface of microglia induces intracellular signaling transduction, which activates the nuclear factor-κB (NF-κB) signaling pathway [7]. NF-κB is a key transcription factor that regulates immune response, inflammation and cell survival. NF-κB signaling activation in microglial cells increases the transcription of proinflammatory enzymes such as inducible nitric oxide synthase (iNOS) and cyclooxygenase-2 (COX-2) and cytokines, eliciting neuroinflammation [2,8,9].

ROS plays a role in cell signaling transmission and homeostasis maintenance. As a secondary messenger, ROS contributes to the modulation of microglial phagocytic activity and inflammatory response [10]. Proinflammatory stimulation induces the interaction between TLR and NADPH oxidase (NOX), causing intracellular ROS generation [11,12]. The induction of intracellular H_2O_2 facilitates inflammatory response by activating NF-κB and MAPK pathways [10,12]. Numerous studies have found that inducing antioxidant proteins such as heme oxygenase-1 (HO-1) by activation of the nuclear factor erythroid-2-related factor 2 (Nrf2), ameliorates inflammatory response in microglia and macrophages [12–14]. Therefore, several medicinal plants and natural products with anti-inflammatory and antioxidative stress properties, have gained interest as therapeutic agents in inflammation-related disorders.

The tree *Quercus acuta* Thunb. (QA) is found in East Asia including Korea, China and Japan [15]. Many species of the genus *Quercus* have been used as medicinal plants to treat septic and gastrointestinal disorders [16]. Previous research on the biological efficacy of QA has shown that the leaf extract of QA has anti-hyperuricemia, antibacterial properties [15,17]. Furthermore, phenolic compounds isolated from QA stem such as catechin, epicatechin and taxifolin have exhibited anti-inflammatory and radical scavenging activity [18]. Moreover, the anti-viral effect of QA on herpes simplex virus-1 (HSV-1) replication via the inhibition of ROS generation and NF-κB signaling has been studied [19]. Despite the fact that QA has antioxidant and anti-inflammatory characteristics, its pharmacological action in activated microglia is unknown. Therefore, the present study investigated the anti-neuroinflammatory activity of QA in lipopolysaccharide (LPS)-stimulated BV2 microglial cells. Furthermore, we explored the regulatory effect of QA on NF-κB, MAPK and Nrf2/HO-1 signaling pathways to determine the molecular mechanisms involved.

2. Materials and Methods

2.1. Materials and Reagents

Methanol extract of dried QA stems (QAE) (Voucher No. PB2418) was provided by the Korea Plant Extract Bank (Cheongju, Republic of Korea), and dissolved in dimethyl sulfoxide (100 mg/mL) for cell treatment. Anti-iNOS (NB300-605) antibody was obtained from Novus Biologicals (Centennial, CO, USA). Anti-COX-2 (#12282), anti-p65 (#8242), anti-phosphorylated p65 (Ser^{536}) (#3033), anti-inhibitory κBα (IκBα) (#4814), anti-phosphorylated IκBα (#9246), anti-TATA-box binding protein (TBP) (#8515), anti-extracellular signal-regulated kinase (ERK) (#4377), anti-phosphorylated ERK (Thr^{202}/Tyr^{204}) (#9102), anti-p38 (#9212), anti-phosphorylated p38 (Thr^{180}/Tyr^{182}) (#9211), anti-c-Jun N-terminal kinase (JNK) (#9252), anti-phosphorylated JNK (Thr^{183}/Tyr^{185}) (#9251), anti-HO-1 (#82206) antibodies were purchased from Cell Signaling Technology (Beverly, MA, USA). Anti-β-actin (sc-81178), anti-Nrf2 (sc-722) antibodies were obtained from Santa Cruz Biotechnology (Santa Cruz, CA, USA). Anti-phosphorylated Nrf2 (Ser^{40}) (ab76026), anti-glutamate-cysteine ligase catalytic subunit (GCLC) (ab190685) antibodies were from Abcam (Cambridge, MA, USA). Anti-Sestrin-2 (SESN2) (10795-1-AP) antibody was provided by Proteintech (Chicago, IL, USA). LPS from *Escherichia coli* and the reference standards of catechin and taxifolin (purity over 97%) used in ultra-high performance liquid chromatography coupled to high-resolution Orbitrap mass spectrometry (UHPLC-UV-HRMS) analysis were obtained from Sigma-Aldrich (St. Louis, MO, USA). MS-grade products, water, acetonitrile, and formic acid used as mobile phase were purchased from Thermo Fisher Scientific (Rockford, IL, USA).

2.2. Cell Culture

BV2, immortalized murine microglial cells, were kindly donated from Professor Kyungho Suk (Kyungpook National University, Daegu, Korea), and were cultured in

Dulbecco's modified Eagle's medium (HyClone Laboratories, Logan, UT, USA) with 10% fetal bovine serum, penicillin (100 U/mL) and streptomycin (100 μg/mL) (HyClone Laboratories), under humidified conditions with 5% CO_2, at 37 °C temperature.

2.3. Cell Viability Assay

BV2 microglial cells were plated in 96-well plates (30,000 cells/well) to determine cell viability. After treatment, cells were further incubated with CCK-8 (Dojindo Laboratories, Kumamoto, Japan) for 2 h. Absorbance at 450 nm was detected using an automated plate reader (SpectraMax i3, Molecular Devices, Sunnyvale, CA, USA). Relative cell viability was calculated using the following equation:

$$\text{Relative cell viability (\%)} = \left(\frac{\text{Absorbance of the treated group}}{\text{Absorbance of the control group}}\right) \times 100$$

2.4. Measurement of Nitric Oxide (NO) Production

To measure NO production, conditioned media was reacted with the same volume of Griess reagent for 10 min. Absorbance at 550 nm was measured using an automated plate reader (Molecular Devices).

2.5. Preparation of Whole Cell Lysates and Nuclear Fraction Extract, and Immunoblot Assay

Cells were lysed with radioimmunoprecipitation (RIPA) buffer containing protease and phosphatase inhibitor cocktail (Roche Molecular Biochemicals, Mannheim, Germany), and incubated on ice for 40 min with vigorous vortex every 10 min. The cell lysate was centrifuged at 15,000× g for 30 min, and the supernatant was collected as whole cell lysates. Nuclear fraction was obtained using a commercial nuclear extraction kit (Thermo Fisher Scientific), according to the manufacturer's instruction. The bicinchoninic acid assay was performed to determine the protein content (Thermo Fisher Scientific). An equal amount of protein was then separated using sodium dodecyl sulfate polyacrylamide gel electrophoresis (SDS-PAGE), and transferred to a polyvinylidene fluoride membrane (Millipore, Bedford, MA, USA). After blocking with 5% skim milk, the membrane was allowed to reacted with primary antibody overnight at 4 °C, and subsequently incubated with horseradish peroxide conjugated-secondary antibody (Cell Signaling Technology). The chemiluminescence density of the protein of interest was detected by the Alliance Q9 chemiluminescence imager (Uvitec, Cambridge, UK), and densitometric analysis was performed using ImageJ (US National Institutes of Health, Bethesda, MD, USA). Equal protein loading was verified by β-actin (whole cell lysates) or TBP (nuclear fraction) immunoblotting.

2.6. RNA Isolation and Real-Time PCR

Total RNA was extracted using QIAzol Lysis Reagent (Qiagen, Hilden, Germany), as directed by the manufacturer. Reverse transcription was conducted using Maxima Reverse Transcriptase (Thermo Fisher Scientific). Real-time PCR was performed using SYBR Green qPCR Master Mix (Thermo Fisher Scientific) and CFX real-time PCR detection system (Bio-Rad, Hercules, CA, USA). Glyceraldehyde-3-phophate dehydrogenase (*GAPDH*) was used as the reference gene. The primer sequences used for the RT-qPCR reaction are given in Table 1. Relative gene expression was calculated using the $2^{-\Delta\Delta CT}$ method [20].

Table 1. Oligonucleotide sequences used in RT-qPCR experiments.

Gene Symbol	Primer Sequence (Sense, Anti-Sense)	Accession Number	Product Size
NOS2	5′-GGCAGCCTGTGAGACCTTTG-3′, 5′-GCATTGGAAGTGAAGCGTTTC-3′	NM_010927.4	72 bp
PTGS2	5′-TGAGTACCGCAAACGCTTCTC-3′, 5′-TGGACGAGGTTTTTCCACCAG-3′	NM_011198.4	151 bp
TNF	5′-ATGAGCACAGAAAGCATGAT-3′, 5′-TACAGGCTTGTCACTCGAAT-3′	NM_013693.3	276 bp
IL1B	5′-ATGGCAACTGTTCCTGAACT-3′, 5′-CAGGACAGGTATAGATTCTT-3′	NM_008361.4	563 bp
IL6	5′-TTCCATCCAGTTGCCTTCTT-3′, 5′-ATTTCCACGATTTCCCAGAG-3′	NM_031168.2	170 bp
CCL2	5′-TGATCCCAATGAGTAGGCTGG-3′, 5′-ATGTCTGGACCCATTCCTTCT-3′	NM_011333.3	132 bp
HOMX1	5′-GGGAATTTATGCCATGTAAA-3′, 5′-AGAACAGCTGCTTTTACAGG-3′	NM_010442.2	294 bp
GAPDH	5′-AACGACCCCTTCATTGAC-3′, 5′-TCCACGACATACTCAGCAC-3′	NM_008084.3	191 bp

2.7. Measurement of Cytokine Release

BV2 microglial cells were preincubated with QAE (25–100 μg/mL) for 3 h, and exposed to LPS for 12 h. After treatment, the level of cytokine in cultured media was measured using commercial enzyme-linked immunosorbent assay (ELISA) kit (Thermo Fisher Scientific) according to the manufacturer's instruction.

2.8. Immunocytochemistry

Cells were seeded onto coverslips (Marienfeld, Lauda-Königshofen, Germany). After treatment, cells were fixed with 4% paraformaldehyde for 15 min, and permeabilized with 0.25% Triton X-100 for 30 min at room temperature. After blocking with 1% bovine serum albumin, cells were incubated with primary antibody (anti-p65) for overnight at 4 °C, and subsequently reacted with Alexa Fluor 488-conjugated secondary antibody (Thermo Fisher Scientific) at room temperature for 2 h. After staining nuclei with Hoechst 33342 (Thermo Fisher Scientific), coverslips were mounted with fluorescence mounting solution (Dako, Glostrup, Denmark), and observed under a Lionheart FX microscope (BioTek, Vermont, Winooski, VT, USA).

2.9. Measurement of H_2O_2 Production

After treatment, BV2 microglial cells were stained with H_2DCF-DA (10 μM) for 30 min. The fluorescence intensity of dichlorofluorescein (DCF) was detected at a wavelength of 485/530 nm (emission/excitation) using an automated plate reader (Molecular Devices).

2.10. UHPLC-UV-HRMS Analysis

Phytochemical identification in QAE was performed by a Q-Exactive quadrupole Orbitrap mass spectrometer coupled with Thermo Dionex UltiMate 3000 system (UHPLC-UV-HRMS, Thermo Fisher Scientific). Chromatographic separation was carried out on a Waters Acquity BEH C18 column (2.1 mm i.d. × 100 mm, 1.7 μm) equipped with VanGuard XBridge BEH C18 pre-column (2.1 mm i.d. × 5 mm, 1.7 μm), with the oven temperature maintained at 40 °C. The mobile phase conditions for analysis were composed of 0.1% formic acid (*v/v*) in water (A) and acetonitrile (B), and the flow rate was maintained at 0.3 mL/min. Full MS and MS/MS spectrums were obtained via UHPLC-UV-HRMS that equipped with a heated electrospray ionization (HESI) interface. The data acquisition and analysis of all data obtained through this study were processed through Xcalibur v.4.2 and Tracefinder v.4.0 software (Thermo Fisher Scientific). In addition, the analysis method used in this study was referred and applied to the previously reported methods [21] and briefly described.

2.11. *Statistical Analysis*

To determine the significance difference among experimental groups, unpaired student's *t*-test or one-way analysis of variance (ANOVA) was performed. According to results from one-way ANOVA, Tukey's honest difference was conducted as post hoc analysis. All numerical data were expressed as bar chart (mean ± standard deviation) or box plot of at least three independent experiments. p Values under 0.05 were considered as statistically significant. The statistical analysis test was accomplished using GraphPad Prism 6.0 software (GraphPad Software, San Diego, CA, USA).

3. Results

3.1. *QAE Inhibited NO Production in LPS-Stimulated BV2 Microglial Cells*

First, the effect of QAE on cell viability was assessed to evaluate the cytotoxicity of QAE and the suitable dose on BV2 microglial cells. The final concentration of QAE up to 100 μg/mL (24 h) had no discernible cytotoxic effect on the viability of BV2 microglial cells (Figure 1a). Furthermore, the NO released in the medium was measured by Griess assay to determine whether QAE possessed any anti-inflammatory activity. Excessive NO production has been linked to the development of neurodegenerative diseases via nitrosative stress, protein misfolding, and neuronal damage [2,8]. After pretreatment with QAE (25–100 μg/mL, 3 h), BV2 microglial cells were exposed to LPS (100 ng/mL) for 24 h. As expected, the LPS treatment significantly increased NO production. Pretreatment with QAE (25–100 μg/mL) reduced NO generation in a dose-dependent manner (Figure 1b). Based on these results, further experiments were carried out at QAE concentrations in the range of 25–100 μg/mL.

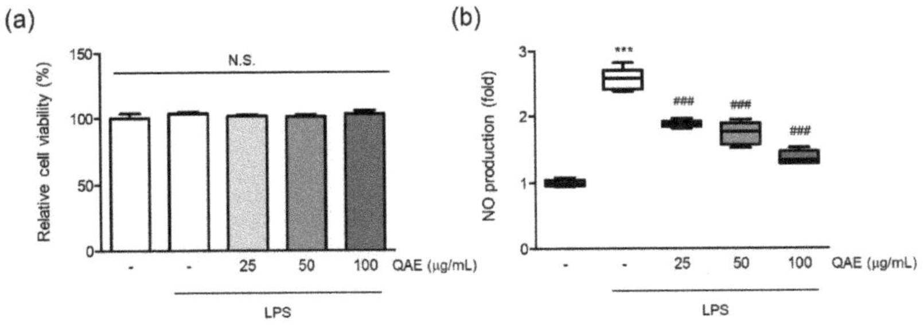

Figure 1. Methanol extract of *Quercus acuta* Thunb. stem (QAE) inhibits nitric oxide (NO) production in lipopolysaccharide (LPS)-stimulated BV2 microglial cells. BV2 microglial cells were pretreated with QAE (25–100 μg/mL) for 3 h, and then exposed to LPS (100 ng/mL) for 24 h. (**a**) Relative cell viability. After treatment, cell viability was measured using CCK-8 (n = 3). (**b**) NO production. Produced NO in media was determined by Griess assay (n = 5). Significant versus control, *** $p < 0.001$; Significant versus LPS alone-treated group, ### $p < 0.001$; N.S., not significant.

3.2. *QAE Suppressed iNOS and COX-2 Expression in LPS-Stimulated BV2 Microglial Cells*

iNOS is a key enzyme responsible for inflammatory response by synthesizing NO from L-arginine [22]. The protein expression of iNOS was evaluated by immunoblot assay if the inhibitory effect of QAE on NO production was mediated by iNOS expression. As expected, LPS stimulation (100 ng/mL, 12 h) significantly increased the expression of iNOS (Figure 2a), and pretreatment with QAE (25–100 μg/mL) significantly suppressed iNOS protein level. According to protein expression, the mRNA level of iNOS was decreased by QAE pretreatment at 50 and 100 μg/mL (Figure 2b). COX-2 is an enzyme catalyzing the conversion of arachidonic acid to prostaglandins which act as an inflammatory mediator [9]. QAE (100 μg/mL) pretreatment decreased COX-2 protein expression (Figure 2a), and significantly suppressed mRNA expression of COX-2 (Figure 2b).

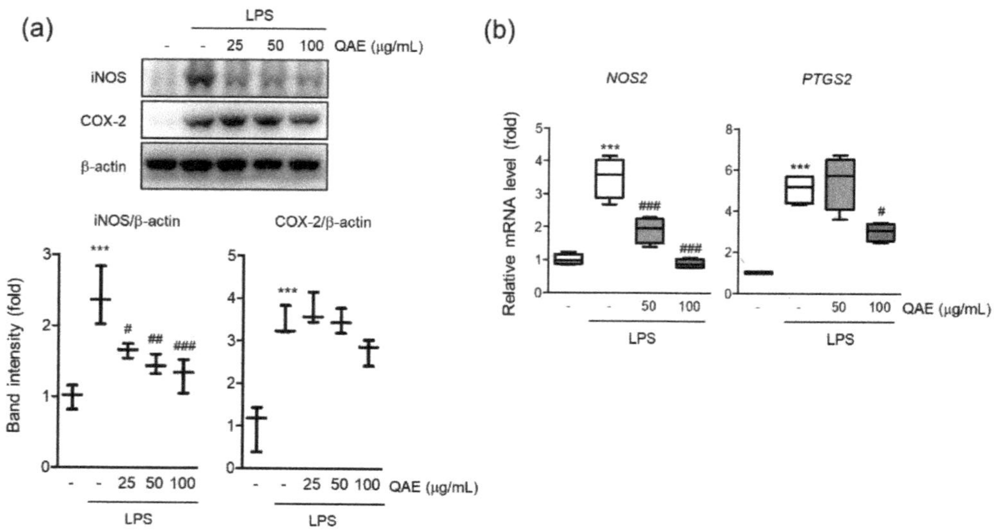

Figure 2. QAE decreases inducible nitric oxide synthase (iNOS) and cyclooxygenase-2 (COX-2) expression in LPS-stimulated BV2 microglial cells. BV2 microglial cells were pretreated with QAE (25–100 µg/mL) for 3 h, and further stimulated with LPS (100 ng/mL) for (**a**) 12 h, (**b**) 8 h. (**a**) Protein expression of iNOS and COX-2. β-actin immunoblotting was used to verify equal protein loading (n = 3). (**b**) mRNA expression of iNOS (*NOS2*) and COX-2 (*PTGS2*). Relative gene expression was quantified by RT-qPCR, and normalized by mouse *GAPDH* expression of each sample (n = 4). Significant versus control, *** $p < 0.001$; Significant versus LPS alone-treated group, # $p < 0.05$, ## $p < 0.01$, ### $p < 0.001$.

3.3. QAE Inhibited Proinflammatory Cytokine Production in LPS-Stimulated BV2 Microglial Cells

Microglia are the major source of proinflammatory cytokines in the brain. Endotoxin-driven TLRs pathway activation exacerbates neuroinflammation by expressing proinflammatory cytokines such as interleukin (IL)-1β, IL-6, and tumor necrosis factor-α (TNF-α) [6]. To determine the effect of QAE on the production of proinflammatory cytokines in LPS-stimulated BV2 microglial cells, the released levels of TNF-α, IL-1β, and IL-6 cytokines in conditioned media were measured by ELISA assay. As is well known, LPS stimulation (100 ng/mL for 12 h) significantly increases the expression of TNF-α, IL-1β and IL-6 in conditioned media. QAE pretreatment (25–100 µg/mL, 3 h) significantly decreased the expression of TNF-α, IL-1β, and IL-6 (Figure 3a). LPS stimulation (100 ng/mL, for 8 h) increased the mRNA levels of TNF-α, IL-1β, IL-6, and monocyte chemoattracted protein-1 (MCP-1), similar to cytokine release in conditioned media, whereas QAE pretreatment (50, 100 µg/mL, 3 h) significantly reduced cytokine mRNA expression (Figure 3b).

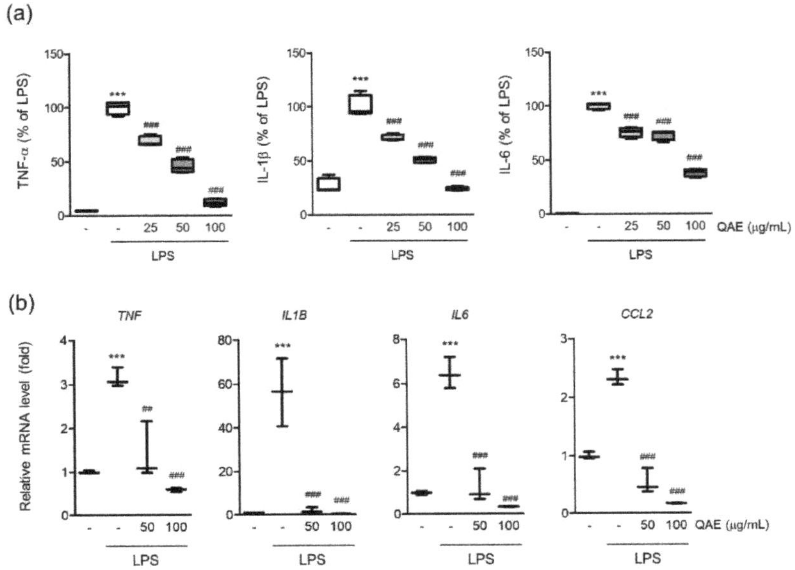

Figure 3. QAE inhibits proinflammatory cytokine expression in LPS-stimulated BV2 microglial cells. BV2 microglial cells were preincubated with QAE (25–100 μg/mL) for 3 h, and subsequently exposed to LPS (100 ng/mL) for (**a**) 12 h or (**b**) 8 h. (**a**) Level of proinflammatory cytokine production. The production of tumor necrosis factor-α (TNF-α), interleukin (IL)-1β, and IL-6 in conditioned media was determined by enzyme-linked immunosorbent assay ($n = 4$). (**b**) mRNA expression of proinflammatory cytokines. Relative mRNA expression of TNF-α (*TNF*), IL-1β (*IL1B*), IL-6 (*IL6*), and monocyte chemoattracted protein-1 (MCP-1) (*CCL2*) was measured by RT-qPCR, and normalized by *GAPDH* expression of each sample ($n = 3$). Significant versus control, *** $p < 0.001$; Significant versus LPS alone-treated group, ## $p < 0.01$, ### $p < 0.001$.

3.4. QAE Inhibited NF-κB Signaling Activation in LPS-Stimulated BV2 Microglial Cells

To determine the molecular mechanisms of the anti-inflammatory effect of QAE, we investigated the effect of QAE on the NF-κB signaling pathway in LPS-stimulated BV2 microglial cells. As expected, LPS treatment (100 ng/mL, 1 h) significantly increased phosphorylation and degradation of IκB. However, QAE pretreatment (50 and 100 μg/mL, 3 h) significantly inhibited IκB phosphorylation and inhibited degradation of IκB (Figure 4a). To investigate the effect of QAE on the nuclear translocation of NF-κB (p65 subunit), nuclear p65 was detected by immunoblotting and immunofluorescence staining using p65 subunit-specific antibody. LPS stimulation (100 ng/mL, 1 h) enhanced the nuclear translocation of p65, which was considerably inhibited by QAE pretreatment (50 or 100 μg/mL, 3 h) (Figure 4b,c). Furthermore, we examined the effect of QAE on phosphorylation at Ser^{536} of the p65 subunit by immunoblotting. It is reported that LPS-induced phosphorylation of the p65 subunit at Ser^{536} is essential to the elevation of NF-κB transcriptional activity [23]. QAE pretreatment (50, 100 μg/mL, 3 h) significantly inhibited phosphorylation at Ser^{536} of p65 (Figure 4d). Overall, these results suggest that QAE exhibited anti-inflammatory activity through NF-κB inhibition.

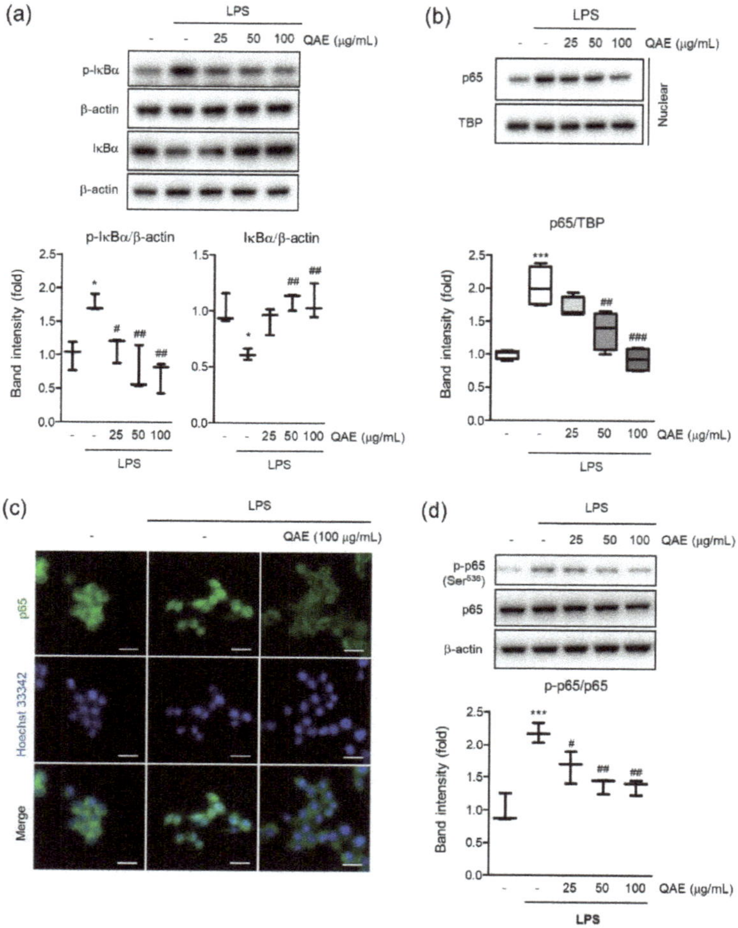

Figure 4. QAE inhibits the nuclear factor-κB (NF-κB) signaling pathway in LPS-stimulated BV2 microglial cells. After QAE pretreatment (25–100 μg/mL) for 3 h, BV2 microglial cells were exposed to LPS (100 ng/mL) for 1 h. (**a**) Phosphorylation and degradation of IκBα. Equal protein loading was verified by β-actin immunoblotting (n = 3). (**b**) p65 expression in nuclear. Equal nuclear protein loading was verified by TATA-box binding protein (TBP) immunoblot (n = 4). (**c**) Immunocytochemistry. After treatment, cells were immunostained with an anti-NF-κB (p65) antibody and Alexa Fluor 488-conjugated secondary antibody. Scale bar represents 25 μm. Hoechst 33342 was used for nuclear counterstaining (n = 3). (**d**) Phosphorylation of p65 at Ser^{536}. Expression of phosphorylated p65 was normalized with expression of total p65 immunoblotting (n = 3). Significant versus control, * $p < 0.05$, *** $p < 0.001$; Significant versus LPS alone-treated group, # $p < 0.05$, ## $p < 0.01$, ### $p < 0.001$.

3.5. *QAE Inhibited Activation of MAPK Signaling Pathway*

MAPK signaling is well known to promote the production of inflammatory mediators [24]. To examine whether the anti-inflammatory activity of QAE is mediated by the abrogation of the MAPK signaling pathway, the phosphorylation of ERK, p38, and JNK was measured by immunoblot analysis. LPS stimulation (100 ng/mL, 1 h) significantly increased the phosphorylation of ERK, p38, and JNK. However, the phosphorylation of ERK, p38, and JNK was remarkably reduced in a dose-dependent manner by QAE pretreatment (25–100 μg/mL, 3 h) (Figure 5). These results suggest that QAE exhibited an inhibitory effect on the inflammatory response via inactivation of MAPK signaling.

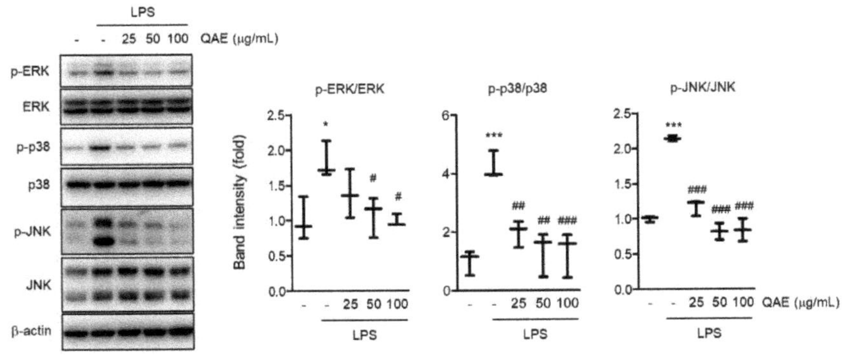

Figure 5. QAE inhibits the mitogen-activated protein kinase (MAPK) signaling pathway in LPS-stimulated BV2 microglial cells. After QAE pretreatment (25–100 μg/mL) for 3 h, BV2 microglial cells were exposed to LPS (100 ng/mL) for 1 h. Expressions of phosphorylated extracellular signal-regulated kinase (ERK), p38, and c-Jun N-terminal kinase (JNK) were normalized by expression of ERK, p38, and JNK protein expression, respectively ($n = 3$). Significant versus control, * $p < 0.05$, *** $p < 0.001$; Significant versus LPS alone-treated group, # $p < 0.05$, ## $p < 0.01$, ### $p < 0.001$.

3.6. QAE Decreased Cellular ROS via Nrf2/HO-1 Activation

Intracellular ROS accumulation by TLR signaling intensifies the inflammatory response [11]. To investigate whether QAE possess inhibitory activity on H_2O_2 production, intracellular H_2O_2 production was measured using H_2DCF-DA in LPS (100 ng/mL, 12 h)-stimulated BV2 microglial cells. As expected, LPS treatment significantly increased the fluorescence intensity of DCF, which reflects the production of H_2O_2 in cells. However, QAE pretreatment (25–100 μg/mL, 3 h) significantly inhibited H_2O_2 production induced by LPS (Figure 6a). Further, the effect of QAE on antioxidant protein expressions such as HO-1, SESN2, and GCLC was examined. QAE treatment (50, 100 μg/mL, 8 h) significantly increased the protein expression of HO-1, SESN2 (Figure 6b). Accordingly, QAE treatment (100, 200 μg/mL, 7 h) significantly increased HO-1 mRNA level (Figure 6c). Because Nrf2 is widely recognized as a master transcriptional factor of antioxidant proteins genes such as HO-1, the phosphorylation of Nrf2 by QAE treatment was monitored. Results showed that the expression of total Nrf2 was constant, whereas the expression of phosphorylated Nrf2 was increased by QAE treatment (50, 100 μg/mL, 3 h) (Figure 6d). In addition, immunoblot analysis using cytosolic and nuclear fraction protein revealed that QAE treatment (50 and 100 μg/mL, 3 h) induced translocation of Nrf2 to the nucleus (Figure 6a). These results suggested that Nrf2 activation by QAE treatment elevates HO-1 expression and decreases H_2O_2 production.

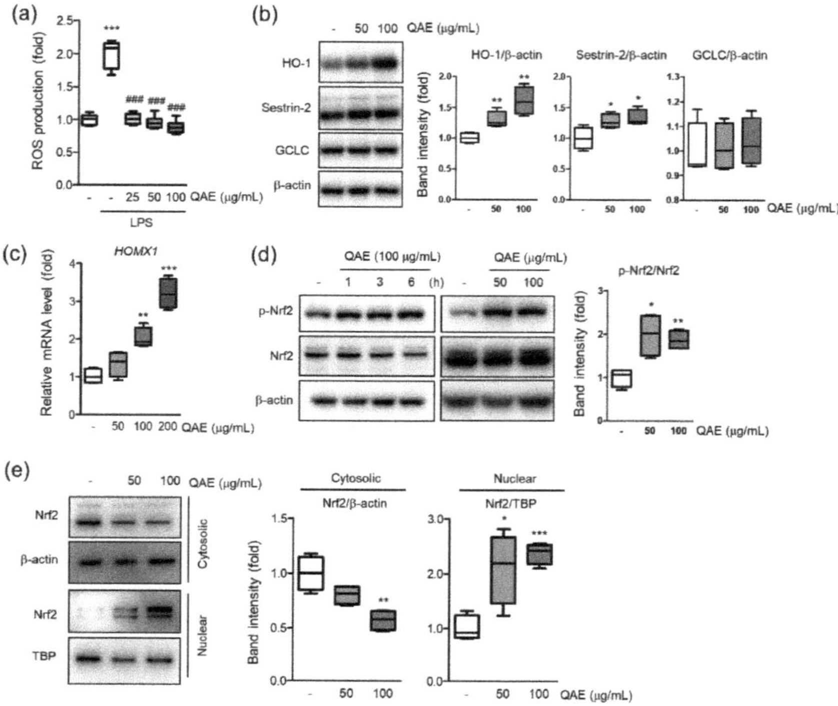

Figure 6. QAE activates the nuclear factor erythroid-2-related factor 2 (Nrf2)/heme oxygenase-1 (HO-1) signaling pathway in BV2 microglial cells. (**a**) H_2O_2 production. Cells were pretreated with QAE (25–100 μg/mL, 3 h), and further exposed to LPS (100 ng/mL, 12 h). After staining with H_2DCF-DA (10 μM, 0.5 h), dichlorofluorescein (DCF) fluorescence intensity was detected using an automated plate reader ($n = 8$). (**b**) Expression of antioxidant protein. Cells were treated with QAE (50, 100 μg/mL) for 8 h. Equal loading of each protein was verified by β-actin immunoblotting ($n = 4$). (**c**) Heme oxygenase-1 (HO-1) (*HOMX1*) mRNA level. Cells were treated with QAE (50–200 μg/mL) for 7 h. Relative mRNA expression of HO-1 was normalized by expression of *GAPDH* ($n = 4$). (**d**) Phosphorylation of Nrf2. Cells were treated with 100 μg/mL of QAE for the indicated time (left), or 50, 100 μg/mL of QAE for 3 h (right) ($n = 4$). (**e**) Expression of Nrf2 in the cytoplasm and nuclear. Expressions of Nrf2 were normalized by β-actin (cytosolic) or TBP (nuclear) immunoblotting ($n = 4$). Significant versus control, * $p < 0.05$, ** $p < 0.01$, *** $p < 0.001$; Significant versus LPS alone-treated group, ### $p < 0.001$.

3.7. Phytochemicals of QAE Inhibited Expression of Proinflammatory Mediators

A previous study revealed five phytochemicals in QAE such as catechin, isoquercitrin, taxifolin, fraxin, and chlorogenic acid [19]. Five phytochemicals showed no obvious cytotoxicity in BV2 microglial cells at the indicated doses in the current investigation (~100 μM, 24 h) (Figure 7a). Next, we investigated whether phytochemicals of QAE possessed anti-inflammatory activities by measuring NO production in LPS (100 ng/mL, 24 h)-stimulated BV2 microglial cells. Catechin and taxifolin exhibited the most potent inhibitory effect on NO production of the five phytochemicals tested (Figure 7b). Furthermore, catechin and taxifolin significantly suppressed iNOS and COX-2 mRNA expression in response to LPS stimulation (100 ng/mL, 8 h) (Figure 7c). Additionally, catechin and taxifolin also significantly reduced mRNA expression of proinflammatory cytokines (TNF-α, IL-1β, IL-6, and MCP-1) induced by LPS stimulation (100 ng/mL, 8 h) (Figure 7d).

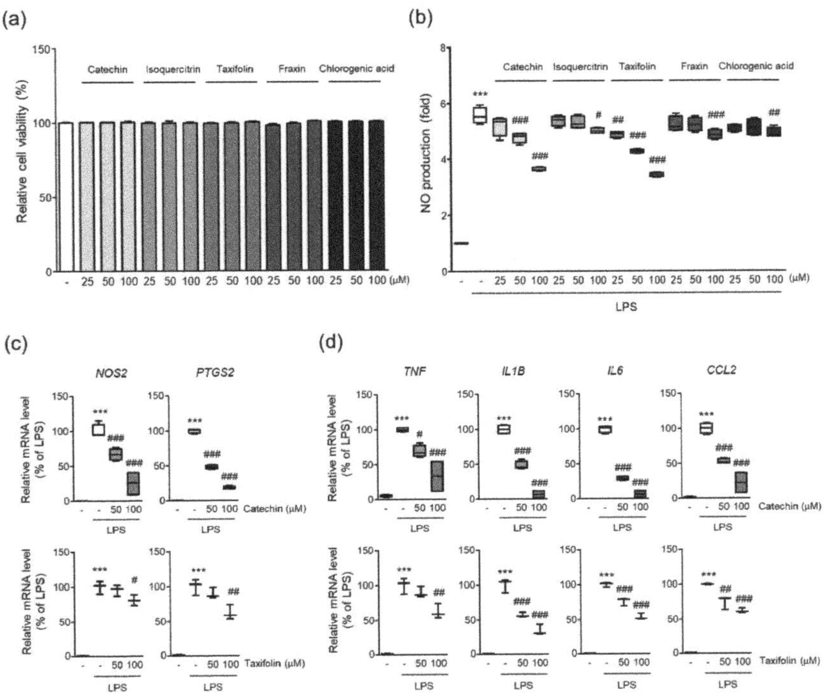

Figure 7. Phytochemicals of QAE inhibited the expression of proinflammatory mediators. (**a**) Relative cell viability. BV2 microglial cells were treated with each phytochemical at indicated dose for 24 h. Cell viability was determined using CCK-8 ($n = 4$). (**b**) NO production. Cells were pretreated with each phytochemical (25–100 μM) for 3 h, and further exposed to LPS (100 ng/mL) for 24 h. Produced NO in media was determined using Griess reagent ($n = 4$). Relative mRNA expression of (**c**) iNOS (*NOS2*), COX-2 (*PTGS2*), and (**d**) TNF-α (*TNF*), IL-1β (*IL1B*), IL-6 (*IL6*), and MCP-1 (*CCL2*). Cells were pretreated with catechin ($n = 4$) or taxifolin ($n = 3$), and subsequently stimulated with LPS (100 ng/mL) for 8 h. Significant versus control, *** $p < 0.001$; Significant versus LPS alone-treated group, # $p < 0.05$, ## $p < 0.01$, ### $p < 0.001$.

3.8. UHPLC-UV-HRMS Analysis of Catechin and Taxifolin in QAE

To identify the phytochemicals in QAE, an analysis was performed using UHPLC-UV-HRMS. As a result of the analysis, two phenolic components, catechin and taxifolin were identified in QAE. Figure 8 shows the extracted ion chromatograms (EIC), which are chromatograms for the precursor ion m/z values of each analyte, and MS/MS spectrums. The component identified at retention time (t_R) of 5.07 min presented precursor ions at m/z 289.0715 $[M\text{-}H]^-$ (Error—0.89 ppm), which was composed of molecular formula $C_{15}H_{14}O_6$ provided by Orbitrap. In addition, the fragment ions of the MS/MS spectrum were m/z 289.0713, 245.0812, 205.0495, and 125.0228. By comparing these results with previous literature, this component was identified as catechin [19,25]. Taxifolin was identified as follows. Taxifolin detected at 6.99 min showed a precursor ion at m/z 303.0508 $[M\text{-}H]^-$ (Error—0.92 ppm), which consists of the molecular formula $C_{15}H_{12}O_7$ provided by Orbitrap. The fragment ions shown in the MS/MS spectrum were m/z 285.0398, 177.0181, and 125.0227. Likewise, it was identified as taxifolin by comparing with the previous literature [19,25,26]. Both catechin and taxifolin identified in QAE were determined by comparing retention time, measured precursor ion, and MS/MS fragments with the values of reference standards, and it was confirmed that the negative ion mode was the optimal condition to for the analysis of both components.

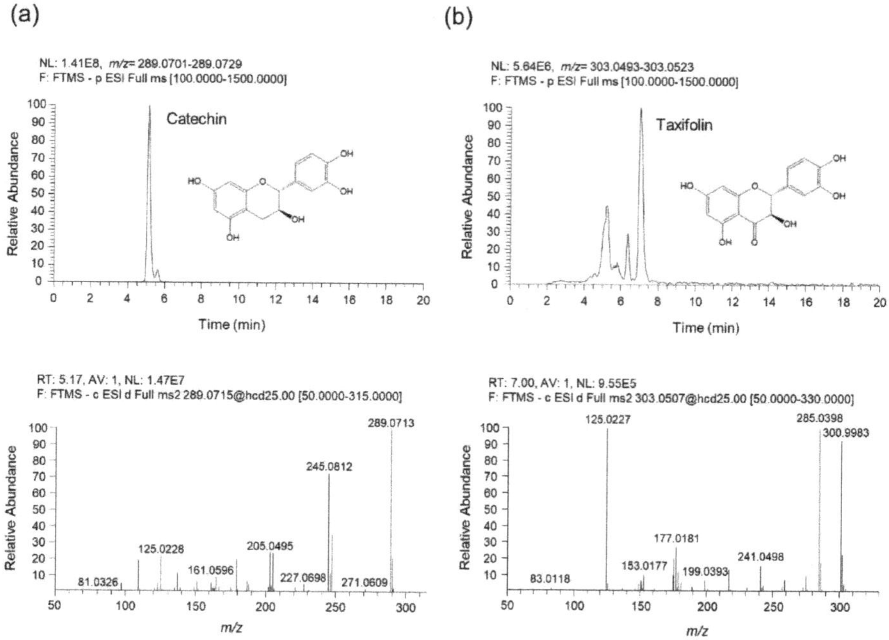

Figure 8. UHPLC-UV-HRMS analysis of catechin and taxifolin in QAE. (**a**) The extracted ion chromatogram (EIC) of catechin identified in QAE (upper chromatogram), and the MS/MS spectrum of catechin (below spectrum). (**b**) The EIC of taxifolin identified in QAE (upper chromatogram), and the MS/MS spectrum of taxifolin (below spectrum). All components were detected in negative ion mode.

4. Discussion

Neuroinflammation has been considered as a significant risk factor for the pathogenesis of neurodegenerative disorders such as Alzheimer's and Parkinson's disease [3,6]. Although the anti-hyperuricemia, antibacterial, antiviral activities of QA have been identified, the beneficial effects of QA on neuroinflammation have not been established [15,17–19]. Therefore, in the present study, we explored the anti-neuroinflammatory properties of QAE using LPS-stimulated BV2 microglial cells.

Prolonged activation of microglia results in the excessive release of cytokines and neurotoxic molecules, which eventually contribute to neurodegeneration [2]. Thus, controlling microglial activation to reduce the emergence of proinflammatory mediators may be advantageous in managing neurodegenerative disease [6]. First, we studied the effect of QAE on the production of proinflammatory mediators in LPS-stimulated BV2 microglial cells. The results revealed that QAE pretreatment significantly inhibited LPS-induced NO production (Figure 1) and iNOS, COX-2 expressions (Figure 2). Furthermore, it reduced mRNA expression and the release of proinflammatory cytokines including TNF-α, IL-1β, and IL-6 by LPS stimulation (Figure 3). These results suggested that QAE exhibited an anti-inflammatory effect via regulating the production of proinflammatory mediators.

Previous research identified the five major phytochemicals of QAE (taxifolin, chlorogenic acid, fraxin, catechin, isoquercitrin), and isoquercitrin exhibiting the most excellent antiviral efficacy against HSV-1 viral infection [19]. In other reports, taxifolin and catechin isolated from QA inhibited LPS-induced NO production in RAW 264.7 macrophage cells [18]. The inhibitory effect of phytochemicals on LPS-induced NO production in BV2 microglial cells was tested to see if the five phytochemicals of QAE had an anti-neuroinflammatory effect. Among the five constituents, catechin and taxifolin showed the most potent anti-inflammatory effect, reducing LPS-induced NO production (Figure 7), and two phenolic compounds in QAE were identified by UHPLC-UV-HRMS analysis (Figure 8).

Furthermore, catechin and taxifolin inhibited LPS-induced mRNA expression of iNOS, COX-2, inflammatory cytokines (TNF-α, IL-1β, IL-6, and MCP-1) (Figure 7b,c). In line with these results, the suppressive effects of catechin and taxifolin on the expression of proinflammatory enzymes in LPS-stimulated BV2 microglial cells were reported previously [27,28]. Based on these findings, the anti-inflammatory activity of QAE is assumed to be connected to the inhibitory efficacy of catechin and taxifolin on the expression of inflammatory mediators.

NF-κB is an essential transcription factor of regulating inflammatory mediators. Signal transduction from TLRs activates IKKα/β, which leads to phosphorylation and degradation of IκB [29]. NF-κB released from IκB translocates to the nucleus and induces the transcription of proinflammatory genes, including iNOS, COX-2, cytokines, and chemokines. In the current study, we revealed that QAE significantly inhibited phosphorylation and degradation of IκB (Figure 4a). Furthermore, immunoblotting results using nuclear protein and immunocytochemistry showed that QAE successfully inhibited the nuclear accumulation of the p65 subunit against LPS stimulation (Figure 4b,c). In addition to nuclear translocation, phosphorylation of p65 on Ser^{536} is essential for transcriptional activation of NF-κB. Evidence showed that phosphorylation of p65 on Ser^{536} by IKKβ enhances NF-κB transcriptional activity, and p65 phosphorylation (Ser^{536}) is independent of the IκBα regulation [23,30]. In our study, phosphorylation of p65 (Ser^{536}) significantly increased by LPS stimulation, whereas QAE pretreatment effectively reduced the level of p-p65 (Ser^{536}) (Figure 4d). We also investigated the effect of QA on the LPS-induced activation of MAPK signaling. QAE pretreatment significantly inhibited LPS-induced phosphorylation of ERK, p38, and JNK (Figure 5). MAPK signaling cascades are involved in regulating cell proliferation, cell survival, and inflammation. It has been known that MAPK activates NF-κB signaling through the degradation of IκB [31]. Moreover, studies using chemical inhibitors showed the abrogation of p38, ERK, and JNK signaling suppressed the production of proinflammatory cytokines such as IL-1β, TNF-α in microglia [32,33]. Thus, the inhibition of the MAPK signaling pathway in microglia is beneficial in resolving inflammation in the CNS. These results suggested that QAE inhibited inflammatory gene expression in the microglial cells by blocking MAPK and NF-κB signaling pathways.

Intracellular ROS acts as key signaling molecules that are implicated in the inflammatory response of microglia [12]. H_2O_2 production by LPS-stimulated TLR4 signaling accelerates the activation of NF-κB signaling via phosphorylation and degradation of IκB [11,34]. On the other hand, HO-1, endogenous antioxidant enzyme, and its by-products attenuate TLR4 signaling transduction by inhibiting ROS generation in macrophages [12]. Accordingly, small molecules that activate the Nrf2/HO-1 pathway simultaneously regulate the inflammatory response in BV2 microglial cells [13]. In our study, QAE treatment significantly increased mRNA and the protein expression of HO-1 in a dose-dependent manner, which ameliorated LPS-induced cellular H_2O_2 generation (Figure 6).

Nrf2 is a major transcriptional factor regulating the expression of antioxidant genes such as HO-1, GCLC, and SESN2. Nrf2 phosphorylation at Ser^{40} results in translocation to the nucleus. Finally, Nrf2 binds to the antioxidant response element (ARE), and transactivates antioxidant genes, thereby serving cellular redox homeostasis [35]. Apart from the antioxidant response, the anti-inflammatory properties of Nrf2 via the inhibition of the NF-κB pathway and blocking transcription of inflammatory mediators have also been suggested [36]. Representatively, chromatin immunoprecipitation-seq (ChIP-seq) study in macrophages revealed that Nrf2 inhibited the transcription of proinflammatory cytokines by binding to its promoter regions [37]. In our results, QAE successfully activated the Nrf2 signaling pathway by inducing its phosphorylation and nuclear translocation (Figure 6). Taken together, our results suggested that the effect of QAE on Nrf2/HO-1 activation contributes to QA's anti-inflammatory efficacy.

Overall, our results showed the anti-neuroinflammatory effect of QAE in LPS-stimulated BV2 microglial cells by reducing the expression of proinflammatory enzymes and mediators. These activities were related to NF-κB, MAPK inhibition, and Nrf2/HO-1 activation.

Although the present study revealed cellular efficacy and molecular mechanisms of QA on the microglial inflammatory response, efficacy evaluation using neurodegenerative disease experimental animal models needs to be further established for clinical application.

5. Conclusions

In the present study, we demonstrated that QAE exhibited anti-inflammatory effects on LPS-stimulated BV2 microglial cells by reducing the expression of proinflammatory enzymes and mediators including iNOS, COX-2, NO, TNF-α, IL-1β, and IL-6. Moreover, we revealed that these inhibitory effects were mediated by inhibiting NF-κB and MAPK signaling pathways. Additionally, we found antioxidative stress efficacy of QAE against LPS-stimulation via the activation of the Nrf2/HO-1 pathway. Thus, QA may be a promising candidate for managing neuroinflammation-mediated neurodegenerative diseases.

Author Contributions: Conceptualization J.K.K. and Y.G.; methodology, J.K.K. and H.J.Y.; validation, J.K.K. and Y.G.; writing—original draft preparation, J.K.K. and H.J.Y.; writing—review and editing, J.K.K. and Y.G.; supervision, Y.G.; and funding acquisition, Y.G. All authors have read and agreed to the published version of the manuscript.

Funding: This work was supported by the Korea Institute of Oriental Medicine (KIOM), grant number KSN2022230, funded by the Ministry of Science and ICT, Republic of Korea.

Institutional Review Board Statement: Not applicable.

Informed Consent Statement: Not applicable.

Data Availability Statement: All the data are available within the article.

Conflicts of Interest: The authors declare no conflict of interest.

References

1. Leng, F.; Edison, P. Neuroinflammation and microglial activation in Alzheimer disease: Where do we go from here? *Nat. Rev. Neurol.* **2021**, *17*, 157–172. [CrossRef] [PubMed]
2. Lyman, M.; Lloyd, D.G.; Ji, X.; Vizcaychipi, M.P.; Ma, D. Neuroinflammation: The role and consequences. *Neurosci. Res.* **2014**, *79*, 1–12. [CrossRef]
3. Heneka, M.T.; Carson, M.J.; Khoury, J.E.; Landreth, G.E.; Brosseron, F.; Feinstein, D.L.; Jacobs, A.H.; Wyss-Coray, T.; Vitorica, J.; Ransohoff, R.M.; et al. Neuroinflammation in Alzheimer's disease. *Lancet Neurol.* **2015**, *14*, 388–405. [CrossRef]
4. Okello, A.; Edison, P.; Archer, H.A.; Turkheimer, F.E.; Kennedy, J.; Bullock, R.; Walker, Z.; Kennedy, A.; Fox, N.; Rossor, M.; et al. Microglial activation and amyloid deposition in mild cognitive impairment: A PET study. *Neurology* **2009**, *72*, 56–62. [CrossRef] [PubMed]
5. Fan, Z.; Aman, Y.; Ahmed, I.; Chetelat, G.; Landeau, B.; Chaudhuri, K.R.; Brooks, D.J.; Edison, P. Influence of microglial activation on neuronal function in Alzheimer's and Parkinson's disease dementia. *Alzheimer's Dement.* **2015**, *11*, 608–621. [CrossRef] [PubMed]
6. Muzio, L.; Viotti, A.; Martino, G. Microglia in Neuroinflammation and Neurodegeneration: From Understanding to Therapy. *Front. Neurosci.* **2021**, *15*, 742065. [CrossRef] [PubMed]
7. Kumar, V. Toll-like receptors in the pathogenesis of neuroinflammation. *J. Neuroimmunol.* **2019**, *332*, 16–30. [CrossRef] [PubMed]
8. Yuste, J.E.; Tarragon, E.; Campuzano, C.M.; Ros-Bernal, F. Implications of glial nitric oxide in neurodegenerative diseases. *Front. Cell. Neurosci.* **2015**, *9*, 322. [CrossRef] [PubMed]
9. Minghetti, L. Cyclooxygenase-2 (COX-2) in inflammatory and degenerative brain diseases. *J. Neuropathol. Exp. Neurol.* **2004**, *63*, 901–910. [CrossRef]
10. Tan, H.Y.; Wang, N.; Li, S.; Hong, M.; Wang, X.; Feng, Y. The Reactive Oxygen Species in Macrophage Polarization: Reflecting Its Dual Role in Progression and Treatment of Human Diseases. *Oxid. Med. Cell. Longev.* **2016**, *2016*, 2795090. [CrossRef]
11. Park, H.S.; Jung, H.Y.; Park, E.Y.; Kim, J.; Lee, W.J.; Bae, Y.S. Cutting edge: Direct interaction of TLR4 with NAD(P)H oxidase 4 isozyme is essential for lipopolysaccharide-induced production of reactive oxygen species and activation of NF-κB. *J. Immunol.* **2004**, *173*, 3589–3593. [CrossRef] [PubMed]
12. Simpson, D.S.A.; Oliver, P.L. ROS Generation in Microglia: Understanding Oxidative Stress and Inflammation in Neurodegenerative Disease. *Antioxidants* **2020**, *9*, 743. [CrossRef] [PubMed]
13. Foresti, R.; Bains, S.K.; Pitchumony, T.S.; de Castro Bras, L.E.; Drago, F.; Dubois-Rande, J.L.; Bucolo, C.; Motterlini, R. Small molecule activators of the Nrf2-HO-1 antioxidant axis modulate heme metabolism and inflammation in BV2 microglia cells. *Pharmacol. Res.* **2013**, *76*, 132–148. [CrossRef] [PubMed]

14. Campbell, N.K.; Fitzgerald, H.K.; Dunne, A. Regulation of inflammation by the antioxidant haem oxygenase 1. *Nat. Rev. Immunol.* **2021**, *21*, 411–425. [CrossRef]
15. Yoon, I.S.; Park, D.H.; Bae, M.S.; Oh, D.S.; Kwon, N.H.; Kim, J.E.; Choi, C.Y.; Cho, S.S. In Vitro and In Vivo Studies on *Quercus acuta* Thunb. (Fagaceae) Extract: Active Constituents, Serum Uric Acid Suppression, and Xanthine Oxidase Inhibitory Activity. *Evid. Based Complement. Altern. Med.* **2017**, *2017*, 4097195. [CrossRef]
16. Taib, M.; Rezzak, Y.; Bouyazza, L.; Lyoussi, B. Medicinal Uses, Phytochemistry, and Pharmacological Activities of *Quercus* Species. *Evid. Based Complement. Altern. Med.* **2020**, *2020*, 1920683. [CrossRef]
17. Kim, M.H.; Park, D.H.; Bae, M.S.; Song, S.H.; Seo, H.J.; Han, D.G.; Oh, D.S.; Jung, S.T.; Cho, Y.C.; Park, K.M.; et al. Analysis of the Active Constituents and Evaluation of the Biological Effects of *Quercus acuta* Thunb. (Fagaceae) Extracts. *Molecules* **2018**, *23*, 1772. [CrossRef]
18. Oh, M.H.; Park, K.H.; Kim, M.H.; Kim, H.H.; Kim, S.R.; Park, K.J.; Heo, J.H.; Lee, M.W. Anti-Oxidative and Anti-Inflammatory Effects of Phenolic Compounds from the Stems of Quercus acuta Thunberg. *Asian J. Chem.* **2014**, *26*, 4582–4586. [CrossRef]
19. Kim, B.; Kim, Y.S.; Hwang, Y.H.; Yang, H.J.; Li, W.; Kwon, E.B.; Kim, T.I.; Go, Y.; Choi, J.G. *Quercus acuta* Thunb. (Fagaceae) and Its Component, Isoquercitrin, Inhibit HSV-1 Replication by Suppressing Virus-Induced ROS Production and NF-κB Activation. *Antioxidants* **2021**, *10*, 1638. [CrossRef]
20. Livak, K.J.; Schmittgen, T.D. Analysis of relative gene expression data using real-time quantitative PCR and the $2^{-\Delta\Delta CT}$ Method. *Methods* **2001**, *25*, 402–408. [CrossRef]
21. Hwang, Y.H.; Ma, J.Y. Preventive Effects of an UPLC-DAD-MS/MS Fingerprinted Hydroalcoholic Extract of Citrus aurantium in a Mouse Model of Ulcerative Colitis. *Planta Med.* **2018**, *84*, 1101–1109. [CrossRef] [PubMed]
22. Coleman, J.W. Nitric oxide in immunity and inflammation. *Int. Immunopharmacol.* **2001**, *1*, 1397–1406. [CrossRef]
23. Yang, F.; Tang, E.; Guan, K.; Wang, C.Y. IKKβ plays an essential role in the phosphorylation of RelA/p65 on serine 536 induced by lipopolysaccharide. *J. Immunol.* **2003**, *170*, 5630–5635. [CrossRef] [PubMed]
24. Kaminska, B. MAPK signalling pathways as molecular targets for anti-inflammatory therapy–from molecular mechanisms to therapeutic benefits. *Biochim. Biophys. Acta* **2005**, *1754*, 253–262. [CrossRef]
25. Saez, V.; Riquelme, S.; Baer, D.V.; Vallverdu-Queralt, A. Phenolic Profile of Grape Canes: Novel Compounds Identified by LC-ESI-LTQ-Orbitrap-MS. *Molecules* **2019**, *24*, 3763. [CrossRef]
26. Chen, G.; Li, X.; Saleri, F.; Guo, M. Analysis of Flavonoids in *Rhamnus davurica* and Its Antiproliferative Activities. *Molecules* **2016**, *21*, 1275. [CrossRef]
27. Syed Hussein, S.S.; Kamarudin, M.N.; Kadir, H.A. (+)-Catechin Attenuates NF-κB Activation Through Regulation of Akt, MAPK, and AMPK Signaling Pathways in LPS-Induced BV-2 Microglial Cells. *Am. J. Chin. Med.* **2015**, *43*, 927–952. [CrossRef]
28. Park, S.Y.; Kim, H.Y.; Park, H.J.; Shin, H.K.; Hong, K.W.; Kim, C.D. Concurrent Treatment with Taxifolin and Cilostazol on the Lowering of β-Amyloid Accumulation and Neurotoxicity via the Suppression of P-JAK2/P-STAT3/NF-κB/BACE1 Signaling Pathways. *PLoS ONE* **2016**, *11*, e0168286. [CrossRef]
29. Liu, T.; Zhang, L.; Joo, D.; Sun, S.C. NF-κB signaling in inflammation. *Signal Transduct. Target. Ther.* **2017**, *2*, 17023. [CrossRef]
30. Sasaki, C.Y.; Barberi, T.J.; Ghosh, P.; Longo, D.L. Phosphorylation of RelA/p65 on serine 536 defines an IκBα-independent NF-κB pathway. *J. Biol. Chem.* **2005**, *280*, 34538–34547. [CrossRef]
31. Spiegelman, V.S.; Stavropoulos, P.; Latres, E.; Pagano, M.; Ronai, Z.; Slaga, T.J.; Fuchs, S.Y. Induction of β-transducin repeat-containing protein by JNK signaling and its role in the activation of NF-κB. *J. Biol. Chem.* **2001**, *276*, 27152–27158. [CrossRef] [PubMed]
32. Lee, Y.B.; Schrader, J.W.; Kim, S.U. p38 map kinase regulates TNF-α production in human astrocytes and microglia by multiple mechanisms. *Cytokine* **2000**, *12*, 874–880. [CrossRef] [PubMed]
33. Kim, S.H.; Smith, C.J.; Van Eldik, L.J. Importance of MAPK pathways for microglial pro-inflammatory cytokine IL-1β production. *Neurobiol. Aging* **2004**, *25*, 431–439. [CrossRef]
34. Takada, Y.; Mukhopadhyay, A.; Kundu, G.C.; Mahabeleshwar, G.H.; Singh, S.; Aggarwal, B.B. Hydrogen peroxide activates NF-κB through tyrosine phosphorylation of IκBα and serine phosphorylation of p65: Evidence for the involvement of IκBα kinase and Syk protein-tyrosine kinase. *J. Biol. Chem.* **2003**, *278*, 24233–24241. [CrossRef] [PubMed]
35. Kaspar, J.W.; Niture, S.K.; Jaiswal, A.K. Nrf2:INrf2 (Keap1) signaling in oxidative stress. *Free Radic Biol. Med.* **2009**, *47*, 1304–1309. [CrossRef] [PubMed]
36. Ahmed, S.M.; Luo, L.; Namani, A.; Wang, X.J.; Tang, X. Nrf2 signaling pathway: Pivotal roles in inflammation. *Biochim. Biophys. Acta Mol. Basis Dis.* **2017**, *1863*, 585–597. [CrossRef]
37. Kobayashi, E.H.; Suzuki, T.; Funayama, R.; Nagashima, T.; Hayashi, M.; Sekine, H.; Tanaka, N.; Moriguchi, T.; Motohashi, H.; Nakayama, K.; et al. Nrf2 suppresses macrophage inflammatory response by blocking proinflammatory cytokine transcription. *Nat. Commun.* **2016**, *7*, 11624. [CrossRef]

antioxidants

MDPI

Article

β-Carotene Increases Activity of Cytochrome P450 2E1 during Ethanol Consumption

Cristian Sandoval [1,2,3,*], Luciana Mella [4], Karina Godoy [5], Khosrow Adeli [6] and Jorge Farías [2,5,*]

1 Escuela de Tecnología Médica, Facultad de Salud, Universidad Santo Tomás, Los Carreras 753, Osorno 5310431, Chile
2 Departamento de Ingeniería Química, Facultad de Ingeniería y Ciencias, Universidad de La Frontera, Temuco 4811230, Chile
3 Departamento de Ciencias Preclínicas, Facultad de Medicina, Universidad de La Frontera, Temuco 4811230, Chile
4 Carrera de Tecnología Médica, Facultad de Medicina, Universidad de La Frontera, Temuco 4811230, Chile; l.mella01@ufromail.cl
5 Núcleo Científico y Tecnológico en Biorecursos (BIOREN), Universidad de La Frontera, Temuco 4811230, Chile; karina.godoy@ufrontera.cl
6 Molecular Medicine, Research Institute The Hospital for Sick Children University of Toronto, Toronto, ON M5G 1X8, Canada; khosrow.adeli@sickkids.ca
* Correspondence: cristian.sandoval@ufrontera.cl (C.S.); jorge.farias@ufrontera.cl (J.F.); Tel.: +56-45-2325720 (C.S.); +56-45-2325956 (J.F.)

Abstract: One of the key routes through which ethanol induces oxidative stress appears to be the activation of cytochrome P450 2E1 at different levels of ethanol intake. Our aim was to determine if oral β-carotene intake had an antioxidant effect on *CYP2E1* gene expression in mice that had previously consumed ethanol. C57BL/6 mice were used and distributed into: control (C), low-dose alcohol (LA), moderate-dose alcohol (MA), β-carotene (B), low-dose alcohol+β-carotene (LA + B), and moderate-dose alcohol+β-carotene (MA + B). Animals were euthanized at the end of the experiment, and liver tissue was taken from each one. *CYP2E1* was measured using qPCR to detect liver damage. The relative expression level of each RNA was estimated using the comparative threshold cycle (Ct) technique ($2^{-\Delta\Delta CT}$ method) by averaging the Ct values from three replicates. The LA+B (2267 $\pm$ 0.707) and MA+B (2.307 $\pm$ 0.384) groups had the highest *CYP2E1* fold change values. On the other hand, the C (1.053 $\pm$ 0.292) and LA (1.240 $\pm$ 0.163) groups had the lowest levels. These results suggest that ethanol feeding produced a fold increase in *CYP2E1* protein in mice as compared to the control group. Increased *CYP2E1* activity was found to support the hypothesis that β-carotene might be dangerous during ethanol exposure in animal models. Our findings imply that β-carotene can increase the hepatic damage caused by low and high doses of alcohol. Therefore, the quantity of alcohol ingested, the exposure period, the regulatory mechanisms of alcoholic liver damage, and the signaling pathways involved in the consumption of both alcohol and antioxidant must all be considered.

Keywords: alcohol intake; alcoholic fatty liver disease; antioxidant treatment; chronic alcohol consumption

1. Introduction

Excessive alcohol drinking has been linked to several deadly illnesses, including cancer, cirrhosis of the liver, vascular disease, neuropsychiatric illness, as well as diabetes [1–3]. In addition, that increased oxidative stress can produce hepatic damage in people has been demonstrated [4,5].

Alcoholic fatty liver, alcoholic hepatitis, and cirrhosis are all caused by ethanol metabolism [6,7]. In the hepatic metabolization, enzymes such *CYP450 2E1* (*CYP2E1*), alcohol dehydrogenase (ADH), and catalase (CAT) are involved in the oxidative pathway, whereas through the non-oxidative pathway the fatty acid ethyl ester (FAEE) synthase creates FAEEs [8,9].

The microsomal respiratory chain and *CYP2E1*-dependent microsomal monooxygenase system are the main sources of ROS during alcohol intake. As to its ability to produce a diversity of hepatotoxic substrates, such as N-nitrosodimethylamine, alcohol, acetaminophen, and carbon tetrachloride, *CYP2E1* is of particular interest [10]. According to this theory, alcohol-induced activation of *CYP2E1* is one of the primary mechanisms by which alcohol produces oxidative stress. Furthermore, *CYP2E1* oxidizes ethanol to form a very reactive particle that might contribute to alcohol's harmfulness, acetaldehyde [11].

The primary contributor to the development of alcohol-mediated liver damage, extracellular matrix changes, and inflammation has been identified as acetaldehyde [12,13]. The generation of reactive oxygen species (ROS) and a redox potential imbalance (NAD/NADH) generate its effects. It also links to DNA, producing oncogenic chemicals like 1,N^2-(3-hydroxypropane)-2′-deoxyguanosine, and creates protein aggregates in hepatocytes, restricting protein synthesis and promoting hepatomegaly. It also forms salsolinol when it reacts with dopamine, which can contribute to alcohol dependency [14,15].

Alcohol-mediated oxidative stress and toxicity have previously been examined in animal models and in vitro studies [16,17]. In consequence, these results have sparked fresh research into new pathophysiological targets that may be used to treat alcoholic liver disease (ALD). In effect, enzymatic mechanisms such as catalase, superoxide dismutase, and glutathione peroxidase and reductase, as well as non-enzymatic mechanisms, might be used to block the hepatocyte's antioxidant defense [18–20]. Many antioxidants, including silymarin, N-acetylcysteine, vitamin E, and S-adenosylmethionine (SAMe), have been examined in recent clinical investigations, although the results have been inconsistent [18,19,21,22]. Therefore, this study aimed to examine the consequences of β-carotene supplementation on *CYP2E1* activity in C57BL/6 mice exposed to alcohol consumption.

2. Materials and Methods

2.1. Animals

Thirty male C57BL/6 mice were used (*Mus musculus*), 50 days old, from the Chilean Public Health Institute. They were kept for 30 days under standardized conditions and a 12 h light/dark cycle (08:00 a.m.–08:00 p.m./08:00 p.m–8:00 a.m.), with a standard laboratory diet (AIN-93M) and water *ad libitum* to help them adjust to their new environment in the Animal Facility of the Center of Excellence in Morphological and Surgical Studies (CEMyQ) at the Universidad de La Frontera. The animals were handled according to the recommendations published by the Institute for Laboratory Animal Research [23]. The Scientific Ethics Committee of the Universidad de La Frontera has approved this project (N°051/2020).

The mice were split into six groups on the first day of the experiment (day 1): 1. control (C); 2. low-dose alcohol (LA): low-dose alcohol consumption (3% *v/v ad libitum*) for 28 days [24]; 3. moderate-dose alcohol (MA): moderate-dose alcohol consumption (7% *v/v ad libitum*) for 28 days [24]; 4. β-carotene (B): administration of 0.52 mg/kg body weight/day of β-carotene for 28 days [25]; 5. low-dose alcohol + β-carotene (LA + B): low-dose alcohol consumption plus administration of 0.52 mg/kg body weight/day of β-carotene for 28 days; and 6. moderate-dose alcohol + β-carotene (MA + B): moderate-dose alcohol consumption plus administration of 0.52 mg/kg body weight/day of β-carotene for 28 days.

Chronic ethanol administration was given according to the modified liquid diet of Lieber-DeCarli [24,26]. β-carotene was orally administered at a dose of 0.52 mg/kg body weight/day [25].

2.2. Euthanasia

On day 28, at the end of the experiment, the animals were fasted for 6 h and euthanized with sodium pentobarbital.

2.3. Liver Tissue

Each animal's liver tissue (n = 30) was obtained after euthanasia. Liver samples were extracted as soon as possible and placed in autoclaved microtubes containing lysis solution (RNeasy Mini Kit, QIAGEN) for RNA stabilization. They were then stored at room temperature for 30 min. The samples were then frozen in liquid nitrogen before being carried to the freezer room. The frozen samples were then transferred to an ultra-freezer and stored at −80 °C until they were utilized.

2.4. Extraction of RNA and cDNA Synthesis

The liver sample was crushed into a fine powder in liquid nitrogen with a prechilled mortar and pestle, then combined with the TRIzol (QIAGEN Diagnostics GmbH, Germany)/lysis buffer given with the kits and extracted according to the methodology previously described [27]. Using High-Capacity cDNA Reverse Transcription Kits, the mRNA strand was reverse transcribed into single-stranded cDNA (Applied Biosystems, Waltham, MA, USA). The cDNA was subsequently amplified using TaqMan™ Universal PCR Master Mix in a quantitative PCR (qPCR) (Applied Biosystems, Waltham, MA, USA).

2.5. Quantification of RNA from Liver Tissue

The amount and integrity of the isolated total RNA were analyzed using the Qubit™ 4.0 Fluorometer (Life Technologies, Thermo Fisher Scientific Inc., Waltham, MA, USA). The RNA IQ# was estimated from the fraction of large and small RNA in the sample. The RNA IQ# is a number that ranges from 1 to 10, where a high number suggests that most of the RNA in the sample is large and/or organized. On the other hand, a low IQ# indicates that the sample is largely small RNA with little tertiary structure. The manufacturer's instructions were followed while using the standard Qubit™ RNA HS Assay Kit (Life Technologies, Thermo Fisher Scientific Inc.). The Qubit™'s functioning solution was developed in accordance with the manufacturer's standards. We added to each assay tube 180 µL of working solution, up to 20 µL of RNA, and enough water to make the final volume 200 µL. For the standard tubes, 10 µL of Qubit™ RNA Standard solutions were placed into the tubes. The assay tubes were vortexed for 2–3 s, centrifuged for 5 s, and then left at room temperature for 2 min before being measured with the Qubit™ Fluorometer. For the Qubit™ Assay, RNA sample concentration was calculated as: [Concentration of your sample] = QF value (the value given by the Qubit® 4.0 Fluorometer) × (200/the number of microliters of sample put into the assay tube). Three different measurements were taken on each sample.

2.6. Quantitative Real-Time PCR

The expression levels of *CYP2E1*, family 2, subfamily e, polypeptide 1, β-actin (*ACTB*), and glyceraldehyde-3-phosphate dehydrogenase (*GAPDH*) were analyzed using qPCR. The data were normalized according to the mRNA expression levels of housekeeping genes, such as *ACTB* and *GAPDH*. *CYP2E1* was the target gene. All qPCR experiments were performed using the QuantStudio3 system (Applied Biosystems, Waltham, MA, USA). All the amplifications were done using TaqMan™ Universal PCR Master Mix (Applied Biosystems, Waltham, MA, USA). The TaqMan gene expression test is a ready-to-use 5′-3′ Taq polymerase assay containing TaqMan® dye-labeled probes (FAM/MGB) and desired primers, as presented in Table 1. In addition, housekeeping genes are presented in Table 2.

Table 1. Primers for gene targeting.

Gene	Gene Symbol	Assay	Chromosome Location	Amplicon Length
cytochrome P450, family 2, subfamily e, polypeptide 1	*Cyp2e1*	Mm00491127_m1	Chr.7: 140763832–140774981	83 bp

Table 2. Housekeeping genes for quantitative PCR.

Gene	Gene Symbol	Assay	Chromosome Location	Amplicon-Length
actx, E430023M04Rik, beta-actin	*Actb*	Mm00607939_s1	Chr.5: 142903116–142906724	115 bp
glyceraldehyde-3-phosphate dehydrogenase	*Gapdh*	Mm99999915_g1	Chr.6: 125161338–125166511	107 bp

After a 10-min denaturation phase at 95 °C, 40 cycles at 95 °C for 30 s, 60 °C for 30 s, and 72 °C for 30 s were performed. A melting curve analysis of each qPCR was carried out after each cycle. The number of times the reporter dye in the PCR reaction crossed a software-defined threshold, which was computed automatically by the QuantStudio™ Design & Analysis Software, is referred to as the 'Ct', or threshold cycle (version 1.3, Applied Biosystems, Waltham, MA, USA). The relative expression level of each RNA was estimated using the comparative threshold cycle (Ct) technique ($2^{-\Delta\Delta Ct}$ method) by averaging the Ct values from three replicates. We utilized the threshold cycle values automatically generated by the qPCR equipment for the $2^{-\Delta\Delta Ct}$ technique.

2.7. Statistical Analysis

The Kolmogorov–Smirnov test (data normality analysis) and Levene's test were used to assess differences in quantitative data (homoscedasticity of the variances). One-way ANOVA was used to analyze the differences among the groups. Tukey's HSD test or Dunnett's T3 test were used to realize such a post-hoc test, as applicable. $p < 0.05$ was considered statistically significant (GraphPad Prism 6, GraphPad Software Inc., San Diego, CA, USA).

3. Results

3.1. Statistical Analysis

RNA was extracted from liver samples and was analyzed for integrity and quality. The assay kit was prepared to be precise concerning initial RNA sample concentrations of 0.5 to 1200 ng/L, yielding a detection range of 10 to 1200 ng, depending on sample volume. A total of 18 specimens had sufficient yield (RNA IQ# >8) to proceed to amplification through a quantitative PCR (qPCR) using TaqMan™ Universal PCR Master Mix (Applied Biosystems, CA, USA). It has been described that RNA is pure and satisfactory for downstream studies if RIN > 7 [28].

3.2. Quantitative Real-Time PCR

In gene quantification analysis, data normalization in qPCR is a critical step [29,30]. Indeed, depending on the experimental settings and pathophysiology of the examined tissue, mRNA levels of the required housekeeping genes, *ACTB* and *GAPDH*, are likely to change to the point where normalization becomes erroneous and/or deceptive.

The expression of the *ACTB* and *GAPDH* genes changes very little between the control and the experimental samples, as seen in Figures 1 and 2; and has a low variability of expression. As a result, the internal control genes *ACTB* and *GAPDH* may be used to provide accurate and consistent findings. *ACTB* and *GAPDH* resolve differences in templates starting with the amount and operational loading errors [31].

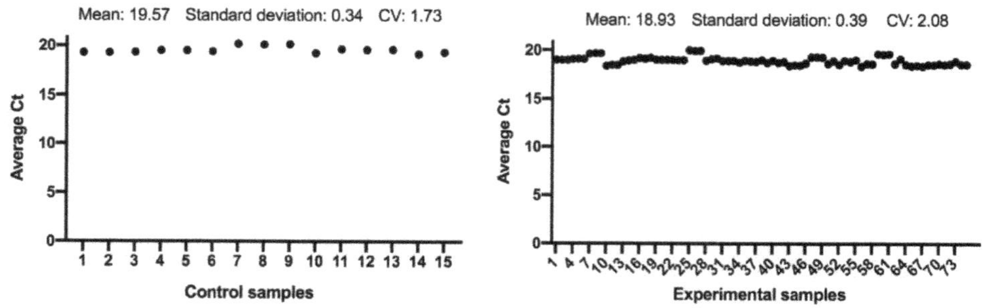

Figure 1. Comparison of *ACTB* gene expression from control and experimental groups, respectively.

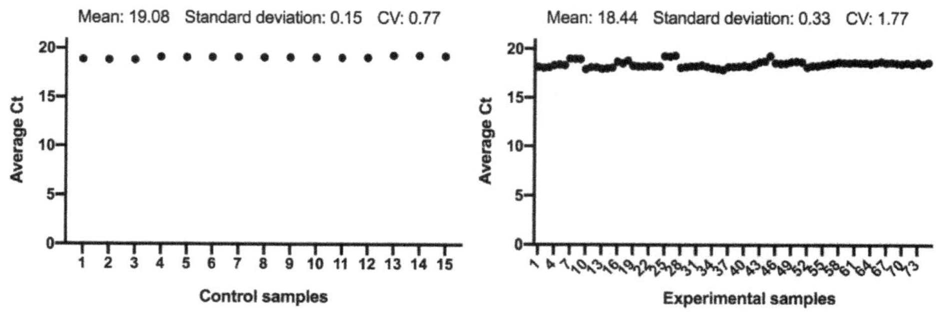

Figure 2. Comparison of *GAPDH* gene expression from control and experimental groups, respectively.

The reference RNA utilized in the standard curve approach is highly efficient and stable. Moreover, this RNA has been useful to determine absolute comparative quantification of target genes by qPCR.

Since chronic ethanol feeding elevates *CYP450 2E1*, the ΔCT values were determined for *CYP2E1* mRNA expression profiles of control and experimental groups after 28 days of ethanol and/or ß-carotene administration (Figure 3). The ΔCT value is the gap among the target gene and housekeeping genes, described as:

$$\Delta CT = \text{average CT (a target gene)} - \text{average CT (housekeeping genes)} \quad (1)$$

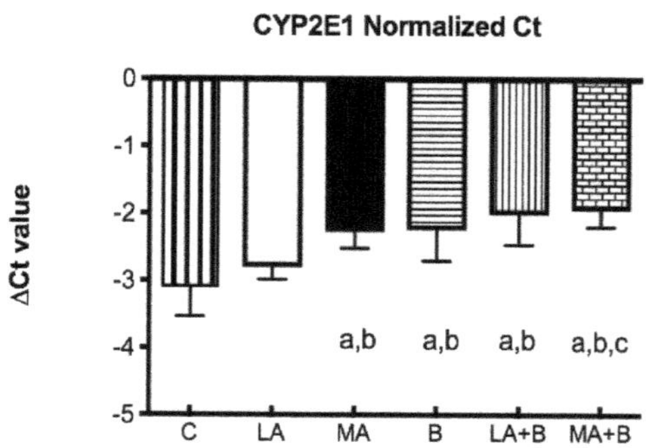

Figure 3. Delta Ct values of *CYP2E1* mRNA of control and experimental groups. Bars represent mean ± SD values of ΔCt per group; a: significant differences ($p < 0.05$) with the C group; b: significant differences ($p < 0.05$) with the LA group; c: significant differences ($p < 0.05$) with the MA group.

The $2^{-\Delta\Delta CT}$ comparative approach was used to estimate relative gene expression. All data were regulated to *ACTB* and *GAPDH* mRNA content. Comparative RNA expression study in experimental versus control groups (calibrator) was performed as follows:

$$\begin{gathered}\text{Experimental groups: } \Delta Ct = Ct\text{ (target)} - Ct\text{ (housekeeping genes)}\\ \text{Control group: } \Delta Ct = Ct\text{ (target)} - Ct\text{ (housekeeping genes)}\\ \Delta\Delta Ct = \Delta Ct\text{ (experimental groups)} - \Delta Ct\text{ (control group)}\\ \text{Ratio} = 2^{-\Delta\Delta Ct}\end{gathered} \quad (2)$$

The average ΔCt value of housekeeping gene RNA was subtracted from the average Ct value of the control and experimental groups, yielding the Ct value. The ΔΔCt value was obtained by subtracting the control group's ΔCt value from the experimental groups' ΔCt value [32]. The ratio $2^{-\Delta\Delta Ct}$ was used to determine the fold of enrichment values.

Figure 4 displays that following 28 days of alcohol intake, levels of *CYP2E1* were bigger 2267 ± 0.707-fold in LA+B and 2.307 ± 0.384-fold in MA+B groups after ethanol and β-carotene exposure. No significant differences were found between C and LA groups.

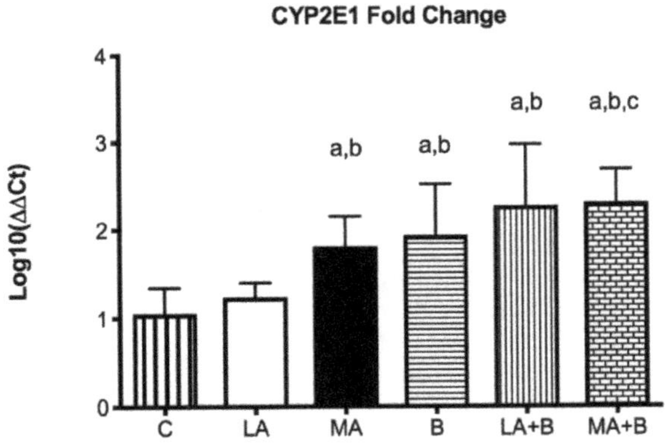

Figure 4. *CYP2E1* mRNA fold change is expressed as fold change using the ΔΔCt method in experimental groups with respect to the control group (calibrator). Bars represent mean ± SD values of fold change per group; a: significant differences ($p < 0.05$) with the C group; b: significant differences ($p < 0.05$) with the LA group; c: significant differences ($p < 0.05$) with the MA group.

The housekeeping genes allow the target gene's gene expression pattern to be normalized against the quantity of input RNA or cDNA. They were adjusted for probable RNA degradation, sample management variations, reverse-transcription efficacy differences, the existence of inhibitors in the RNA sample and RNA content, and sample handling differences. The comparative approach (ratio $2^{-\Delta\Delta CT}$) was applied, with *ACTB* and *GAPDH* serving as housekeeping genes and the control group serving as a calibrator [33].

4. Discussion

4.1. Summary of Key Findings and Interpretation

For reference samples, the relative gene expression is commonly adjusted to 1 since CT equals 0 and hence 20 equals 1. The $2^{-\Delta\Delta CT}$ method approach assumes that all samples have a consistent PCR amplification efficiency of 100 percent [33,34]. The number 2 is 1 plus the PCR amplification efficiency (100 %). This assumption simplifies the method and ensures that it is valid in ideal circumstances. However, since there are variables such as the existence of PCR inhibitors or enhancers, extraction of RNA, and various primers, probes, and enzymes, PCR efficiency cannot be guaranteed.

Ethanol dependence is a disease that progresses nearly five years after the primary alcohol use starts and it takes nearly 15–20 years for the alcohol addict to request medical care [35,36]. While much of the effort on alcohol metabolism has been on ADH, chronic alcohol consumption might raise levels of other enzymes such as *CYP2E1* [15,30,37–39]. In effect, elevated levels of *CYP2E1* in the liver of patients with alcoholic and nonalcoholic liver diseases have been shown [40]. As expected, ethanol has shown that feeding mice using the Lieber-DeCarli liquid diet results in a fold increase in *CYP2E1* levels compared to control (Figures 3 and 4). Under these conditions, previous studies using the same protocol have shown alcohol feeding produces fatty liver and raises LDL-C levels [39].

Following alcohol intake, the microsomal monooxygenase system, and the microsomal respiratory chain, both of which rely on *CYP2E1*, are the major generators of ROS in hepatocytes. Because of its propensity to process and stimulate diverse hepatotoxic substrates in the liver, cytochrome P450 2E1 is particularly essential in the prevention of carbon tetrachloride, alcohol, N-nitroso dimethylamine, and acetaminophen turning into more toxic compounds [10]. The activation of *CYP2E1* by ethanol seems to be one of the keyways by which alcohol produces oxidative stress. In effect, *CYP2E1* also converts ethanol to acetaldehyde, a highly reactive molecule that contributes to ethanol toxicity [11]. Furthermore, heavy alcohol use appears to be associated with *CYP2E1* blood expression [41]. Due to enhanced NADPH oxidase activity and strong production of O_2 and H_2O_2 radicals even in the absence of substrate, *CYP2E1* is a potent ROS generator [42].

Previous research has established the alcohol and β-carotene dosages, as well as the treatment times [24,25]. In this regard, low doses of oral β-carotene supplementation have been shown to protect against liver damage caused by the antioxidant pathway [20,43]. Our findings have reported greater levels of *CYP2E1* fold-change in the MA, LA+B, and MA+B groups in comparison to the C group and even the LA group (Figure 4, $p < 0.05$). The C and LA groups had the lowest levels (1.053 ± 0.292 and 1.240 ± 0.163, respectively). Hence, our findings show that ß-carotene increases the activity of *CYP2E1* during ethanol consumption in low and moderate doses. Conversely, previous studies using the same protocol found it could prevent alcoholic liver disease and improve health when biochemical markers and histopathological and transmission electron microscopy are used in the evaluation [20,39]. In effect, reduced oxidative stress and reduced *CYP2E1* activity have been found to support the protection provided by induced-β-carotene in a moderate alcohol intake. These discrepancies demonstrate the need to identify the action pathway of β-carotene during ethanol metabolism, which could be directly related to acetaldehyde and even acetate. Despite these findings, it is necessary to understand that these discrepancies with other studies are mainly due to the amount of alcohol drunk, the exposure period, the regulatory mechanisms of alcoholic liver damage, and the signaling pathways involved in the consumption of both alcohol and antioxidants. In fact, previous reviews have described these discrepancies using vitamins and supplements against alcoholic liver disease [44].

Although increases in *CYP2E1* mRNA can occur at very high blood ethanol levels [45], ethanol induction of *CYP2E1* is mostly posttranscriptional, suggesting the stability of *CYP2E1* against proteosome-mediated destruction. Ethanol is both a ligand and a substrate for *CYP2E1*, which explains its ability to stabilize and prolong the half-life of the enzyme [46,47]. This study also shown that combining ethanol with p-carotene causes a more severe hepatic damage in C57BL/6 mice than either drug alone, raising concerns about the use of B-carotene as a source of retinol and as an anticancer agent when substantial alcohol use or abuse is present. It has been frequently suggested that a lack of carotenoids in a diet is related to an increased risk of cancer [48–50], while several studies have failed to show such a link [51,52].

4.2. Scope and Limitations

The aim of this research was to evaluate the consequences of β-carotene on *CYP2E1* activity of C57BL/6 mice exposed to alcohol exposure. In addition, our data update the existing linkage between *CYP2E1* and β-carotene therapy. However, the absence of information linking alcohol dehydrogenase and aldehyde dehydrogenase expressions with antioxidant treatments remains a limitation of this research and must be attended to in future studies. Unfortunately, previous research results provide little support for this notion. Considering these data, seems that β-carotene exposure did not improve the hepatotoxic damage induced by alcohol exposure in C57BL/6 mice.

5. Conclusions

Because *CYP2E1* activity increases after moderate alcohol consumption and β-carotene, our findings imply that antioxidant therapies might be dangerous during ethanol exposure in animal models. Despite all the progress made in understanding the effects of antioxidant supplementation, future studies should use specific cell lines and clinical trials to better understand the relationship between alcohol consumption, antioxidant therapies, the signaling pathways involved, and enzymatic and non-enzymatic mechanisms.

Author Contributions: C.S., L.M., J.F., K.G. and K.A. carried out the conception and design of the research. C.S. and K.G. have participated in the experimental phase, and they did the molecular analysis using qPCR. L.M. and J.F. did the quantification of RNA from liver tissue. C.S. and K.A. participated in obtaining funding. All authors have read and agreed to the published version of the manuscript.

Funding: Universidad de La Frontera, DIUFRO Project DI22–0007 and Programa de Formación de Investigadores Postdoctorales 2022.

Institutional Review Board Statement: The animal study protocol was approved by the Scientific Ethics Committee of the Universidad de La Frontera (N°051/2020).

Informed Consent Statement: Not applicable.

Data Availability Statement: The data that support the findings of this study are openly available in "figshare" at https://doi.org/10.6084/m9.figshare.19766218.v1, accessed on 4 April 2022.

Conflicts of Interest: The authors have declared no conflict of interest.

References

1. Haber, P.S.; Apte, M.V.; Moran, C.; Applegate, T.L.; Pirola, R.C.; Korsten, M.A.; McCaughan, G.W.; Wilson, J.S. Non-oxidative Metabolism of Ethanol by Rat Pancreatic Acini. *Pancreatology* **2004**, *4*, 82–89. [CrossRef] [PubMed]
2. Brust, J.C.M. Ethanol and Cognition: Indirect Effects, Neurotoxicity, and Neuroprotection: A Review. *Int. J. Environ. Res. Public Health* **2010**, *7*, 1540–1557. [CrossRef] [PubMed]
3. World Health Organization. *Alcohol. Descriptive Note N° 349*; World Health Organization: New York, NY, USA, 2011.
4. Cederbaum, A.I. Cytochrome P450 2E1-dependent Oxidant Stress and Upregulation of Anti-oxidant Defense in Liver Cells. *J. Gastroenterol. Hepatol.* **2006**, *21*, S22–S25. [CrossRef] [PubMed]
5. Cohen, J.I.; Roychowdhury, S.; DiBello, P.M.; Jacobsen, D.W.; Nagy, L.E. Exogenous Thioredoxin Prevents Ethanol-induced Oxidative Damage and Apoptosis in Mouse Liver. *Hepatology* **2009**, *49*, 1709–1717. [CrossRef]
6. García Gutiérrez, E.; Lima Mompó, G.; Aldana Vilas, L.; Casanova Carrillo, P.; Feliciano Álvarez, V. Alcoholismo y sociedad, tendencias actuales. *Rev. Cub. Med. Mil.* **2004**, *33*, 3.
7. Arias, R. Reacciones fisiológicas y neuroquímicas del alcoholismo. *Diversitas* **2005**, *1*, 138–147. [CrossRef]
8. Lakshman, R.; Cederbaum, A.I.; Hoek, J.B.; Konishi, M.; Koop, D.; Donohue, T.M. Use of CYP2E1-transfected Human Liver Cell Lines in Elucidating the Actions of Ethanol. *Alcohol Clin. Exp. Res.* **2006**, *29*, 1726–1734. [CrossRef]
9. Wu, H.; Cai, P.; Clemens, D.L.; Jerrells, T.R.; Shakeel Ansari, G.A.; Kaphalia, B.S. Metabolic Basis of Ethanol-induced Cytotoxicity in Recombinant HepG2 Cells: Role of Nonoxidative Metabolism. *Toxicol. Appl. Pharmacol.* **2006**, *216*, 238–247. [CrossRef]
10. Lu, Y.; Cederbaum, A.I. CYP2E1 and Oxidative Liver Injury by Alcohol. *Free Rad. Biol. Med.* **2008**, *44*, 723–738. [CrossRef]
11. Yang, Z.; Klionsky, D.J. Eaten Alive: A History of Macroautophagy. *Nat. Cell Biol.* **2010**, *12*, 814–822. [CrossRef]
12. Bailey, S.M.; Cunningham, C.C. Contribution of Mitochondria to Oxidative Stress Associated with Alcoholic Liver Disease. *Free Rad. Biol. Med.* **2002**, *32*, 11–16. [CrossRef]
13. Hoek, J.B.; Cahill, A.; Pastorino, J.G. Alcohol and Mitochondria: A Dysfunctional Relationship. *Gastroenterology* **2002**, *122*, 2049–2063. [CrossRef]
14. Tuma, D.J.; Casey, C.A. Dangerous Byproducts of Alcohol Breakdown: Focus on Adducts. *Alcohol Res. Health* **2003**, *27*, 285–290.
15. Sandoval, C.; Vaásquez, B.; Mandarim-de-Lacerda, C.; del Sol, M. Ethanol Intake and Toxicity: In Search of New Treatments. *Int. J. Morphol.* **2017**, *35*, 942–949. [CrossRef]
16. Schattenberg, J.M.; Czaja, M.J. Regulation of the Effects of CYP2E1-induced Oxidative Stress by JNK Signaling. *Redox Biol.* **2014**, *3*, 7–15. [CrossRef]

17. Diesinger, T.; Buko, V.; Lautwein, A.; Dvorsky, R.; Belonovskaya, E.; Lukivskaya, O.; Naruta, E.; Kirko, S.; Andreev, V.; Buckert, D.; et al. Drug Targeting CYP2E1 for the Treatment of Early-stage Alcoholic Steatohepatitis. *PLoS ONE* **2020**, *15*, e0235990. [CrossRef]
18. Zelko, I.N.; Mariani, T.J.; Folz, R.J. Superoxide Dismutase Multigene Family: A Comparison of the CuZn-SOD (SOD1), Mn-SOD (SOD2), and EC-SOD (SOD3) Gene Structures, Evolution, and Expression. *Free Rad. Biol. Med.* **2002**, *33*, 337–349. [CrossRef]
19. Chang, P.; Cheng, E.; Brooke, S.; Sapolsky, R. Marked Differences in the Efficacy of Post-insult Gene Therapy with Catalase versus Glutathione Peroxidase. *Brain Res.* **2005**, *1063*, 27–31. [CrossRef]
20. Sandoval, C.; Vásquez, B.; Souza-Mello, V.; Adeli, K.; Mandarim-de-Lacerda, C.; del Sol, M. Morphoquantitative Effects of Oral β-carotene Supplementation on Liver of C57BL/6 Mice Exposed to Ethanol Consumption. *Int. J. Clin. Exp. Pathol.* **2019**, *12*, 1713–1722.
21. Schott, M.B.; Rasineni, K.; Weller, S.G.; Schulze, R.J.; Sletten, A.C.; Casey, C.A.; McNiven, M.A. β-Adrenergic Induction of Lipolysis in Hepatocytes Is Inhibited by Ethanol Exposure. *J. Biol. Chem.* **2017**, *292*, 11815–11828. [CrossRef]
22. Werling, K. A Májbetegségek Kialakulásának Új Szempontjai—Különös Tekintettel Az Autophagiára És A mikro-RNS Szerepére. *Orvosi Hetilap* **2020**, *161*, 1449–1455. [CrossRef]
23. Committee for the Update of the Guide for the Care and Use of Laboratory Animals; Institute for Laboratory Animal Research; Division on Earth and Life Studies. *Guide for the Care and Use of Laboratory Animals*, 8th ed.; The National Academies Press: Washington, DC, USA, 2011.
24. Furuya, D.T.; Binsack, R.; Machado, U.F. Low ethanol consumption increases insulin sensitivity in Wistar rats. *Braz. J. Med. Biol. Res.* **2003**, *36*, 125–130. [CrossRef]
25. Peng, H.C.; Chen, Y.L.; Yang, S.Y.; Ho, P.Y.; Yang, S.S.; Hu, J.T.; Yang, S.C. The antiapoptotic effects of different doses of β-carotene in chronic ethanol-fed rats. *Hepatobiliary Surg. Nutr.* **2013**, *2*, 132–141. [CrossRef]
26. Diao, Y.; Nie, J.; Tan, P.; Zhao, Y.; Zhao, T.; Tu, J.; Ji, H.; Cao, Y.; Wu, Z.; Liang, H.; et al. Long-term low-dose ethanol intake improves healthspan and resists high-fat diet-induced obesity in mice. *Aging* **2020**, *12*, 13128–13146. [CrossRef]
27. Rio, D.C.; Ares, M.; Hannon, G.J.; Nilsen, T.W. Purification of RNA Using TRIzol (TRI Reagent). *Cold Spring Harb. Protoc.* **2010**, *2010*, pdb.prot5439. [CrossRef]
28. Sheng, Q.; Vickers, K.; Zhao, S.; Wang, J.; Samuels, D.C.; Koues, O.; Shyr, Y.; Guo, Y. Multi-perspective quality control of Illumina RNA sequencing data analysis. *Brief. Funct. Genom.* **2017**, *16*, 194–204. [CrossRef]
29. Pfaffl, M.W. A new mathematical model for relative quantification in real-time RT-PCR. *Nucleic Acids Res.* **2001**, *29*, e45. [CrossRef]
30. Bustin, S.A. Quantification of mRNA using real-time reverse transcription PCR (RT-PCR): Trends and problems. *J. Mol. Endocrinol.* **2002**, *29*, 23–29. [CrossRef]
31. Pfaffl, M.W. Quantification Strategies in Real-Time PCR. In *A-Z of Quantitative PCR*; International University Line: La Jolla, CA, USA, 2004.
32. Finis, K.; Sültmann, H.; Ruschhaupt, M.; Buness, A.; Helmchen, B.; Ruprecht, K.; Gross, M.L.; Fink, B.; Schirmacher, P.; Poustka, A. Analysis of pigmented villonodular synovitis with genome-wide complementary DNA microarray and tissue array technology reveals insight into potential novel therapeutic approaches. *Arthritis Rheum.* **2006**, *54*, 1009–1019. [CrossRef]
33. Livak, K.J.; Schmittgen, T.D. Analysis of relative gene expression data using real time quantification PCR and the $2^{-\Delta\Delta CT}$ method. *Methods* **2001**, *25*, 402–408. [CrossRef]
34. Arocho, A.; Chen, B.; Ladanyi, M.; Pan, Q. Validation of the 2-DeltaDeltaCt calculation as an alternate method of data analysis for quantitative PCR of BCR-ABL P210 transcripts. *Diagn. Mol. Pathol.* **2006**, *15*, 56–61. [CrossRef] [PubMed]
35. Silvestri, L.; Sonzogni, L.; De Silvestri, A. CYP Enzyme Polymorphısms and Susceptıbılıty to HCV-Related Chronıc Liver Disease and Liver Cancer. *Int. J. Cancer* **2003**, *104*, 310–317. [CrossRef] [PubMed]
36. Lieber, C.S. The Discovery of the Microsomal Ethanol Oxidizing System and Its Physiologic and Pathologic Role. *Drug Metab. Rev.* **2004**, *36*, 511–529. [CrossRef] [PubMed]
37. Sandoval, C.; Vásquez, B.; Souza-Mello, V.; Mandarim-de-Lacerda, C.A.; del Sol, M. Rol del consumo de alcohol y antioxidantes sobre la metilación global del ADN y cáncer. *Int. J. Morphol.* **2018**, *36*, 367–372. [CrossRef]
38. Carrasco, C.; Carrasco, C.; Souza-Mello, V.; Sandoval, C. Effectiveness of antioxidant treatments on cytochrome P450 2E1 (CYP2E1) activity after alcohol exposure in humans and in vitro models: A systematic review. *Int. J. Food Prop.* **2021**, *24*, 1300–1317. [CrossRef]
39. Sandoval, C.; Vásquez, B.; Vasconcellos, A.; Souza-Mello, V.; Adeli, K.; Mandarim-de-Lacerda, C.; del Sol, M. Oral supplementation of b-carotene benefits the hepatic structure and metabolism in mice exposed to chronic ethanol consumption. *Sains Malays.* **2022**, *51*, 285–296. [CrossRef]
40. Niemelä, O.; Parkkila, S.; Pasanen, M.; Viitala, K.; Villanueva, J.A.; Halsted, C.H. Induction of cytochrome P450 enzymes and generation of protein-aldehyde adducts are associated with sex-dependent sensitivity to alcohol-induced liver disease in micropigs. *Hepatology* **1999**, *30*, 1011–1017. [CrossRef]
41. Liangpunsakul, S.; Kolwankar, D.; Pinto, A.; Gotski, J.C.; Hall, S.D.; Chalasani, N. Activity of CYP2E1 and CYP3A enzymes in adults with moderate alcohol consumption: A comparison with nonalcoholics. *Hepatology* **2005**, *41*, 1144–1150. [CrossRef]
42. Lu, Y.; Cederbaum, A.I. Cisplatin-Induced Hepatotoxicity Is Enhanced by Elevated Expression of Cytochrome P450 2E1. *Toxicol. Sci.* **2006**, *89*, 515–523. [CrossRef]
43. Lin, W.T.; Huang, C.C.; Lin, T.J.; Chen, J.R.; Shieh, M.J.; Peng, H.C.; Yang, S.C.; Huang, C.Y. Effects of beta-carotene on antioxidant status in rats with chronic alcohol consumption. *Cell Biochem. Funct.* **2009**, *27*, 344–350. [CrossRef]

44. Sandoval, C.; Farías, J.; Zamorano, M.; Herrera, C. Vitamin Supplements as a Nutritional Strategy against Chronic Alcohol Consumption? An Updated Review. *Antioxidants* **2022**, *11*, 564. [CrossRef]
45. Ronis, M.J.; Huang, J.; Crouch, J.; Mercado, C.; Irby, D.; Valentine, C.R.; Lumpkin, C.K.; Ingelman-Sundberg, M.; Badger, T.M. Cytochrome P450 CYP 2E1 induction during chronic alcohol exposure occurs by a two-step mechanism associated with blood alcohol concentrations in rats. *J. Pharmacol. Exp. Ther.* **1993**, *264*, 944–950.
46. Eliasson, E.; Johansson, I.; Ingelman-Sundberg, M. Ligand-dependent maintenance of ethanol-inducible cytochrome P-450 in primary rat hepatocyte cell cultures. *Biochem. Biophys. Res. Commun.* **1988**, *150*, 436–443. [CrossRef]
47. Lieber, C.S. Microsomal ethanol-oxidizing system (MEOS): The first 30 years (1968–1998)—A review. *Alcohol. Clin. Exp. Res.* **1999**, *23*, 991–1007. [CrossRef]
48. Ziegler, R.G. A review of epidemiologic evidence that carotenoids reduce the risk of cancer. *J. Nutr.* **1989**, *119*, 116–122. [CrossRef]
49. Ziegler, R.G. Vegetables, fruits, and carotenoids and the risk of cancer. *Am. J. Clin. Nutr.* **1991**, *53* (Suppl. S1), 251S–259S. [CrossRef]
50. Stahelin, H.B.; Gey, K.F.; Eichholzer, M.; Ludin, E. B-carotene and cancer prevention: The Basel Study. *Am. J. Clin. Nutr.* **1991**, *53* (Suppl. S1), 265S–269S. [CrossRef]
51. Paganini-Hill, A.; Chao, A.; Ross, R.K.; Henderson, B.E. Vitamin A, @-carotene, and the risk of cancer: A prospective study. *J. Natl. Cancer Inst.* **1987**, *79*, 443–448.
52. Leo, M.A.; Kim, C.; Lowe, N.; Lieber, C.S. Interaction of ethanol with beta-carotene: Delayed blood clearance and enhanced hepatotoxicity. *Hepatology* **1992**, *15*, 883–891. [CrossRef]

antioxidants

MDPI

Article

Menthae Herba Attenuates Neuroinflammation by Regulating CREB/Nrf2/HO-1 Pathway in BV2 Microglial Cells

Yeo Jin Park [1,2], Hye Jin Yang [2], Wei Li [2], You-Chang Oh [2] and Younghoon Go [2,*]

[1] Korean Medicine Life Science, University of Science and Technology, Daejeon 34054, Korea; pyjin5526@kiom.re.kr

[2] Korean Medicine (KM)-Application Center, Korea Institute of Oriental Medicine (KIOM), Daegu 41062, Korea; hjyang@kiom.re.kr (H.J.Y.); liwei1986@kiom.re.kr (W.L.); ulivuli@kiom.re.kr (Y.-C.O.)

* Correspondence: gotra827@kiom.re.kr

Abstract: Chronic inflammation and oxidative stress cause microglia to be abnormally activated in the brain, resulting in neurodegenerative diseases such as Alzheimer's disease (AD). Menthae Herba (MH) has been widely used as a medicinal plant with antimicrobial, anti-inflammatory, and antioxidant properties. In this study, we sought to evaluate the effects of MH on the inflammatory response and possible molecular mechanisms in microglia stimulated with lipopolysaccharide (LPS). Transcriptional and translational expression levels of the proinflammatory factors were measured using ELISA, RT-qPCR, and Western blot analysis. MH extract inhibited the production of proinflammatory enzymes and mediators nitric oxide (NO), NO synthase, cyclooxygenase-2, tumor necrosis factor-α, and interleukin-6 in LPS-stimulated cells. Our molecular mechanism study showed that MH inhibited the production of reactive oxygen species (ROS) and the phosphorylation of mitogen-activated protein kinase and nuclear factor (NF)-κB. In contrast, MH activated HO-1 and its transcriptional factors, cAMP response element-binding protein (CREB), and the nuclear factor erythroid 2-related factor 2 (Nrf2) signaling pathways. Thus, MH reduces ROS and NF-κB-mediated inflammatory signaling and induces CREB/Nrf2/HO-1-related antioxidant signaling in microglia. Together, these results may provide specific prospects for the therapeutic use of MH in the context of neuroinflammatory diseases, including AD.

Keywords: BV2; anti-neuroinflammation; ROS; NF-κB; HO-1

1. Introduction

Along with an increase in lifespan, interest in neurodegenerative diseases and neurological health is growing, leading to a better quality of life. Accumulating evidence indicates that chronic neurological diseases, including Alzheimer's disease (AD) and Parkinson's disease, are closely associated with neuroinflammation [1]. Since proinflammatory responses induced by pathological factors, including reactive oxygen species (ROS) and the production and secretion of cytokines, may be harmful to normal neurons, leading to synaptic dysfunction, neuronal death, and loss of synapses [2], it is important to control the inflammatory response to prevent the progression of chronic neurodegenerative disease.

Although microglia are a primary constituent of the dedicated immune system in the brain, it has been reported that the abnormal activation of microglia is involved in multiple pathological signaling pathways [3]. As glial cells provide multiple physiological functions under basal or disease conditions, inflammatory stimuli, such as lipopolysaccharide (LPS), can induce microglia to aggravate inflammation by over-releasing proinflammatory factors, such as interleukin (IL)-1β, IL-6, tumor necrosis factor-α (TNF-α), nitric oxide (NO), and ROS, eventually causing neuronal dysfunction and cell death [4].

Increasing evidence shows that ROS are secondary messengers in microglia that develop progressive inflammation processes and may contribute to dysregulation of the immune response [5]. In pathological situations and exposure to stressful environmental stimuli, ROS production is excessively increased [6], leading to persistent and inappropriate inflammation. However, the brain defends itself against oxidative stress owing to elevated ROS generation, modest antioxidant defenses, and insufficient regeneration ability [7]. Antioxidants are both endogenous and exogenous molecules that limit the detrimental effects of free radicals in neuronal cells through detoxification [8]. Heme oxygenase (HO-1) is an antioxidant enzyme that plays a key role in protecting against oxidative damage and inflammation [9]. HO-1 ultimately induces bilirubin from heme, and its product is a powerful antioxidant [10]. Therefore, the induction of HO-1 is generally considered a defense response against neuronal damage in various inflammatory neurological conditions owing to its enzymatic function [11].

We aimed to identify anti-inflammatory drugs from natural products owing to their fewer side effects on the human body. We screened over 300 natural extracts to identify the most effective product with anti-inflammatory effects and, finally, selected one for being effective on inflammation in BV2 cells (data not shown). Menthae Herba (MH), a perennial herbaceous plant of the Lamiaceae family, is widely used in food, cosmetics, and medicines [12]. Over several millennia, MH has been used traditionally in Korea and China to treat various diseases, including fever, headache, sore throat, thrush, and rubella [13]. In addition, pharmacological studies have shown that MH possesses various biological activities, including antimicrobial, anti-inflammatory, antioxidant, antitumor, gastrointestinal protective, and hepatoprotective activities [14]. MH consists of various volatile compounds that mainly contain monoterpenoids, menthone, and menthol, which may contribute to its therapeutic activities [15,16]. Although several pharmacological studies of MH have revealed its anti-inflammatory activities, its effects on neuroinflammation have not previously been reported in detail. In this study, we evaluated the anti-inflammatory and antioxidant effects of an ethanolic extract of MH in LPS-stimulated BV2 microglial cells.

2. Materials and Methods

2.1. Materials and Reagents

Dulbecco's modified Eagle's medium (DMEM) was obtained from Welgne Inc. (Gyeongsan, Korea). Fetal bovine serum and antibiotics were purchased from Hyclone (Logan, UT, USA). LPS, dexamethasone (DEX), and dimethyl sulfoxide (DMSO) were purchased from Sigma-Aldrich Co. (St. Louis, MO, USA). Bovine serum albumin (BSA) was purchased from GenDEPOT (Katy, TX, USA). Cell counting kits (CCK) were obtained from Dojindo Molecular Technologies, Inc. (Kumamoto, Japan). A RNeasy Mini Kit for RNA extraction was obtained from Qiagen (Hilden, Germany). A Maxima First Strand cDNA Synthesis Kit for RT-qPCR and PowerTrack SYBR Green Master Mix were obtained from Thermo Scientific (Waltham, MA, USA). Oligonucleotide primers for real-time quantitative PCR were synthesized by Cosmogenetech (Seoul, Korea). Enzyme-linked immunosorbent assay (ELISA) kits were obtained from R&D Systems (Minneapolis, MN, USA). Primary antibodies were purchased from Cell Signaling Technology, Inc. (Boston, MA, USA). Horseradish peroxidase (HRP)-conjugated secondary antibodies were purchased from Bethyl Laboratories Inc. (Montgomery, TX, USA). Cell culture dishes and plates were purchased from Sarstedt (Nümbrecht, Germany). MH extract (freeze-dried powder) was provided from the Korea Plant Extract Bank (Ochang, Korea), which was extracted in 95% ethanol by reflux and dissolved in DMSO and stored at -20 °C for bioassay use. The standard compound (±)menthol used for analysis was purchased from Sigma-Aldrich Co. (St. Louis, MO, USA), and the purity was above 98%. High-Performance Liquid Chromatography (HPLC)-grade acetonitrile and methanol were obtained from Merck KGaA (Darmstadt, Germany). ACS reagent-grade formic acid was purchased from Sigma-Aldrich Co. (St. Louis, MO, USA). The ultrapure water used for HPLC analysis was purified by UP Quality 18.2 MΩ cm^{-1} using a Puris-Evo UP Water System equipped with Evo-UP Dio VFT and Evo-ROP Dico20 (Mirae ST Co., Ltd., Anyang, Gyeonggi-do, Korea).

2.2. High-Performance Liquid Chromatography-Charged Aerosol Detector (HPLC-CAD) Analysis

A stock solution of menthol was prepared at a concentration of 1000 μg/mL in methanol. Then, it was diluted to a final concentration of 500 μg/mL to prepare a working solution. The analytical sample (MH extract) was prepared in methanol at a concentration of 10 mg/mL. Then, menthol and the sample solution were filtered through a 0.22 μm PTFE membrane filter (Whatman International Ltd., Maidstone, UK) prior to injection into the HPLC-CAD system. A chromatographic analysis was performed using a charged aerosol detector (CAD) coupled with HPLC to identify menthol, the main component of MH. The HPLC system Thermo Dionex UltiMate 3000 was equipped with a binary pump, an autosampler, a column oven, a diode array UV/VIS detector, and a CAD (Thermo Fisher Scientific, San Jose, CA, USA). The analysis was conducted according to a previously reported method [17]. The separation was carried out using a Gemini C18 column (4.6 × 150 mm, 5 μm, Phenomenex, Torrance, CA, USA). The temperature of the column oven was maintained at 35 °C, and the injection volume of each sample was 10 μL. The mobile phase consisted of 0.1% formic acid (*v*/*v*) in water (A) and acetonitrile (B), and to improve chromatographic separation capacity, the gradient elution system was programmed at a flow rate of 0.8 mL/min as follows: 80% B, 0–5 min; 80–100% B, 5–15 min; 100% B, 15–25 min; 100–80% B, 25–26 min; and 80% B, 26–35 min. The CAD channel was optimized using the following parameters: power function, 1.00; data collection rate, 10 Hz; filter, 0.1; peak width, 0.02 min; control evaporator temperature, low; and wait ready, ±5.0 K. All data acquisition and analyses were performed using Dionex Chromelon software.

2.3. Cell Culture and Drug Treatment

BV2 microglial cells were provided by Professor Kyoungho Suk of Kyungpook National University (Daegu, Korea). The cells were cultured in DMEM containing 1% antibiotics and 10% FBS in a humidified atmosphere of 5% CO_2 at 37 °C. The cells were pretreated with MH extract (10, 25, or 50 μg/mL) or DEX (10 μM) for 1 h and then stimulated with 100 ng/mL LPS for the indicated periods.

2.4. Cell Viability Test

BV2 microglial cells were plated in 96-well plates at a density of 5×10^4 cells/well for 18 h. To determine the cytotoxic effect of the drug, the cells were incubated with the indicated concentrations of the MH extract for 24 h. After treatment with 10 μL of CCK solution per well for 1 h, the cell viability was analyzed using an ELISA reader at 450 nm.

2.5. Determination of NO Production

NO production in the culture supernatants of the BV2 cells was evaluated by measuring the absorbance of nitrite at the 550 nm wavelength. Cells (5×10^4 cells/well) were seeded in 96-well plates, pretreated with the indicated concentrations of MH extract for 1 h, and then stimulated with LPS for 24 h. Conditioned media were collected, added to the same volume of Griess reagent, and measured spectrophotometrically. Sodium nitrite was used as the standard to calculate the concentration of nitrite.

2.6. ELISA for the Analysis of Inflammatory Cytokine Secretion

To evaluate the effects of the MH extract on the secretion of inflammatory cytokines in the culture supernatants of BV2 cells, ELISA assays for TNF-α and IL-6 were performed. The cells (2.5×10^5 cells/well) were plated in 24-well plates, processed, and treated with MH and LPS under the same conditions as those used in the other experiments. After 6 h of stimulation, the culture media were retrieved according to the manufacturer's standard protocols for the ELISA assay. The amount of each inflammatory cytokine was calculated using absorbance at a wavelength of 450 nm.

2.7. Western Blotting for Protein Expression Measurement

For Western blot analysis, BV2 cells were plated on a six-well plate at a density of 1.5×10^5 cells/well and cultured for 24 h. After treatment with MH extract and LPS, cultured cells were collected at predetermined times. The cells were lysed in radioimmunoprecipitation assay lysis buffer (Millipore, Burlington, MA, USA) and incubated on ice for 30 min. After centrifugation at 15,000× *g* rpm for 10 min at 4 °C, the total proteins were collected and normalized using the BCA protein assay kit (#23225, Thermo Fisher Scientific). Protein samples were separated using 10% sodium dodecyl sulfate–polyacrylamide gel electrophoresis and transferred to polyvinylidene fluoride membranes (Millipore, Burlington, MA, USA). Following blocking with 3% bovine serum albumin for 1 h at room temperature, the samples were incubated with specific primary antibodies at 4 °C. The membranes were then rinsed with Tris-buffered saline (TBS)-T containing 0.1% Tween 20 and incubated with horseradish peroxidase-conjugated secondary antibodies. Protein bands were detected using Alliance Q9 mini (UVITEC, Cambridge, UK) and quantified using ImageJ software.

2.8. RNA Isolation and RT-qPCR

Total cellular RNA was isolated using the QIAzol Lysis Reagent (Qiagen, Hilden, Germany). The purified RNA was reverse-transcribed into cDNA using the Maxima First Strand cDNA Synthesis Kit for RT-qPCR (Thermo Fisher Scientific, Waltham, MA, USA). The amplification reactions for qPCR were performed in 10-µL reaction volumes containing the primers and Power SYBR® Green Master mix (Thermo Fisher Scientific) and detected using a C1000 Touch™ Thermal Cycler (Bio-Rad, Hercules, CA, USA). The samples were compared using the relative CT methods. The expression of 36B4 was used for normalization and quantified using the $2^{-\Delta\Delta Cq}$ method. The primer sequences used are listed in Table 1.

Table 1. Primers used for RT-qPCR.

Target Gene	Primer Sequence (5′-3′)	Gene Bank Accession Number
COX-2	F: TGAGTACCGCAAACGCTTCTC R: TGGACGAGGTTTTTCCACCAG	NM_011198.4
HO-1	F: TGAAGGAGGCCACCAAGGAGG R: AGAGGTCACCCAGGTAGCGGG	NM_010442.2
IL-1β	F: ATGGCAACTGTTCCTGAACTCAACT R: CAGGACAGGTATAGATTCTTTCCTTT	NM_008361.4
IL-6	F: TCCAGTTGCCTTCTTGGGAC R: GTGTAATTAAGCCTCCGACTTG	NM_001314054.1
iNOS	F: GGCAGCCTGTGAGACCTTTG R: GCATTGGAAGTGAAGCGTTTC	NM_001313922.1
MCP1	F: ACCTGGATCGGAACCAAATG R: CCTTAGGGCAGATGCAGTTT	NM_011333.3
TNF-α	F: TTCTGTCTACTGAACTTCGGGGTGATCGGTCC R: GTATGAGATAGCAAATCGGCTGACGGTGTGGG	NM_001278601.1
36B4	F: ACCTCCTTCTTCCAGGCTTT R: CTCCAGTCTTTATCAGCTGC	NM_007475.5

F—forward and R—reverse.

2.9. Immunofluorescence

Coverslips were coated with poly-D-lysine (Gibco, CA, USA) in 24-well plates for 15 min, and then, BV2 cells were seeded onto glass coverslips for 24 h. Cells were pretreated with MH or DEX in serum-free medium for 1 h, stimulated with LPS (100 ng/mL) for 1 h, washed with phosphate-buffered saline (PBS), and fixed in cold 4% paraformaldehyde in distilled water for 30 min. Then, the cells were rinsed, permeabilized with 0.5% Triton X-100 in PBS, incubated in blocking solution for 30 min, and then probed with an NF-κB p65 antibody (Cell Signaling Technology, Inc., Danvers, MA, USA) at 4 °C for 24 h. A secondary antibody, anti-rabbit IgG (Bethyl Laboratories, Montgomery, TX, USA) labeled with Alexa 488 (1:1000) was used for the visualization of fluorescence. To stain the nuclei, the cells were treated with Hoechst 33258 (Thermo Fisher Scientific) in PBS (1:10,000) for 5 min at room temperature in the dark. After washing and mounting, stained cells were visualized using a fluorescence microscope (Lionheart FX automated microscope, BioTek, Winooski, VT, USA).

2.10. ROS Measurement

BV2 cells were cultured in 96-well plates (5×10^4 cells/well). The cells were pretreated with the indicated concentration of MH for 1 h and then stimulated with 100 ng/mL LPS for 24 h. Intracellular ROS levels were detected using ROS-sensitive probe 2′,7′-dichlorodihydrofluorescein diacetate (DCFDA, Invitrogen, D-399). DCFDA was dissolved in DMSO and further diluted before use. BV2 cells were incubated with 5-μM DCFDA staining solution in PBS in the dark for 1 h at 37 °C, then washed and imaged with a fluorescence microscope (Nikon ECLIPSE Ti-U, Nikon Co., Tokyo, Japan). Signal quantification was determined by measuring the fluorescence using a GloMax® Explorer Multimode Microplate Reader (Promega). The results were presented as the relative ratio compared with untreated control cells.

2.11. Statistical Analysis

Data were presented as the mean ± SEM obtained from three individual experiments, and the experiments were performed at least in triplicate ($n = 3$). All data were analyzed by one-way analysis of variance (ANOVA) followed by Tukey's honest significant difference test using GraphPad Prism 5.0 (GraphPad Software, San Diego, CA, USA). Differences were considered statistical significant at p-values < 0.05.

3. Results

3.1. HPLC-CAD Analysis of MH

Menthol, the main component in MH, was determined by comparing the retention time (tR), UV spectra, and chromatogram patterns with those of the standard using a HPLC system. In addition, the detection sensitivity of the menthol component was improved using a detector to measure the amount of chemicals in the sample by generating charged aerosol particles. As shown in Figure 1, menthol was well-separated within 26 min and showed good resolution while minimizing the interference by other analytes. The retention time of the standard compound was analyzed at 18.13 min in the chromatogram. Under the same conditions, the retention times of the observed analytes in MH were 18.12 min.

3.2. Effects of MH on Cell Viability of BV2 Cells

As a preliminary assessment, an appropriate MH concentration was established to attenuate LPS-stimulated neuroinflammation in the absence of overt toxicity. CCK and Griess reagent assays were conducted on BV2 microglial cells in the presence or absence of LPS and the extract. As shown in the CCK assay for the cell viability test, MH was not toxic to the cells at concentrations up to 50 μg/mL when activated separately or together with LPS (Figure 2A,B). Therefore, MH was used at concentrations below 50 μg/mL in subsequent studies to eliminate the possibility of toxic effects of MH on cells.

3.3. MH Reduces NO Production in LPS-Stimulated BV2 Cells

BV2 cells were administered LPS (100 ng/mL), which is extensively used in inflammation studies. The Griess reagent assay showed that BV2 cells produced high concentrations of NO (a mediator of neurotoxic reactions) in the presence of LPS. When comparing the effect of MH between drug-treated groups and the LPS-treated control, NO release increased with LPS stimulation and was significantly decreased at concentrations ≥25 μg/mL MH in a dose-dependent manner. DEX, a representative anti-inflammatory drug, also slightly decreased the NO release of LPS-induced BV2 cells (Figure 2C).

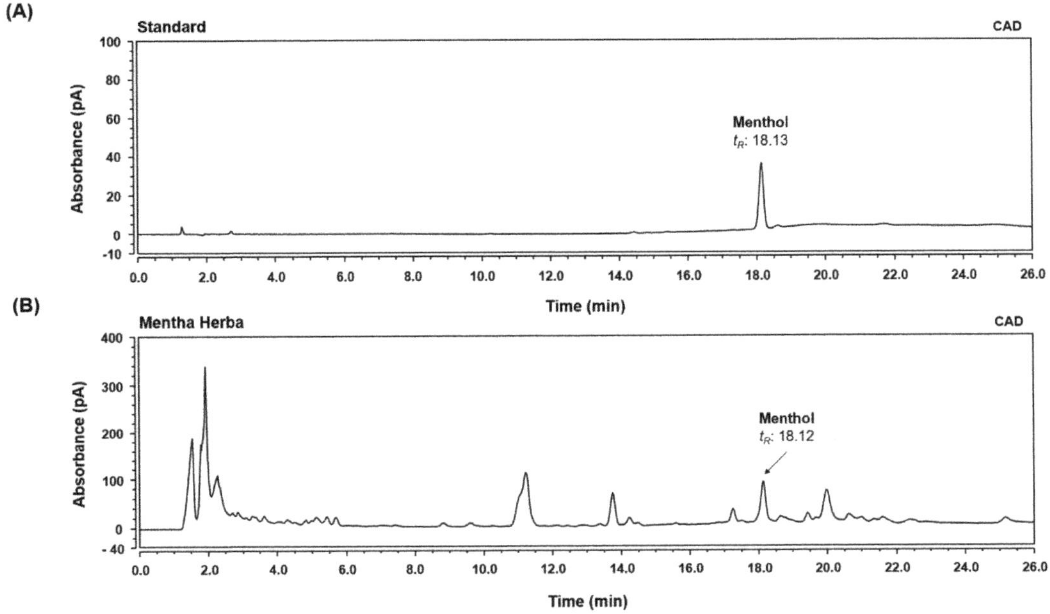

Figure 1. HPLC-CAD analysis of menthol in Mentha Herba (MH). CAD chromatogram of standard solution (**A**) and MH (**B**).

3.4. MH Decreases Inflammatory Cytokines and Chemokine in LPS-Stimulated BV2 Cells

To evaluate the efficacy of MH extract treatment in mitigating the proinflammatory processes triggered by LPS in microglia, we measured the production of proinflammatory cytokines, including TNF-α and IL-6. ELISA showed that LPS markedly produced and secreted TNF-α and IL-6 cytokines, while pretreatment with MH reduced their secretion in a dose-dependent manner. DEX significantly decreased the secretion of TNF-α and failed to induce IL-6 secretion (Figure 2D,E).

To evaluate whether MH extract also altered the transcriptional changes in the mRNA expression of the cytokines, we analyzed the relative gene expression of the inflammatory cytokines in BV2 cells by RT-qPCR. LPS stimulation resulted in higher mRNA expression levels of TNF-α and IL-6 in BV2 cells than those in the nontreated controls. At all concentrations tested (10, 25, and 50 μg/mL), pretreatment with the MH extract significantly decreased the expression of cytokines (Figure 2F,G). Furthermore, we examined additional proinflammatory cytokines (IL-1β and MCP-1) in BV2 cells, where LPS showed a statistically significant increase in both cytokine transcripts compared to those in the untreated control (Figure 2H,I). However, treatment with MH rescued cytokine gene expression in the cells. Specifically, it reduced IL-1β mRNA expression in a dose-dependent manner, reaching levels at 50 μg/mL comparable to those observed in the control (Figure 2H). The increased mRNA expression of MCP-1 triggered by LPS was also significantly reduced at a concentration of 50 μg/mL (Figure 2I).

3.5. MH Represses LPS-Mediated Expression of iNOS and COX-2

Considering the effects of MH treatment on the attenuation of NO induction in LPS-exposed BV2 cells, we investigated whether such profitable effects were also related to transcriptional changes in the mRNA expression of inducible NOS, an enzyme that synthesizes NO. In contrast, COX-2 is an enzyme that catalyzes the production of prostaglandins

(PG), resulting in the exacerbation of neuroinflammatory processes [18]. As shown in Figure 3A,B, the mRNA and protein expression levels of iNOS and COX-2 were dramatically increased in BV2 cells stimulated with LPS, demonstrating that these genes have emerged as major players in brain inflammation, and increased iNOS and COX-2 expression was associated with neurodegeneration. Furthermore, inhibitory effects on both the mRNA and protein expression of iNOS and COX-2 were observed at the indicated concentrations of MH, resulting in levels at 50 μg/mL not significantly different from the control. DEX exposure also attenuated the mRNA and protein expression of LPS-mediated iNOS and COX-2.

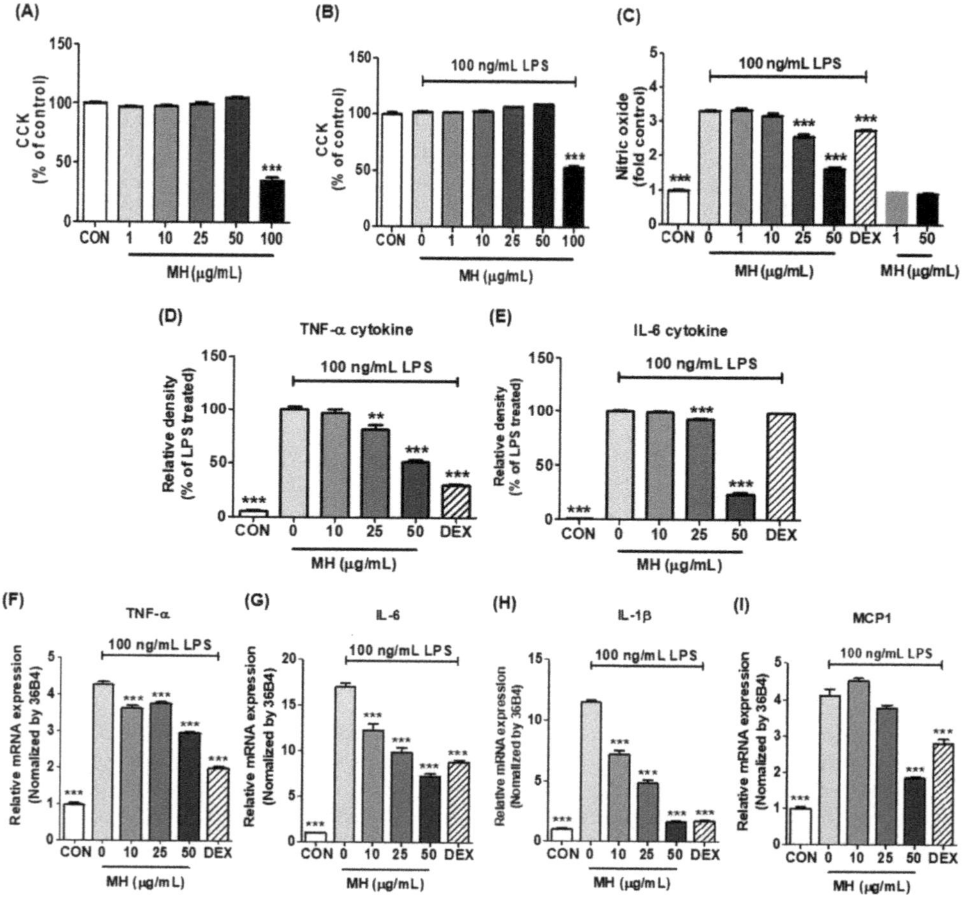

Figure 2. Effects of MH on the production of NO and inflammatory cytokines in LPS-stimulated BV2 cells. (**A**) Cell viability in BV2 cells treated with MH at the indicated concentrations for 24 h. (**B**) BV2 cells were treated with MH at the indicated concentrations for 1 h and then stimulated with LPS for 24 h. Cell viability was measured by the CCK assay. (**C**) BV2 cells were treated with MH at the indicated concentrations or 10 μM DEX for 1 h before stimulation with LPS for 24 h. NO production was measured by Griess reagent assays. (**D**,**E**) Inflammatory cytokines secretion and (**F–I**) mRNA expression of inflammatory cytokines in BV2 cells treated with MH at the indicated concentrations for 1 h and then stimulated with LPS for 6 h. Data are expressed as the mean ± SEM. ** $p < 0.01$ and *** $p < 0.001$ vs. LPS control.

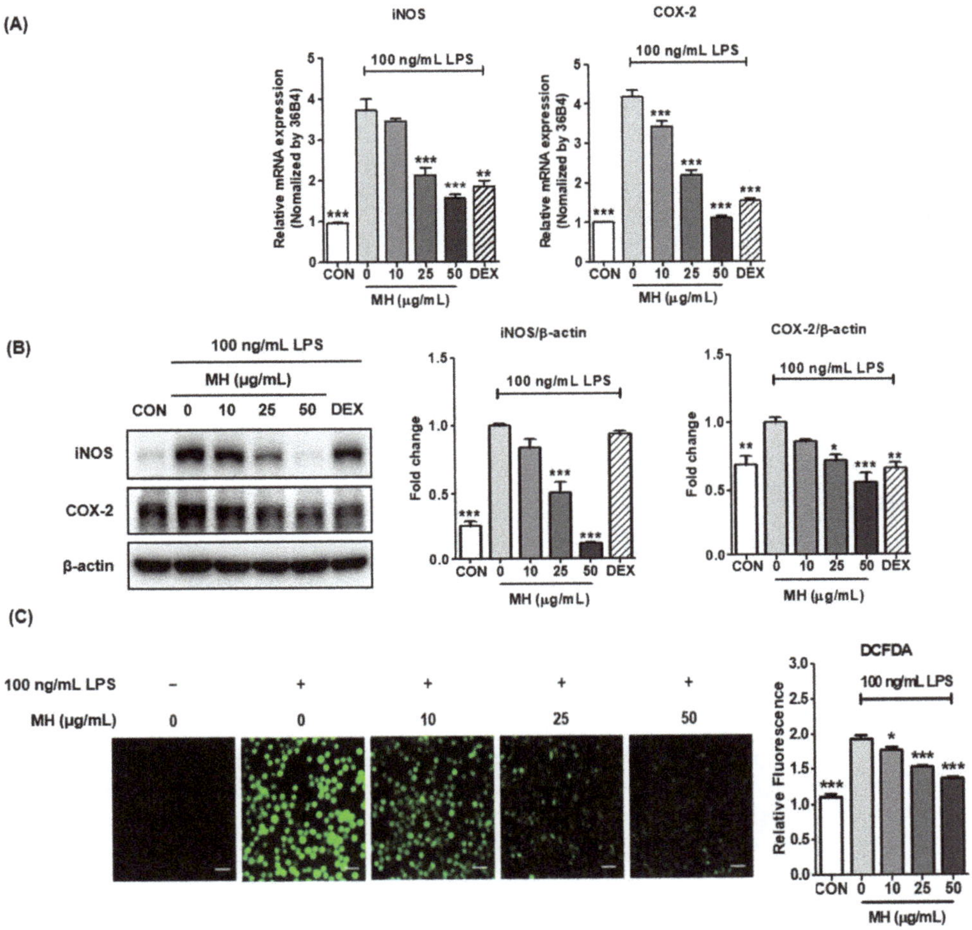

Figure 3. Effects of MH on (**A**) mRNA expressions and (**B**) protein levels of iNOS, COX-2, and (**C**) ROS production. The cells were preincubated with MH at the indicated concentrations or 10 μM DEX for 1 h before stimulation with LPS. The cells were stimulated with LPS for 12 h (mRNA and protein) or 18 h (ROS). The ROS levels were measured by DCFDA; the relative amount of cellular ROS was represented as a fold change over BV2 cells. Data are expressed as the mean ± SEM. * $p < 0.05$, ** $p < 0.01$, and *** $p < 0.001$ vs. LPS control. Scale bar: 100 μm.

3.6. MH Inhibits ROS Production in LPS-Treated BV2 Cells

Based on the data showing that MH treatment decreased the NO, cytokine, and enzyme levels related to neuroinflammatory processes, we further tested whether MH regulates ROS production in BV2 cells. As expected, MH treatment notably decreased ROS production, which was significantly increased compared to that in the control after LPS treatment (Figure 3C). These data indicate that MH significantly downregulated the neuroinflammatory response in LPS-induced BV2 cells.

3.7. MH Suppresses the Transcriptional Activity of NF-κB

Since the activation of nuclear factor (NF)-κB during the inflammatory response promotes the transcription of proinflammatory mediators and cytokines, we further evaluated whether MH inhibits LPS-induced NF-κB activation. To determine this, immunofluorescence staining was performed on BV2 cells treated with MH with or without stimulation with LPS for 1 h. As shown in Figure 4A, LPS stimulation increased the fluorescence intensity of NF-κB p65 in the nuclei, which was reversed in a dose-dependent manner in the presence of MH, indicating that MH inhibited the nuclear translocation of NF-κB p65. Consistent with the immunofluorescence images, the expression of NF-κB p65 in the nuclei of LPS-treated cells was markedly increased. Furthermore, the phosphorylation and degradation of IκBα were elevated by LPS treatment, indicating that NF-κB p65 was activated. However, pretreatment with MH reduced the nuclear translocation of NF-κB p65 and phosphorylation and degradation of IκBα in LPS-treated cells (Figure 4B). These findings suggest that MH may prevent neuroinflammation through inactivation of the NF-κB signaling pathway.

3.8. MH Inactivates MAPK Signaling

It has been reported that NF-κB is activated by mitogen-activated protein kinases (MAPKs), including ERK, p38, and JNK [19,20]. To assess the inhibitory effect of MH on the causative molecular mechanisms of NF-κB activation in LPS-induced inflammatory conditions, we tested whether MH affected the MAPK signaling pathways. As shown in Figure 5, LPS stimulation significantly phosphorylates the main MAPK extracellular signal-regulated kinase (ERK), c-Jun NH_2-terminal kinase (JNK), and p38, and pretreatment with MH represses the phosphorylation of MAPK in a dose-dependent manner without altering the total form. The LPS-treated cells also responded significantly to DEX exposure.

3.9. MH Increases HO-1 Expression through Nrf2/CREB Signaling

To evaluate the other molecular mechanisms responsible for the anti-inflammatory effects of MH, we tested whether MH mediates the antioxidant signaling pathway. Transcriptional and translational upregulation of HO-1 was observed in BV2 cells treated with MH following stimulation with LPS (Figure 6A,B). To investigate the upstream mechanism of HO-1, we assessed the activation of the transcriptional factors of HO-1 by Western blot. According to the results of the Western blot analysis, phosphorylation of the cAMP response element-binding protein (CREB) was increased by MH in a dose-dependent manner. In addition, the degradation of Kelch-like ECH-associated protein (Keap)1 was increased by MH, indicating that nuclear factor erythroid 2-related factor 2 (Nrf2) was translocated from the cytoplasm to the nucleus (Figure 6C). These results suggest that the anti-inflammatory effects of MH are mediated by the antioxidant effect of HO-1 through the activation of CREB and Nrf2.

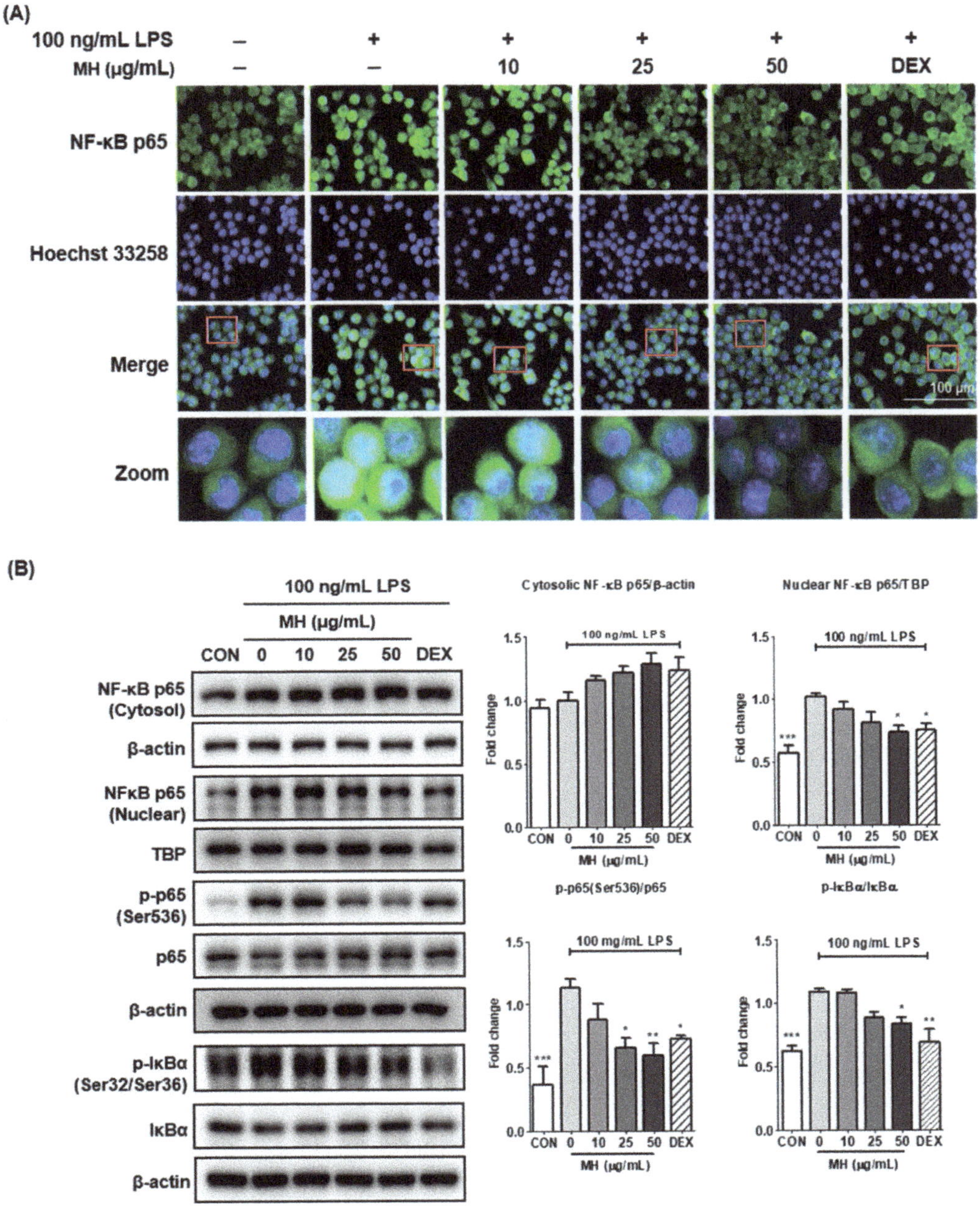

Figure 4. Effects of MH on the (**A**) translocation, (**B**) cytosol, nuclear protein expression, phosphorylation of NF-κB p65, and phosphorylation and degradation of IκBα. The cells were preincubated with the indicated concentration of MH or 10 μM DEX for 1 h before stimulation with LPS. The cells were stimulated with LPS for 1 h (NF-κB p65) or 30 min (IκBα). Immunofluorescence staining was used for detecting NF-κB p65 translocation. The cells were counterstained with Hoechst 33258. (**B**) β-actin was used as the cytosolic endogenous control. TBP was used as the nuclear endogenous control. Data are expressed as the mean ± SEM. * $p < 0.05$, ** $p < 0.01$, and *** $p < 0.001$ vs. LPS control.

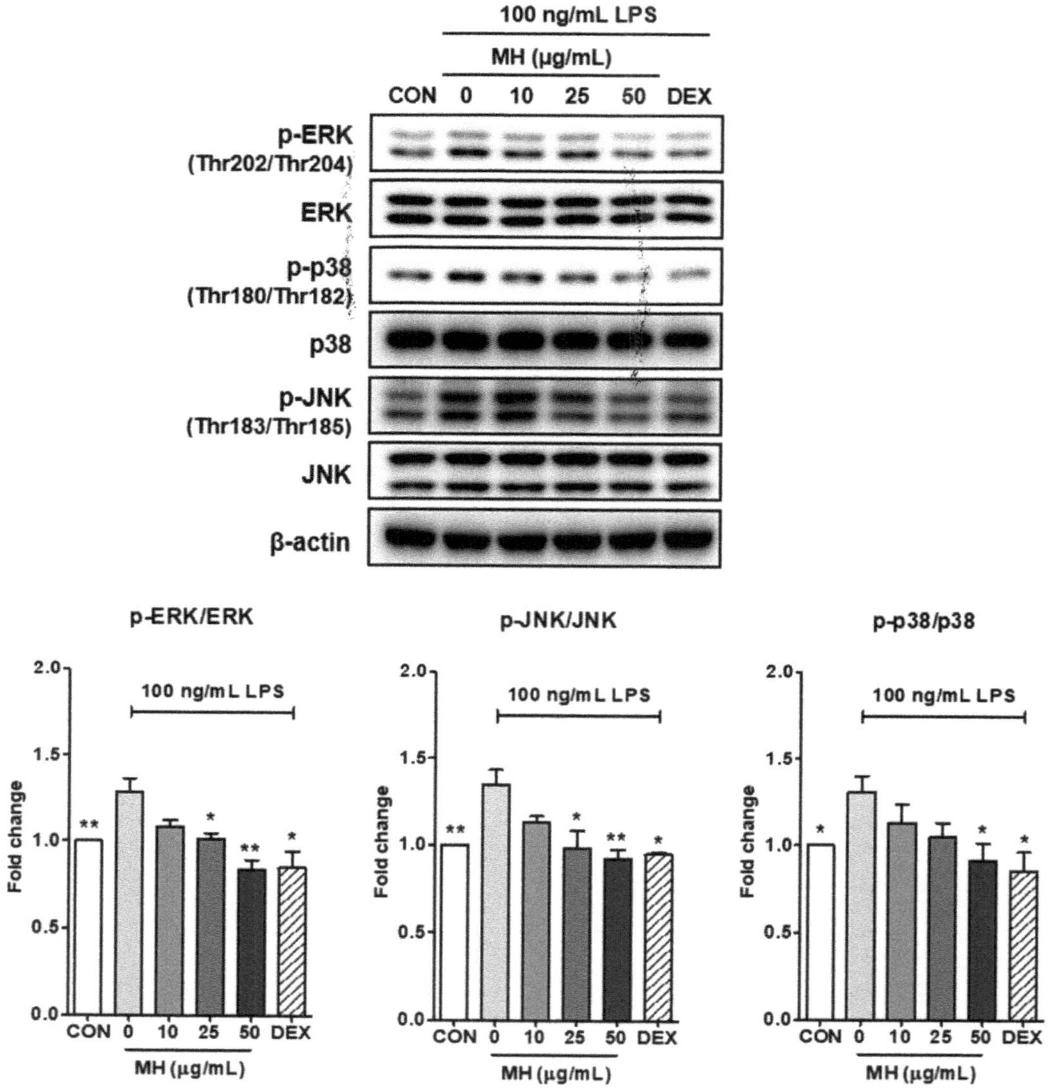

Figure 5. Effects of MH on the phosphorylation of ERK, p38, and JNK MAPK. Representative Western blots and densitometric quantification of p-ERK, ERK, p-p38, p38, p-JNK, and JNK expression in BV2 cells treated with LPS (100 ng/mL) alone or in combination with 3 doses of MH or left untreated (CON). The cells were preincubated with the indicated concentration of MH or 10 μM DEX for 1 h before stimulation with LPS. The cells were stimulated with LPS for 30 min. β-actin was used as an endogenous control. Data are expressed as the mean ± SEM. * $p < 0.05$ and ** $p < 0.01$ vs. LPS control.

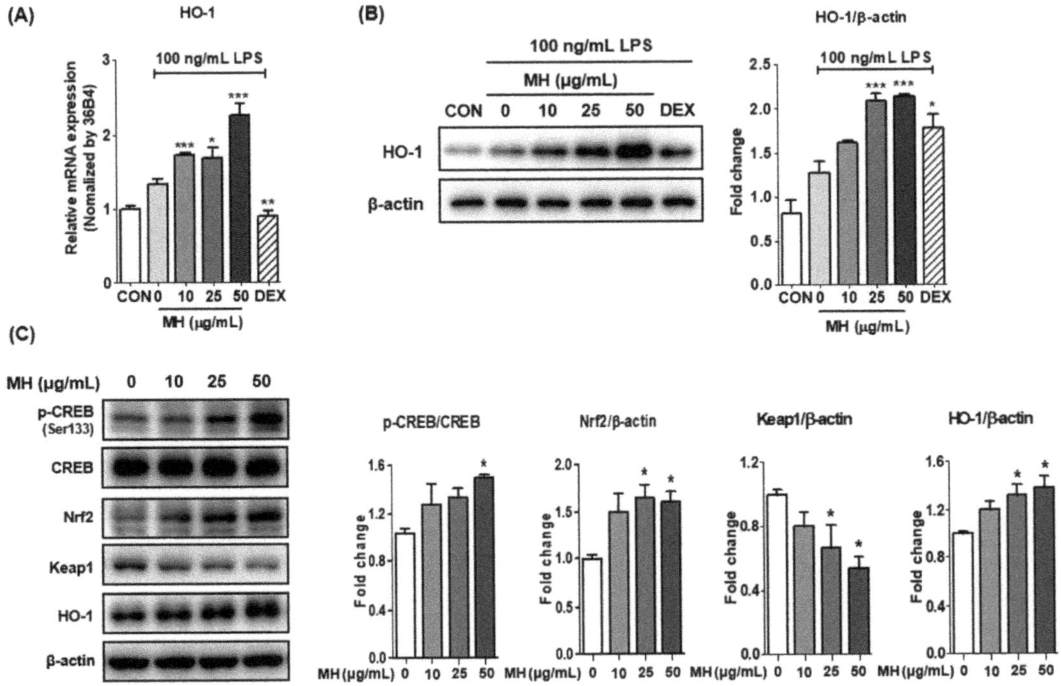

Figure 6. Effects of MH on antioxidative genes. (**A**) mRNA expression levels of HO-1 in BV2 cells treated with LPS and MH. (**B**) Representative Western blots and densitometric quantification of the HO-1 protein levels in BV2 cells treated with LPS and MH. The cells were preincubated with the indicated concentration of MH or 10 μM DEX for 1 h before stimulation with LPS. (**C**) Representative Western blots and densitometric quantification of CREB, phosphorylated CREB, Nrf2, and Keap1 in BV2 cells only treated with MH. β-actin was used as an endogenous control. Data are expressed as the mean ± SEM. * $p < 0.05$, ** $p < 0.01$, and *** $p < 0.001$ vs. LPS control.

4. Discussion

It is widely accepted that inflammation is involved in numerous diseases of the whole body, especially in the brain, and contributes to neurodegenerative diseases, including AD. Under inflammatory stimuli, microglia are progressively and abnormally activated and exacerbate inflammation worsened by the overexpression of proinflammatory molecules such as IL-1β, IL-6, TNF-α, NO, and ROS, eventually resulting in neuronal dysfunction and cell death [4]. Thus, a drug that reduces neuroinflammation via microglial activation is a promising therapeutic candidate for treating neurodegenerative diseases.

Previous studies have shown the anti-inflammatory effects of *Mentha* species in activated RAW264.7 cells [21] and an asthma mouse model [22]. However, it is still unclear whether MH inhibits neuronal inflammation in the central nervous system. We demonstrated the ability of MH to inhibit LPS-stimulated microglial activation, which contributes to neurodegenerative processes. In this study, the underlying mechanisms included, but were not limited to, the downregulation of proinflammatory genes associated with the NF-κB and HO-related antioxidant signaling pathways.

Microglial cells are abnormally activated by inflammatory stimulation, such as LPS, and secrete inflammatory factors, such as NO and PG and the proinflammatory cytokines IL-6, IL-1β, and TNF-α [18]. Under oxidative stress conditions, NO participates in oxidative-reductive processes to form reactive nitrogen species, which are noxious compounds for cognitive function in the CNS. NO is synthesized by members of the NO synthase family, including iNOS, neuronal NOS (nNOS), and endothelial NOS (eNOS) [23]. The current study showed that MH decreased the NO secretion and iNOS expression induced by LPS stimulation. Our results suggest that MH decreases NO production by inhibiting iNOS mRNA and protein expression. In contrast, the downregulation or inhibition of COX-2 is associated with a decrease in PG production, because COX-2 is an enzyme that synthesizes PGs [18]. In this study, MH decreased the expression of COX-2 mRNA and protein. This result suggests that MH may modulate PG production.

Next, we evaluated the mechanisms underlying the anti-inflammatory effects of MH in LPS-treated BV2 cells. LPS stimulation activates Toll-like receptor (TLR) 4/MyD88, which substantially triggers the NF-κB and MAPK signaling pathways in microglia. Upon phosphorylation and degradation of IκBα, the released NF-κB translocates from the cytosol into the nucleus and induces an upsurge in the transcription of inflammatory genes [24]. In addition, among the transactivation domains of NF-κB p65 subunits, Ser536 is a well-known residue that contributes to NF-κB activity [25,26]. In our study, the LPS-activated MAPK signaling pathway was inhibited by MH treatment in microglial cells. Moreover, Western blot results showed that p-IκBα (Ser32/Ser36), p-p65 (Ser536), and nuclear p65 were detected at high levels and that IκBα was low in LPS-induced microglia, which decreased the degradation of IκBα and the translocation and phosphorylation of p65 in a dose-dependent manner. Furthermore, we confirmed the suppressive effect of MH treatment on LPS-induced NF-κB translocation using fluorescence imaging. These results suggest that MH can reduce neuroinflammatory gene expression by inhibiting the degradation of IκBα and activation of MAPK signaling, thus preventing translocation and phosphorylation of the p65 subunit.

In our study, intracellular ROS levels were lower in microglial cells treated with MH than those in the LPS control. Many previous studies have shown that ROS mediates the generation of proinflammatory molecules under stressful conditions and plays a key role in aging and the development of a variety of inflammatory diseases, such as neurodegenerative diseases. Furthermore, these molecules and their metabolites and intracellular ROS overproduction can also induce mitochondrial oxidative damage, eventually resulting in cell death [27,28].

To prevent oxidative damage during inflammation, vertebrate cells have evolved an array of antioxidant defense systems that remove ROS. Among the antioxidant genes, HO-1 is a representative antioxidant and anti-inflammatory gene that produces bilirubin and CO through heme group degradation [29,30]. The expression of HO-1 is closely associated with Nrf2, which is a key transcription factor of many cytoprotective genes, including HO-1, NAD(P)H quinone oxidoreductase (NQO)-1, and the glutamate-cysteine ligase modifier subunit (GCLM). The activity of Nrf2 is essentially involved in Keap1. Under normal conditions, Keap1 binds to Nrf2, resulting in the ubiquitination and degradation of Nrf2 [31]. Under stimulated conditions, Nrf2 liberated from cytoplasmic Keap1 is phosphorylated and subsequently transported into the nucleus, where it induces the production of HO-1 [32]. In the current study, MH treatment increased HO-1 expression. Furthermore, Nrf2 expression and Keap1 degradation were significantly induced by MH treatment for 3 h. Our results suggest that Nrf2 translocation may be accompanied by the phosphorylation of Nrf2 following MH induced-Keap1 degradation that masks the nuclear localization signal of Nrf2. Since translocated Nrf2 has transcriptional activity, it may increase the expression of HO-1, as well as of other cytoprotective genes. In contrast, we showed that the phosphorylation of CREB was induced in a dose-dependent manner by treatment with MH for 3 h. Previous studies have demonstrated that the PKA/CREB pathway is an upstream modulator as a transcription factor that binds the HO-1 promoter directly or as an inducer of Nrf2-binding proteins and activator protein-1 proteins indirectly in microglial cells [33,34]. Interestingly, a previous study showed that the abolition of CREB1 decreased PKCε-induced Nrf2 DNA binding, suggesting that CREB1 may be a transcriptional cooperator of Nrf2 [35]. Furthermore, Lee et al. reported that CREB plays a role in ROS detoxification [34]. Therefore, MH increased CREB/Nrf2/HO-1 signaling, and the activation of this signaling can reduce LPS-induced ROS production in BV2 cells.

Menthol, a main chemical constituent extracted from *Mentha* species, has been reported to display a defense effect against inflammatory disease conditions in various tissues, including the stomach [36], lungs [37], and brain [38]. A previous study showed that menthol, administered during the initiation phase, attenuates forestomach carcinogenesis by suppressing early peripheral leukocyte blood genotoxicity, epithelium cell proliferation, and apoptosis induced by benzo(a)pyrene [36]. On the other hand, it also reported that L-menthol significantly alleviated cigarette smoke extract-induced lung injury in rats with its anti-inflammatory and antioxidative properties. L-menthol suppressed the total white blood cells and macrophages influx, as well as NF-κB and p38 MAPK pathways. The antioxidative activity of L-menthol can be indicated as a decrease of oxidative stress markers (malondialdehyde and myeloperoxidase) and encouraging the Nrf2 pathway [37]. Jian et al. reported that menthol prevented the neuroinflammatory response through MAPK, NF-κB, and AKT. Additionally, these effects of menthol led to improvement of impaired dopaminergic neurons and motor dysfunction in an LPS-induced PD rat model [38]. Based on these previous studies, the effects of menthol may be closely related to the protective effects of MH on neuroinflammatory conditions.

Taken together, MH is a promising anti-inflammatory agent for preventing the inflammatory activation of BV2 cells through the inhibition of ROS generation and HO-1 induction. Nevertheless, our study has some limitations. Since this study evaluated the anti-inflammatory effect of MH in BV2 cells, further studies on AD drugs are required to investigate the inhibitory effect of MH on neurodegeneration in other cell lines. Furthermore, it is necessary to evaluate its anti neuroinflammatory effects in vivo. However, it should be further explored to elucidate the active compounds in MH and to carry out more specific mechanistic studies with the compounds to consolidate evidence that MH could regulate neuroinflammation by increasing the antioxidative pathway.

5. Conclusions

In this study, the inflammatory response of MH and possible molecular mechanisms in microglia stimulated with LPS were evaluated. The results showed that MH inhibited the production of proinflammatory enzymes and mediators. Moreover, MH inhibited the production of ROS and phosphorylation of MAPK and NF-κB. Additionally, MH activated HO-1 and its transcriptional factors, the CREB and Nrf2 signaling pathways. Our results suggest that MH could be used as a potential therapeutic agent for neuroinflammatory diseases due to the attenuation of ROS and inflammation signaling related to MAPK and NF-κB through the induction of CREB/Nrf2/HO-1-mediated antioxidant signaling in LPS-stimulated BV2 microglial cells.

Author Contributions: Conceptualization, Y.J.P., W.L., Y.-C.O. and Y.G.; methodology, Y.J.P., Y.-C.O. and H.J.Y.; validation, Y.J.P. and Y.G.; writing—original draft preparation, Y.J.P. and H.J.Y.; writing—review and editing, Y.J.P., W.L. and Y.G.; supervision, Y.G.; and funding acquisition, Y.G. All authors have read and agreed to the published version of the manuscript.

Funding: This work was supported by the Korea Institute of Oriental Medicine (KIOM), grant number KSN2022230 provided by the Ministry of Science and ICT, Korea.

Institutional Review Board Statement: Not applicable.

Informed Consent Statement: Not applicable.

Data Availability Statement: All the data are available within the article.

Conflicts of Interest: The authors declare no conflict of interest.

References

1. Dansokho, C.; Heneka, M.T. Neuroinflammatory responses in Alzheimer's disease. *J. Neural Transm.* **2018**, *125*, 771–779. [CrossRef] [PubMed]
2. Yang, S.; Yang, Y.; Chen, C.; Wang, H.; Ai, Q.; Lin, M.; Zeng, Q.; Zhang, Y.; Gao, Y.; Li, X.; et al. The Anti-Neuroinflammatory Effect of Fuzi and Ganjiang Extraction on LPS-Induced BV2 Microglia and Its Intervention Function on Depression-Like Behavior of Cancer-Related Fatigue Model Mice. *Front. Pharmacol.* **2021**, *12*, 670586. [CrossRef] [PubMed]
3. Kanegawa, N.; Collste, K.; Forsberg, A.; Schain, M.; Arakawa, R.; Jucaite, A.; Lekander, M.; Olgart Hoglund, C.; Kosek, E.; Lampa, J.; et al. In vivo evidence of a functional association between immune cells in blood and brain in healthy human subjects. *Brain Behav. Immun.* **2016**, *54*, 149–157. [CrossRef]
4. Guzman-Martinez, L.; Maccioni, R.B.; Andrade, V.; Navarrete, L.P.; Pastor, M.G.; Ramos-Escobar, N. Neuroinflammation as a Common Feature of Neurodegenerative Disorders. *Front. Pharmacol.* **2019**, *10*, 1008. [CrossRef]
5. Simpson, D.S.A.; Oliver, P.L. ROS Generation in Microglia: Understanding Oxidative Stress and Inflammation in Neurodegenerative Disease. *Antioxidants* **2020**, *9*, 743. [CrossRef] [PubMed]
6. Brieger, K.; Schiavone, S.; Miller, F.J., Jr.; Krause, K.H. Reactive oxygen species: From health to disease. *Swiss Med. Wkly.* **2012**, *142*, w13659. [CrossRef]
7. Cobley, J.N.; Fiorello, M.L.; Bailey, D.M. 13 reasons why the brain is susceptible to oxidative stress. *Redox Biol.* **2018**, *15*, 490–503. [CrossRef]
8. Sinyor, B.; Mineo, J.; Ochner, C. Alzheimer's Disease, Inflammation, and the Role of Antioxidants. *J. Alzheimers Dis. Rep.* **2020**, *4*, 175–183. [CrossRef]
9. Nakahira, K.; Takahashi, T.; Shimizu, H.; Maeshima, K.; Uehara, K.; Fujii, H.; Nakatsuka, H.; Yokoyama, M.; Akagi, R.; Morita, K. Protective role of heme oxygenase-1 induction in carbon tetrachloride-induced hepatotoxicity. *Biochem. Pharmacol.* **2003**, *66*, 1091–1105. [CrossRef]
10. Siow, R.C.M.; Sato, H.; Mann, G.E. Heme oxygenase–carbon monoxide signalling pathway in atherosclerosis: Anti-atherogenic actions of bilirubin and carbon monoxide? *Cardiovasc. Res.* **1999**, *41*, 385–394. [CrossRef]
11. Yang, J.; Kim, C.-S.; Tu, T.H.; Kim, M.-S.; Goto, T.; Kawada, T.; Choi, M.-S.; Park, T.; Sung, M.-K.; Yun, J.W.; et al. Quercetin Protects Obesity-Induced Hypothalamic Inflammation by Reducing Microglia-Mediated Inflammatory Responses via HO-1 Induction. *Nutrients* **2017**, *9*, 650. [CrossRef] [PubMed]
12. He, X.F.; Geng, C.A.; Huang, X.Y.; Ma, Y.B.; Zhang, X.M.; Chen, J.J. Chemical Constituents from *Mentha haplocalyx* Briq. (*Mentha canadensis* L.) and Their alpha-Glucosidase Inhibitory Activities. *Nat. Prod. Bioprospect.* **2019**, *9*, 223–229. [CrossRef] [PubMed]
13. Kim, J.-H.; Seo, C.-S.; Shin, H.-K. Simultaneous Determination of (-)-Menthone and (-)-Menthol in Menthae Herba by Gas Chromatography and Principal Component Analysis. *Nat. Prod. Sci.* **2010**, *16*, 180–184.
14. Cao, G.; Shan, Q.; Li, X.; Cong, X.; Zhang, Y.; Cai, H.; Cai, B. Analysis of fresh *Mentha haplocalyx* volatile components by comprehensive two-dimensional gas chromatography and high-resolution time-of-flight mass spectrometry. *Analyst* **2011**, *136*, 4653–4661. [CrossRef]
15. She, G.; Xu, C.; Liu, B.; Shi, R. Two New Monoterpenes from *Mentha haplocalyx* Briq. *Helv. Chim. Acta* **2010**, *93*, 2495–2498. [CrossRef]
16. Singh, A.; Raina, V.; Naqvi, A.; Patra, N.; Kumar, B.; Ram, P.; Khanuja, S. Essential oil composition and chemoarrays of menthol mint (*Mentha arvensis* L. *F. piperascens* Malinvaud ex. Holmes) cultivars. *Flavour Fragr. J.* **2005**, *20*, 302–305. [CrossRef]
17. Shaikh, K.; Patil, S.D. Sensitive and selective method for the analysis of menthol from pharmaceutical products by RP-HPLC with refractive index detector. *J. Pharm. Bioallied Sci.* **2010**, *2*, 360. [CrossRef]
18. Ganesan, P.; Kim, B.; Ramalingam, P.; Karthivashan, G.; Revuri, V.; Park, S.; Kim, J.S.; Ko, Y.T.; Choi, D.-K. Antineuroinflammatory Activities and Neurotoxicological Assessment of Curcumin Loaded Solid Lipid Nanoparticles on LPS-Stimulated BV-2 Microglia Cell Models. *Molecules* **2019**, *24*, 1170. [CrossRef]
19. Pan, X.; Cao, X.; Li, N.; Xu, Y.; Wu, Q.; Bai, J.; Yin, Z.; Luo, L.; Lan, L. Forsythin inhibits lipopolysaccharide-induced inflammation by suppressing JAK-STAT and p38 MAPK signalings and ROS production. *Inflamm. Res.* **2014**, *63*, 597–608. [CrossRef]
20. Li, J.; Shui, X.; Sun, R.; Wan, L.; Zhang, B.; Xiao, B.; Luo, Z. Microglial Phenotypic Transition: Signaling Pathways and Influencing Modulators Involved in Regulation in Central Nervous System Diseases. *Front. Cell. Neurosci.* **2021**, *15*, 736710. [CrossRef]
21. Chen, X.; Zhang, S.; Xuan, Z.; Ge, D.; Chen, X.; Zhang, J.; Wang, Q.; Wu, Y.; Liu, B. The Phenolic Fraction of *Mentha haplocalyx* and Its Constituent Linarin Ameliorate Inflammatory Response through Inactivation of NF-κB and MAPKs in Lipopolysaccharide-Induced RAW264.7 Cells. *Molecules* **2017**, *22*, 811. [CrossRef] [PubMed]
22. Lee, M.-Y.; Lee, J.-A.; Seo, C.-S.; Ha, H.; Lee, N.-H.; Shin, H.-K. Protective Effects of *Mentha haplocalyx* Ethanol Extract (MH) in a Mouse Model of Allergic Asthma. *Phytother. Res.* **2011**, *25*, 863–869. [CrossRef] [PubMed]
23. Calabrese, V.; Mancuso, C.; Calvani, M.; Rizzarelli, E.; Butterfield, D.A.; Giuffrida Stella, A.M. Nitric oxide in the central nervous system: Neuroprotection versus neurotoxicity. *Nat. Rev. Neurosci.* **2007**, *8*, 766–775. [CrossRef] [PubMed]
24. Shimizu, H.; Mitomo, K.; Watanabe, T.; Okamoto, S.; Yamamoto, K. Involvement of a NF-kappa B-like transcription factor in the activation of the interleukin-6 gene by inflammatory lymphokines. *Mol. Cell. Biol.* **1990**, *10*, 561–568. [PubMed]
25. Morgan, M.J.; Liu, Z.-g. Crosstalk of reactive oxygen species and NF-κB signaling. *Cell Res.* **2011**, *21*, 103–115. [CrossRef] [PubMed]

26. Huante-Mendoza, A.; Silva-García, O.; Oviedo-Boyso, J.; Hancock, R.; Baizabal Aguirre, V.M. Peptide IDR-1002 Inhibits NF-κB Nuclear Translocation by Inhibition of IκBα Degradation and Activates p38/ERK1/2–MSK1-Dependent CREB Phosphorylation in Macrophages Stimulated with Lipopolysaccharide. *Front. Immunol.* **2016**, *7*, 533. [CrossRef] [PubMed]
27. Shevtsova, E.F.; Vinogradova, D.V.; Neganova, M.E.; Avila-Rodriguez, M.; Ashraf, G.M.; Barreto, G.E.; Bachurin, S.O.; Aliev, G. Mitochondrial Permeability Transition Pore as a Suitable Targ e t for Neuroprotective Agents Against Alzheimer's Disease. *CNS Neurol. Disord. Drug Targets* **2017**, *16*, 677–685. [CrossRef]
28. Alexiou, A.; Soursou, G.; Chatzichronis, S.; Gasparatos, E.; Kamal, M.A.; Yarla, N.S.; Perveen, A.; Barreto, G.E.; Ashraf, G.M. Role of GTPases in the Regulation of Mitochondrial Dynamics in Alzheimer's Disease and CNS-Related Disorders. *Mol. Neurobiol.* **2019**, *56*, 4530–4538. [CrossRef]
29. Nitti, M.; Piras, S.; Brondolo, L.; Marinari, U.M.; Pronzato, M.A.; Furfaro, A.L. Heme oxygenase 1 in the nervous system: Does it favor neuronal cell survival or induce neurodegeneration? *Int. J. Mol. Sci.* **2018**, *19*, 2260. [CrossRef]
30. Paine, A.; Eiz-Vesper, B.; Blasczyk, R.; Immenschuh, S. Signaling to heme oxygenase-1 and its anti-inflammatory therapeutic potential. *Biochem. Pharmacol.* **2010**, *80*, 1895–1903. [CrossRef]
31. Taguchi, K.; Motohashi, H.; Yamamoto, M. Molecular mechanisms of the Keap1–Nrf2 pathway in stress response and cancer evolution. *Genes Cells* **2011**, *16*, 123–140. [CrossRef] [PubMed]
32. He, F.; Ru, X.; Wen, T. NRF2, a transcription factor for stress response and beyond. *Int. J. Mol. Sci.* **2020**, *21*, 4777. [CrossRef] [PubMed]
33. Lee, E.-J.; Ko, H.-M.; Jeong, Y.-H.; Park, E.-M.; Kim, H.-S. β-Lapachone suppresses neuroinflammation by modulating the expression of cytokines and matrix metalloproteinases in activated microglia. *J. Neuroinflamm.* **2015**, *12*, 133. [CrossRef] [PubMed]
34. Jung, J.-S.; Shin, J.A.; Park, E.-M.; Lee, J.-E.; Kang, Y.-S.; Min, S.-W.; Kim, D.-H.; Hyun, J.-W.; Shin, C.-Y.; Kim, H.-S. Anti-inflammatory mechanism of ginsenoside Rh1 in lipopolysaccharide-stimulated microglia: Critical role of the protein kinase A pathway and hemeoxygenase-1 expression. *J. Neurochem.* **2010**, *115*, 1668–1680. [CrossRef]
35. Mylroie, H.; Dumont, O.; Bauer, A.; Thornton, C.C.; Mackey, J.; Calay, D.; Hamdulay, S.S.; Choo, J.R.; Boyle, J.J.; Samarel, A.M.; et al. PKCε-CREB-Nrf2 signalling induces HO-1 in the vascular endothelium and enhances resistance to inflammation and apoptosis. *Cardiovasc. Res.* **2015**, *106*, 509–519. [CrossRef]
36. Santo, S.G.E.; Romualdo, G.R.; Santos, L.A.d.; Grassi, T.F.; Barbisan, L.F. Modifying effects of menthol against benzo (a) pyrene-induced forestomach carcinogenesis in female Swiss mice. *Environ. Toxicol.* **2021**, *36*, 2245–2255. [CrossRef]
37. Liu, Y.; Li, A.; Feng, X.; Jiang, X.; Sun, X.; Huang, W.; Zhu, X.; Zhao, Z. l-Menthol alleviates cigarette smoke extract induced lung injury in rats by inhibiting oxidative stress and inflammation via nuclear factor kappa B, p38 MAPK and Nrf2 signalling pathways. *RSC Adv.* **2018**, *8*, 9353–9363. [CrossRef]
38. Du, J.; Liu, D.; Zhang, X.; Zhou, A.; Su, Y.; He, D.; Fu, S.; Gao, F. Menthol protects dopaminergic neurons against inflamma-tion-mediated damage in lipopolysaccharide (LPS)-Evoked model of Parkinson's disease. *Int. Immunopharmacol.* **2020**, *85*, 106679. [CrossRef]

antioxidants

MDPI

Article

Synergistic Anti-Inflammatory Effects of Lipophilic Grape Seed Proanthocyanidin and Camellia Oil Combination in LPS-Stimulated RAW264.7 Cells

Linli Zhang [1], Juan Chen [2], Ruihong Liang [1], Chengmei Liu [1], Mingshun Chen [1,*] and Jun Chen [1]

[1] State Key Laboratory of Food Science and Technology, Nanchang University, Nanchang 330047, China; 357900210032@email.ncu.edu.cn (L.Z.); liangruihong@ncu.edu.cn (R.L.); liuchengmei@ncu.edu.cn (C.L.); chen-jun@ncu.edu.cn (J.C.)

[2] Moutai Institute, Renhuai 564501, China; kilooy@163.com

* Correspondence: chenshun1221@ncu.edu.cn; Tel.: +86-0791-88305871

Abstract: Combination drug therapy has become an effective strategy to control inflammation. Lipophilic grape seed proanthocyanidin (LGSP) and camellia oil (CO) have been independently investigated to show anti-inflammatory effects, but their synergistic anti-inflammatory effects are unknown. The aim of this study was to investigate the synergistic anti-inflammatory effects of LGSP and CO. The anti-inflammatory activity of LGSP and CO individual or in combination on RAW264.7 cells was detected by MTT assay, Griess reagent, RT-PCR, 2′,7′-dichlorfluoroescein diacetate and Western blot analysis. The combined treatment of LGSP with CO (20 μg/mL and 1 mg/mL) synergistically suppressed the production of NO, TNF-α, IL-6 and ROS. Further studies showed that the synergistic effect was attributed to their suppression of the activation of NF-κB and MAPK signaling pathways. Overall, our findings demonstrate the potential synergistic effect between LGSP and CO in LPS-induced inflammation.

Keywords: Lipophilic grape seed proanthocyanidin; camellia oil; anti-inflammatory; synergistic effect; RAW264.7 cells

1. Introduction

Inflammation is an intricate defensive response of the body to achieve self-protection against various irritation and infection. Excess inflammation is a favorable condition for the formation of various chronic diseases, including diabetes, colitis, obesity and even cancer [1,2]. NO is an inflammatory mediator, and excess NO in immune-inflammatory diseases can induce the production of some pro-inflammatory factors, such as TNF-α, IL-6 and IL-1β, which are the main indicators of the severity of inflammation in vivo [3]. IL-10 is a cytokine with strong anti-inflammatory properties that play an important role in the immune response against pathogens [4]. With the rapid pace of modern life and the deterioration of the living environment, people are easily exposed to the stimulation factors, such as stress, staying up late, unhealthy diet and bacteria, leading to inflammation [5,6]. Nonsteroidal anti-inflammatory drugs and corticosteroids are the most commonly used anti-inflammatory drugs, but they have some side effects [7,8]. Thus, it is necessary to develop novel and more effective anti-inflammatory drugs continuously.

Phytochemicals from various foods such as fruits, vegetables, grains, nuts and cocoa/chocolate, including phenols, terpenoids and organic sulfur compounds, have been reported to have anti-inflammatory activity [9,10]. Grape seed proanthocyanidin (GSP), a natural antioxidant of polyphenols in grape seed, was proven to have anti-inflammatory activity [11]. Lipophilic grape seed proanthocyanidins (LGSP), synthesized by enzymatic esterification of GSP, showed stronger antioxidant [12], anticancer [13] and anti-inflammatory activities [14], compared with GSP. Based on the complexity of digestion, absorption, metabolism and interaction between bioactive phytochemicals and food, combined drug therapy has become an effective strategy to control inflammation, which can improve pharmacological activity and decrease side effects by acting on multiple targets of inflammation. Camellia oil (CO), edible oil with extremely high nutritional value, contains abundant unsaturated fatty acids and phytochemicals, including squalene, sterols, tocopherols and polyphenols [15], exhibiting a variety of bioactivities, such as antioxidant, antitumor and anti-inflammatory [16]. Pallares et al. [17] reported that proanthocyanidins and polyunsaturated fatty acids showed synergistic anti-inflammatory effects in vitro. Thus, we hypothesize that the combination of LGSP and CO might amplify their anti-inflammatory effect.

This study aimed to investigate whether the combination of LGSP and CO had a synergistic anti-inflammatory effect. The underlying mechanism of the synergistic anti-inflammatory effect was also explored. The results might provide the basis for the combined utilization of LGSP and CO as an effective anti-inflammatory agent in the future.

2. Materials and Methods

2.1. Materials and Reagents

CO was obtained from Jiangxi Qiyunshan Food Co., Ltd. (Ganzhou, China). DMEM, penicillin/streptomycin and FBS were purchased from Gibco (Grand Island, NY, USA). Lipopolysaccharides (LPS) and MTT were provided by Sigma-Aldrich (St. Louis, MO, USA). Griess reagent NO assay kit and BCA kit were purchased from Beyotime Biotechnology Co. (Shanghai, China). The antibodies were purchased from Cell Signaling Technology (Beverly, MA, USA). Trizol reagent was purchased from Tiangen Biochemical Technology Co. (Beijing, China).

2.2. Preparation of LGSP

LGSP was prepared as described in our previous literature [12]. GSP and lauric acid were added into a screw-capped glass bottle containing ethanol solvent, using Lipozyme TLIM as the catalyst. The reaction was carried out at 45 °C for 12 h and stopped by removing the enzyme. LGSP was obtained by concentrating the product (Figure S1).

2.3. Cell Culture

RAW264.7 cells were purchased from ATCC (Rockville, MD, USA) and cultured in DMEM supplemented with 10% FBS at 37 °C under 5% CO_2.

2.4. Cell Viability Assay

The cytotoxicity of LGSP and CO was determined by MTT assay [18]. Cells (1×10^4 cells/mL) were seeded on 96 well-plates. After adherence, cells were treated with LGSP (5–80 μg/mL) and/or CO (0.125–2 mg/mL) for 24 h. An amount of 200 μL of MTT reagent (0.5 mg/mL) was added and incubated for 4 h. Then, DMSO was added and measured at 490 nm by a microplate reader (KHB ST-360, Shanghai, China).

2.5. NO Production Measurement

NO production was detected by Griess reagent [19]. Cells (1×10^5 cells/mL) were seeded into 96 well-plates. After attaching, cells were pretreated with different concentrations of LGSP and/or CO for 2 h, and then the LPS (1 μg/mL) was added and co-treated for 24 h. The supernatant was then mixed with Griess reagent in the ratio of 1:1 and incubated for 10 min in the dark. The absorbance was measured at 540 nm by a microplate reader (KHB ST-360, Shanghai, China).

2.6. Synergistic Effect Analysis

The synergistic effect of LGSP and CO on the NO level was analyzed by CompuSyn software 2.0, as described in the literature [20]. The NO inhibition data collected after treatment with LGSP and/or CO were entered into the CompuSyn software [21].

2.7. Quantitative Real-Time PCR Analysis

The Trizol reagent was used to extract the total cellular RNA, and the concentration of RNA was measured using NanoDrop 1000 Spectrophotometer (DeNovix). An amount of 2 μg of total RNA was converted to single-stranded cDNA from each sample, which was then amplified by Brilliant II SYBR Green QRT-PCR Master Mix Kit. The gene expression was quantitatively detected by the CFX96 Real-Time PCR Detection System (Bio-Rad, Hercules, CA, USA). The primer pairs were synthesized by Integrated DNA Technologies, Inc. (Coralville, IA, USA), and the following sense and antisense primer sequences used for RT-PCR analysis were: TNF-α, 5′-CACCACCATCAAGGACTCAAAT-3′ (forward) and 5′-CAGGGAAGAATCTGGAAAGGT-3′ (reverse); IL-6, 5′-CTGGGAAATCGTGGAAATGAG-3′ (forward) and 5′-GACTCTGGCTTGTCTTTCTTGTTA-3′ (reverse); IL-1β, 5′-AGATAGA AGTCAAGAGCAAAGTGGA-3′ (forward) and 5′-TGGGGAAGGCATTAGAAACAG-3′ (reverse); IL-10, 5′-TGGACAACATACTGCTAACCGAC-3′ (forward) and 5′-ATGCTCCTT GATTTCTGGGC-3′ (reverse); GAPDH, 5′-AGGTCGGTGTGAACGGATTTG-3′ (forward) and 5′-TGTAGACCATGTAGTTGAGGTCA-3′ (reverse). The $2^{-\Delta\Delta Ct}$ method was used to calculate the copy number of each transcript relative to the GADPH [22].

2.8. Measurement of Reactive Oxygen Species (ROS) Production

ROS were detected using 2′,7′-dichlorfluoroescein diacetate (DCFHDA) [23]. Cells (1×10^5 cells/well) were seeded into 6-well plates for 12 h. After the adherent, cells were pretreated with various concentrations of LGSP and/or CO for 2 h and then exposed to LPS (1 μg/mL) for 12 h. The cells were then treated with 10 μM DCFH-DA for 30 min in the dark and detected by a fluorescence microplate reader (Nikon Eclipse Ti, Tokyo, Japan).

2.9. Western Blot Analysis

Cells (1×10^5 cells/well) were plated into 6-well plates. After incubated for 12 h, cells were pretreated with various concentrations of LGSP and/or CO for 2 h, and LPS (1 μg/mL) was subsequently added and incubated for 6 h. The protein was obtained by lysing cells and quantified using the BCA protein analysis kit. Cell lysates were denatured in SDS loading buffer, separated by SDS-PAGE gel and transferred onto immunoblot membranes. After being blocked at room temperature for 1 h, the membranes were incubated at 4 °C overnight with a specific primary antibody, followed by a peroxidase-conjugated secondary antibody at room temperature for 1 h. Protein bands were detected using the ChemiDoc XRS+ system (Bio-Rad).

2.10. Statistical Analysis

All experiments were performed at least in triplicate and results were presented as mean ±standard deviation. Significant differences between means were determined using Duncan's multiple range test ($p < 0.05$).

3. Results and Discussion

3.1. Effect of LGSP and CO on Cell Viability of RAW264.7 Cells

The effect of LGSP and CO on RAW264.7 cell viability was detected by MTT assay to eliminate the influence of cytotoxicity on their anti-inflammatory activity. As shown in Figure 1, LGSP at doses of 5–80 μg/mL and CO at doses of 0.125–2 mg/mL had no effect on the viability of RAW264.7 cells. Meanwhile, the combination of LGSP and CO in the ratio of 1:25, 1:50, 1:100 (LGSP: CO) also had no effect on the viability of RAW264.7 cells. The results suggested that LGSP and/or CO were not cytotoxic to RAW264.7 cells at the above doses.

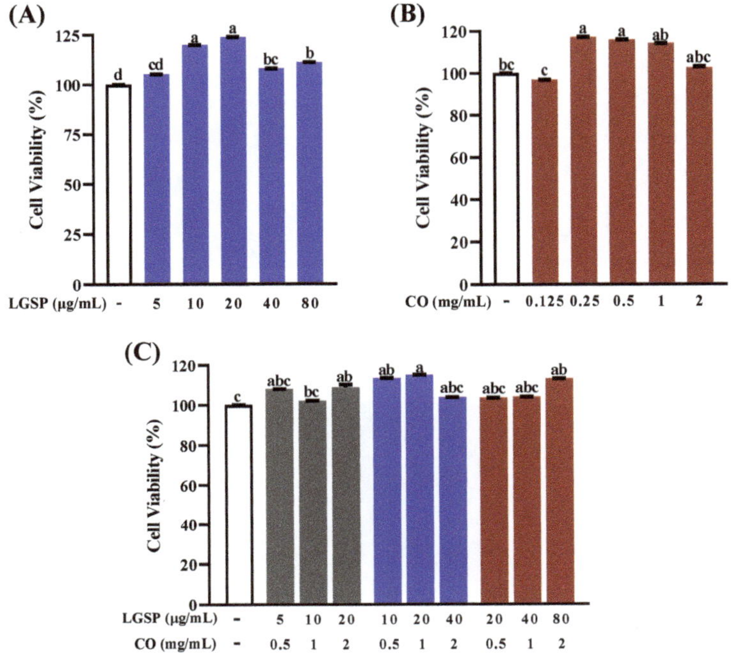

Figure 1. Effect of LGSP (**A**), CO (**B**) and their combination (**C**) on viability of RAW264.7 cells. All values are means ± SD, $n = 3$. Different letters above the bars indicate significant differences ($p < 0.05$).

3.2. *Effect of LGSP and CO on Inhibition of NO Production*

NO is an endogenous gaseous signal molecule, which can be produced as an inflammatory mediator in LPS-induced macrophages [24]. The effect of LGSP and CO on NO levels in LPS-induced RAW264.7 cells was displayed in Figure 2. LPS stimulation led to a sharp increase in NO production ($p < 0.05$). LGSP inhibited LPS-induced NO in a dose-dependent manner by 2.43, 4.71, 12.03, 20.17 and 29.16% at doses of 5, 10, 20, 40 and 80 μg/mL, respectively (Figure 2A). While treatment with CO resulted in a dose-dependent inhibition on NO production by 2.19, 11.22, 20.49, 21.47 and 37.41% at 0.125, 0.25, 0.5, 1 and 2 mg/mL, respectively (Figure 2B). Notably, the combination of LGSP and CO showed a stronger inhibitory effect on NO production than LGSP or CO individual ($p < 0.05$), and the inhibitory effect depended on the concentration and ratio of LGSP and CO (Figure 2C). The inhibition rate of each concentration gradient with LGSP and CO ratio of 1:50 was higher than that of 1:25 and 1:100. Notably, treatment with LGSP (40 μg/mL) and CO (2 mg/mL) caused a 20.17% and 37.41% reduction in NO production, respectively, but the combined treatment with LGSP and CO resulted in a 53.94% inhibition rate of NO production.

3.3. *Synergistic Effect Analysis*

Based on the results of LGSP and CO treatment on NO production, the mode of interaction between LGSP and CO in inhibiting NO production was further determined. Figure 3 showed the isobologram analysis for the combination of LGSP and CO. As shown in Figure 3A–C, the combination of LGSP and CO in the ratio of 1:50 to achieve 90% inhibition, 75% inhibition and 50% inhibition were all below the corresponding lines, suggesting that there was a synergistic anti-inflammatory effect between LGSP and CO.

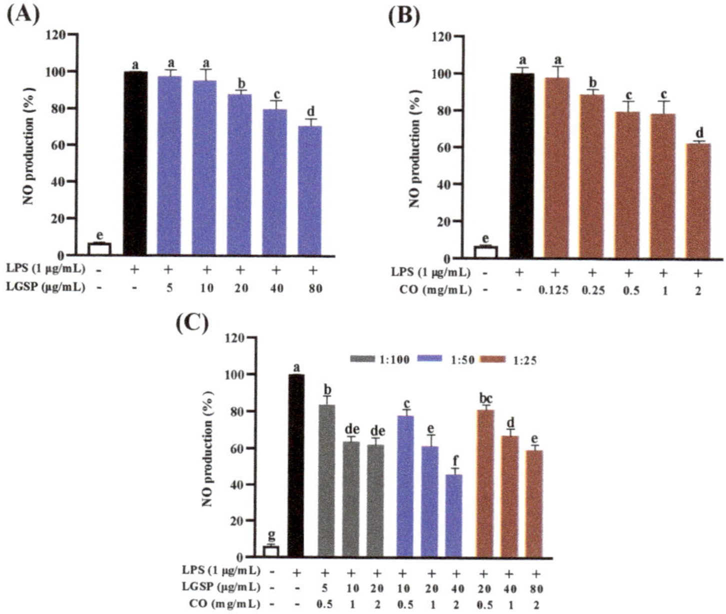

Figure 2. Effect of LGSP (**A**), CO (**B**) and their combination (**C**) on NO production in RAW264.7 cells. All values are means ± SD, $n = 3$. Different letters above the bars indicate significant differences ($p < 0.05$).

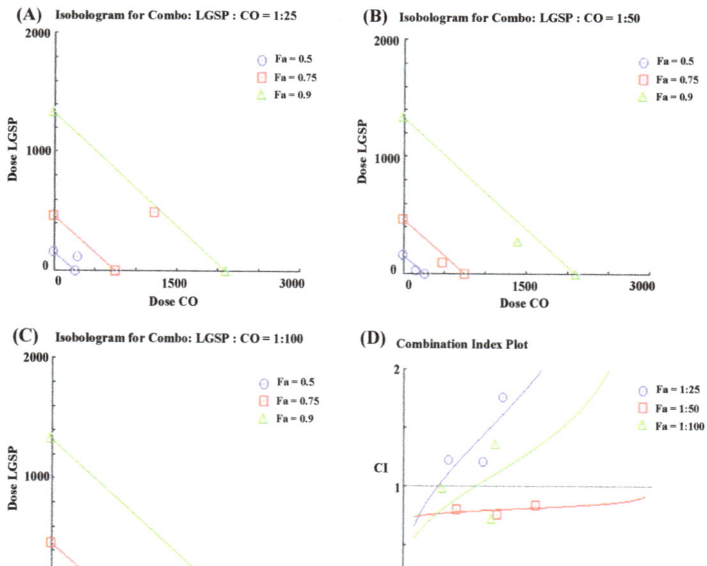

Figure 3. Isobologram curves for the combination of LGSP and CO in ratios of 1:25 (**A**), 1:50 (**B**) and 1:100 (**C**) on inhibition of NO production. Combination index (CI) for LGSP and CO on inhibition of NO production in LPS-stimulated RAW264.7 cells (**D**).

The combination index (CI), a quantitative indicator of the extent of drug interaction, is used to determine the presence of synergy or antagonism between drugs. A lower CI value represents a stronger synergistic effect. Figure 3D displayed that CI values of LGSP and CO (ratio of 1:50) were between 0.76 and 0.84. LGSP (20 μg/mL) combined with CO (1 mg/mL) exhibited the lowest CI value (0.76), indicating the best synergistic effect. Therefore, the combination of LGSP (20 μg/mL) and CO (1 mg/mL) was selected for further study.

3.4. Effect of LGSP and CO on the mRNA Expression of Inflammatory Cytokines

In inflammatory diseases, pro-inflammatory cytokines are over-expressed in LPS-induced RAW264.7 cells, and these inflammatory cytokines were verified as the major indicators of the severity of inflammation in vivo [25,26]. Then, the influence of LGSP and CO on the expression of TNF-α, IL-6 and IL-1β mRNA was detected by RT-PCR. As shown in Figure 4, LPS dramatically enhanced the expression of TNF-α, IL-6, IL-1β and IL-10 mRNA, compared with control ($p < 0.05$). LGSP treatment inhibited the expression of TNF-α and IL-6 mRNA by 36.8% and 14.6%, which was in line with our previous study [14]. CO caused 42.4% and 46.08% inhibition of TNF-α and IL-6 mRNA. Cheng et al. [27] reported that CO could decrease the level of IL-6 in the rat intestinal mucosal injury model. The combination of LGSP and CO strongly inhibited TNF-α and IL-6 mRNA levels by 90.5% and 74.8%, respectively (Figure 4A,B). However, no significant difference was observed in the mRNA expression of IL-1β and IL-10 between their combination treatment and LGSP or CO individual treatment (Figure 4C,D). These results clearly demonstrated that LGSP combined with CO had a synergistic higher inhibitory effect on TNF-α and IL-6 mRNA expression than LGSP or CO individual.

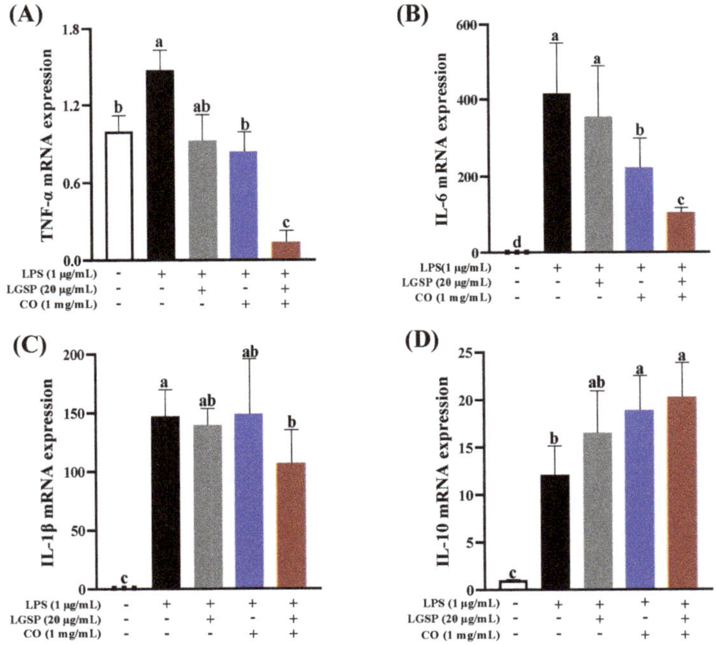

Figure 4. Effects of LGSP, CO and their combination on LPS-induced production of inflammatory cytokines TNF-α (**A**), IL-6 (**B**), IL-1β (**C**) and IL-10 (**D**) in RAW264.7 cells. All values are means ± SD, $n = 3$. Different letters above the bars indicate significant differences ($p < 0.05$).

3.5. Effect of LGSP and CO on LPS-Induced ROS

Macrophages release ROS after being stimulated, leading to cell and tissue damage, thereby further stimulating the inflammatory condition [28]. ROS is one of the major factors that cause the enhancement of oxidative stress, which plays a vital role in the pathogenesis of chronic inflammation [29]. Thus, the effect of LGSP (20 μg/mL) and CO (1 mg/mL) on LPS-induced ROS was evaluated. As shown in Figure 5, LPS stimulation resulted in a remarkable increase in the fluorescence intensity of ROS in RAW264.7 cells, compared with control ($p < 0.05$). This intensity was remarkably decreased after treatment with LGSP and/or CO. The combination of LGSP and CO exhibited a stronger inhibitory effect (93.5%) than LGSP (38.3%) or CO (58.6%) individually on LPS-induced ROS. Bumrungpert et al. [30] reported that the CO-enriched diet could decrease the biomarkers of oxidative stress caused by ROS in hypercholesterolemia. Combined with our experimental data, the synergistic anti-inflammatory activity of LGSP and CO might be attributed to their ability to inhibit oxidative stress-induced damage by scavenging ROS, at least partly.

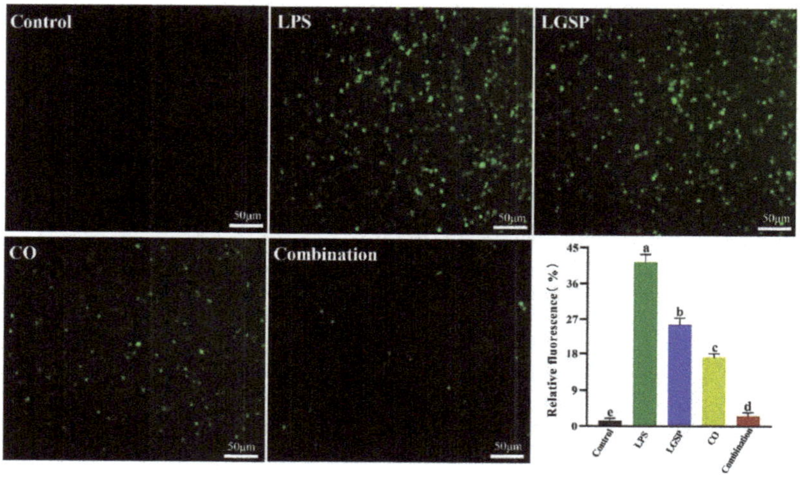

Figure 5. Effect of LGSP (20 μg/mL), CO (1 mg/mL) and their combination on ROS in LPS-induced RAW264.7 cells. All values are means ± SD, $n = 3$. Different letters above the bars indicate significant differences ($p < 0.05$).

3.6. *Effect of LGSP and CO on LPS-Induced iNOS, NF-κB and MAPK Pathway*

In order to further explore the potential synergistic anti-inflammatory mechanism of LGSP and CO, iNOS and two typical inflammatory pathways, NF-κB and MAPK, were detected in LPS-stimulated RAW264.7 cells. iNOS can be activated by endotoxin and pro-inflammatory mediators in macrophages and regulate the production of NO. As shown in Figure 6A, LGSP at 20 μg/mL and CO at 1 mg/mL significantly decreased iNOS levels compared to the LPS-treated group. However, no significant difference in iNOS expression was found between LGSP or CO individual treatment and their combination treatment. NO regulation is a complex and intensive process. When cells are stimulated by cytokines or immune microorganisms, iNOS catalyzes the production of a large number of NO, and the variation in NO content in cells is affected by substrate availability, iNOS protease activity, expression level or other subtypes of NOS enzyme activity [31]. The synergistic inhibitory effect of LGSP and CO on NO was not entirely determined by their effect on iNOS activity.

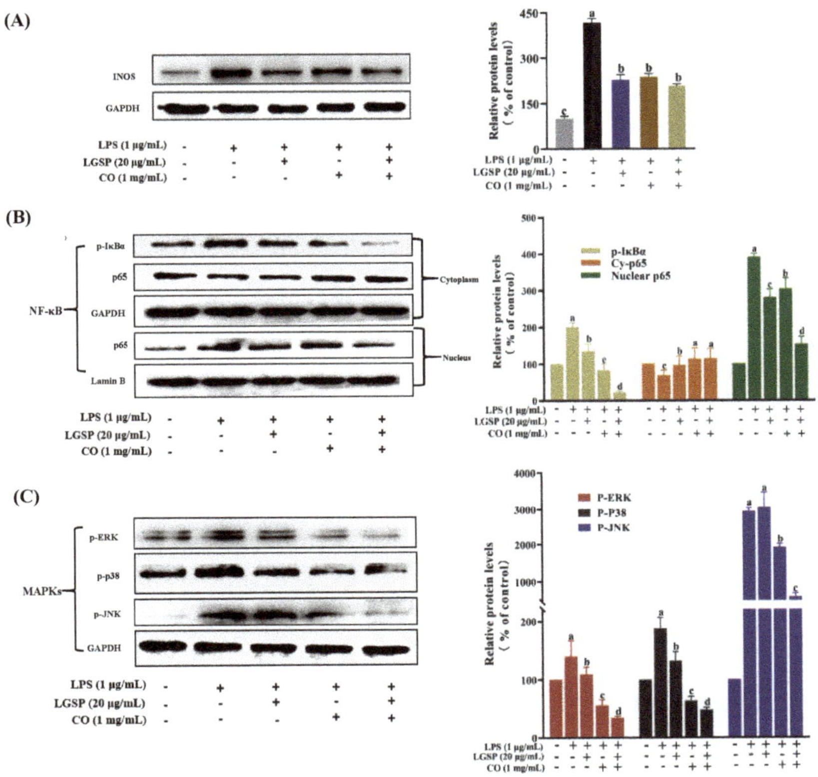

Figure 6. Effect of LGSP, CO and their combination on iNOS (**A**), NF-κB (**B**) and MAPKs (**C**) signaling pathways. All values are means ± SD, $n = 3$. Different letters above the bars indicate significant differences ($p < 0.05$).

As a nuclear transcriptional factor, NF-κB regulates the expression of various inflammation and immune response-related genes [32]. It can be activated in LPS-stimulated macrophages, followed by the phosphorylation of the repressor protein IκBα and dissociated from NF-κB, leading to NF-κB translocation into the nucleus to induce the expression of inflammatory cytokines genes [33]. In order to clarify whether LGSP and CO could regulate NF-κB translocation, the nuclear and cytoplasmic fractions were determined by Western blotting analysis. As shown in Figure 6B, LPS treatment significantly promoted the nuclear translocation of NF-κB (p65), and the IκBα was obviously phosphorylated, compared with control ($p < 0.05$). However, treatment with LGSP and CO individual or in combination (20 μg/mL and 1 mg/mL) could attenuate the activation of NF-κB induced by LPS. Particularly, the inhibitory effect of LGSP combined with CO on nuclear translocation of NF-κB (p65) and phosphorylation of IκBα was markedly stronger than that of LGSP and CO individuals. Sanchez-Fidalgo et al. [34] reported that squalene prevented the continuous increase in inflammatory cytokine levels by inhibiting NF-κB in DSS-induced acute colitis. Meanwhile, tea polyphenol could inhibit the release of pro-inflammatory cytokines and the activation of NF-κB in macrophages [35]. Squalene and tea polyphenols are the main active phytochemicals in CO; Thus, LGSP combined with CO synergistically inhibited the activation of NF-κB, which may be attributed to the complementary and overlapping effects between LGSP and these active substances in CO.

Additionally, we examined the effects of LGSP and CO on LPS-stimulated MAPK by monitoring the expression of p-p38, p-JNK and p-ERK. MAPK plays a critical role in the initiation of the inflammatory response by participating in the regulation of the synthesis of inflammatory cytokines at the levels of transcription and translation [36]. LPS treatment resulted in excessive expression of p-p38, p-JNK and p-ERK (Figure 6C). Treatment with LGSP and CO individuals obviously suppressed the expression of p-p38, p-JNK and p-ERK, while treated with LGSP and CO in combination (20 μg/mL and 1 mg/mL) caused a stronger inhibition. Strikingly, the combination of LGSP and CO displayed the most remarkable synergistic effect on the down-regulation of p-JNK. Eriocitrin combined with resveratrol could strongly inhibit the expression of NO, TNF-α and IL-1β by reducing the levels of p-P38 and p-JNK in LPS-induced RAW264.7 cells, but p-ERK was not affected [37]. Tocopherols, a rich phytochemical in CO, was shown to reduce ERK and p38 phosphorylation in the inflammatory response of bronchial and alveolar epithelial cells [38]. In addition, LGSP showed a stronger inhibitory effect on p-JNK than GSP in our previous study [14]. The results indicated that the combination of LGSP and CO (20 μg/mL and 1 mg/mL) could inhibit the pro-inflammatory mediators (NO, TNF-α and IL-6) and the ROS mediated through the suppression of NF-κB and MAPK signaling pathways (Figure 7).

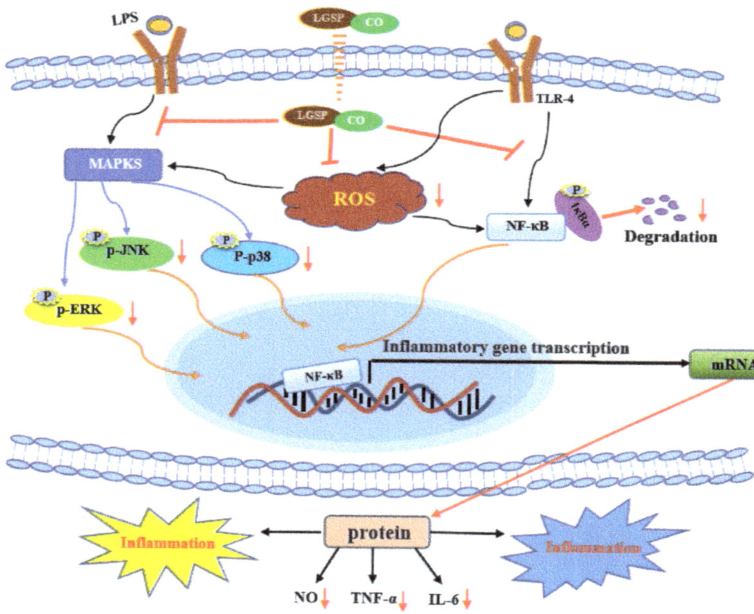

Figure 7. The underlying synergistic anti-inflammatory mechanism of LGSP and CO.

4. Conclusions

The present study demonstrated for the first time that LGSP and CO possessed synergistically anti-inflammatory effects. LGSP and CO in combination suppressed the production of NO, TNF-α and IL-6 and ROS in LPS-stimulated RAW264.7 cells by blocking NF-κB and MAPK signaling pathways. These results may provide a novel and effective agent for the treatment of inflammatory.

Supplementary Materials: The following supporting information can be downloaded at: https://www.mdpi.com/article/10.3390/antiox11020289/s1, Figure S1: Synthesis of LGSP.

Author Contributions: L.Z.: Investigation, Methodology, Writing—original draft, Writing—review and editing. J.C. (Juan Chen): Writing—review and editing. R.L.: Conceptualization, Supervision, Writing—review and editing. C.L.: Data curation, Software, Writing—review and editing. M.C.: Methodology, Supervision, Funding acquisition, Investigation, Writing—review and editing. J.C.

(Jun Chen): Funding acquisition, Project administration. All authors have read and agreed to the published version of the manuscript.

Funding: This research was funded by the first batch of high-end talents (Youth) projects of science and technology innovation in Jiangxi Province (2019201076) and the Natural Science Foundation of Jiangxi Province (20202BAB215014).

Institutional Review Board Statement: Not applicable.

Informed Consent Statement: Not applicable.

Data Availability Statement: Data is contained within the article and supplementary material.

Acknowledgments: The authors would like to thank the Center of Analysis and Testing of Nanchang University and State Key Laboratory of Food Science and Technology for their expert technical assistance.

Conflicts of Interest: The authors declare no conflict of interest.

References

1. Ono, M. Molecular links between tumor angiogenesis and inflammation: Inflammatory stimuli of macrophages and cancer cells as targets for therapeutic strategy. *Cancer Sci.* **2008**, *99*, 1501–1506. [CrossRef] [PubMed]
2. Xie, W.; Du, L. Diabetes is an inflammatory disease: Evidence from traditional chinese medicines. *Diabetes Obes. Metab.* **2011**, *13*, 289–301. [CrossRef] [PubMed]
3. Chae, U.; Kim, H.-S.; Kim, K.-M.; Lee, H.; Lee, H.-S.; Park, J.-W.; Lee, D.-S. IDH_2 deficiency in microglia decreases the pro-inflammatory response via the ERK and NF-κB pathways. *Inflammation* **2018**, *41*, 1965–1973. [CrossRef] [PubMed]
4. Ip, W.K.E.; Hoshi, N.; Shouval, D.S.; Snapper, S.; Medzhitov, R. Anti-inflammatory effect of IL-10 mediated by metabolic reprogramming of macrophages. *Science* **2017**, *356*, 513–519. [CrossRef]
5. Kiecolt-Glaser, J.K. Stress, food, and inflammation: Psychoneuroimmunology and nutrition at the cutting edge. *Psychosom. Med.* **2010**, *72*, 365–369. [CrossRef]
6. Kim, Y.-S.; Lee, W.-H.; Choi, E.-J.; Choi, J.-P.; Heo, Y.-J.; Gho, Y.-S.; Jee, Y.-K.; Oh, Y.-M.; Kim, Y.-K. Extracellular vesicles derived from gram-negative bacteria, such as *Escherichia coli*, induce emphysema mainly via IL-17A-mediated neutrophilic inflammation. *J. Immunol.* **2015**, *194*, 3361–3368. [CrossRef]
7. Huang, M.Y.; Lin, J.; Lu, K.; Xu, H.G.; Geng, Z.Z.; Sun, P.H.; Chen, W.M. Anti-inflammatory effects of cajaninstilbene acid and its derivatives. *J. Agr. Food Chem.* **2016**, *64*, 2893–2900. [CrossRef]
8. Cardinal, S.; Azelmat, J.; Grenier, D.; Voyer, N. Anti-inflammatory properties of quebecol and its derivatives. *Bioorg. Med. Chem. Lett.* **2016**, *26*, 440–444. [CrossRef]
9. Zhang, L.; Virgous, C.; Si, H. Synergistic anti-inflammatory effects and mechanisms of combined phytochemicals. *J. Nutr. Biochem.* **2019**, *69*, 19–30. [CrossRef]
10. de Cassia da Silveira e Sa, R.; Andrade, L.N.; de Sousa, D.P. A review on anti-inflammatory activity of monoterpenes. *Molecules* **2013**, *18*, 1227–1254. [CrossRef]
11. Kim, S.H.; Bang, J.; Son, C.N.; Baek, W.K.; Kim, J.M. Grape seed proanthocyanidin extract ameliorates murine autoimmune arthritis through regulation of TLR4/MyD88/NF-κB signaling pathway. *Korean J. Intern. Med.* **2018**, *33*, 612–621. [CrossRef] [PubMed]
12. Chen, M.; Yu, S. Lipophilized grape seed proanthocyanidin derivatives as novel antioxidants. *J. Agr. Food Chem.* **2017**, *65*, 1598–1605. [CrossRef] [PubMed]
13. Chen, M.; Yu, S. Lipophilic grape seed proanthocyanidin exerts anti-proliferative and pro-apoptotic effects on PC3 human prostate cancer cells and suppresses PC3 xenograft tumor growth in vivo. *J. Agr. Food Chem.* **2019**, *67*, 229–235. [CrossRef] [PubMed]
14. Loo, Y.T.; Howell, K.; Chan, M.; Zhang, P.; Ng, K. Modulation of the human gut microbiota by phenolics and phenolic fiber-rich foods. *Compr. Rev. Food Sci. F.* **2020**, *19*, 1268–1298. [CrossRef] [PubMed]
15. Chang, M.; Qiu, F.; Lan, N.; Zhang, T.; Guo, X.; Jin, Q.; Liu, R.; Wang, X. Analysis of phytochemical composition of *Camellia oleifera* oil and evaluation of its anti-inflammatory effect in lipopolysaccharide-stimulated RAW 264.7 macrophages. *Lipids* **2020**, *55*, 353–363. [CrossRef]
16. Xiao, X.; He, L.; Chen, Y.; Wu, L.; Wang, L.; Liu, Z. Anti-inflammatory and antioxidative effects of *Camellia oleifera* abel components. *Future Med. Chem.* **2017**, fmc-2017-0109. [CrossRef]
17. Pallares, V.; Calay, D.; Cedo, L.; Castell-Auvi, A.; Raes, M.; Pinent, M.; Ardevol, A.; Arola, L.; Blay, M. Additive, antagonistic, and synergistic effects of procyanidins and polyunsaturated fatty acids over inflammation in RAW 264.7 macrophages activated by lipopolysaccharide. *Nutrition* **2012**, *28*, 447–457. [CrossRef]
18. Tao, J.-Y.; Zheng, G.-H.; Zhao, L.; Wu, J.-G.; Zhang, X.-Y.; Zhang, S.-L.; Huang, Z.-J.; Xiong, F.-L.; Li, C.-M. Anti-inflammatory effects of ethyl acetate fraction from melilotus suaveolens ledeb on LPS-stimulated RAW 264.7 cells. *J. Ethnopharmacol.* **2009**, *123*, 97–105. [CrossRef]

19. Kim, H.-Y.; Kim, J.-H.; So, Y.; Kang, S.-Y.; Jeong, H.-G.; Jin, C.-H. Anti-inflammatory effect of lupinalbin a isolated from *Apios americna* on lipopolysaccharide-treated RAW264.7 cells. *Molecules* **2018**, *23*, 583. [CrossRef]
20. Zhang, N.; Fu, J.-N.; Chou, T.-C. Synergistic combination of microtubule targeting anticancer fludelone with cytoprotective panaxytriol derived from panax ginseng against MX-1 cells in vitro: Experimental design and data analysis using the combination index method. *Am. J. Cancer Res.* **2016**, *6*, 97–104.
21. Huang, H.; Chen, X.; Li, D.; He, Y.; Li, Y.; Du, Z.; Zhang, K.; DiPaola, R.; Goodin, S.; Zheng, X. Combination of α-tomatine and curcumin inhibits growth and induces apoptosis in human prostate cancer cells. *PLoS ONE* **2015**, *10*, e0144293. [CrossRef] [PubMed]
22. Gao, Y.; Tollefsbol, T.O. Combinational proanthocyanidins and resveratrol synergistically inhibit human breast cancer cells and impact epigenetic–mediating machinery. *Int. J. Mol. Sci.* **2018**, *19*, 2204. [CrossRef] [PubMed]
23. Lee, D.S. Antioxidant and antibacterial activities of chitosan-phloroglucinol conjugate. *Fish. Aquat. Sci.* **2013**, *16*, 229–235. [CrossRef]
24. Lechner, M.; Lirk, P.; Rieder, J. Inducible nitric oxide synthase (iNOS) in tumor biology: The two sides of the same coin. *Semin. Cancer Biol.* **2005**, *15*, 277–289. [CrossRef] [PubMed]
25. Yahfoufi, N.; Alsadi, N.; Jambi, M.; Matar, C. The immunomodulatory and anti-inflammatory role of polyphenols. *Nutrients* **2018**, *10*, 1618. [CrossRef] [PubMed]
26. Wang, K.; Hu, L.; Jin, X.-L.; Ma, Q.-X.; Marcucci, M.C.; Netto, A.A.L.; Sawaya, A.C.H.F.; Huang, S.; Ren, W.-K.; Conlon, M.A.; et al. Polyphenol-rich propolis extracts from china and brazil exert anti-inflammatory effects by modulating ubiquitination of TRAF6 during the activation of NF-κB. *J. Funct. Foods* **2015**, *19*, 464–478. [CrossRef]
27. Cheng, Y.-T.; Wu, S.-L.; Ho, C.-Y.; Huang, S.-M.; Cheng, C.-L.; Yen, G.-C. Beneficial effects of camellia oil (*Camellia oleifera* abel.) on ketoprofen-induced gastrointestinal mucosal damage through upregulation of HO-1 and VEGF. *J. Agr. Food Chem.* **2014**, *62*, 642–650. [CrossRef]
28. Vasarri, M.; Leri, M.; Barletta, E.; Ramazzotti, M.; Marzocchini, R.; Degl'Innocenti, D. Anti-inflammatory properties of the marine plant *Posidonia* oceanica (L.) delile. *J. Ethnopharmacol.* **2020**, *247*, 112252. [CrossRef]
29. Miller, M.W.; Lin, A.P.; Wolf, E.J.; Miller, D.R. Oxidative stress, inflammation, and neuroprogression in chronic PTSD. *Harvard Rev. Psychiat.* **2018**, *26*, 57–69. [CrossRef]
30. Bumrungpert, A.; Pavadhgul, P.; Kalpravidh, R.W. Camellia oil-enriched diet attenuates oxidative stress and inflammatory markers in hypercholesterolemic subjects. *J. Med. Food* **2016**, *19*, 895–898. [CrossRef]
31. Parhiz, H.; Roohbakhsh, A.; Soltani, F.; Rezaee, R.; Iranshahi, M. Antioxidant and anti-inflammatory properties of the citrus flavonoids hesperidin and hesperetin: An updated review of their molecular mechanisms and experimental models. *Phytother. Res.* **2015**, *29*, 323–331. [CrossRef] [PubMed]
32. Sen, R.; Baltimore, D. Multiple nuclear factors interact with the immunoglobulin enhancer sequences (reprinted from cell, vol 46, pg 705–716, 1986). *J. Immunol.* **2006**, *177*, 7485–7496. [PubMed]
33. Hamadou, M.H.; Kerkatou, M.; Zucal, C.; Bisio, A.; Provenzani, A.; Inga, A.; Menad, A.; Benayache, S.; Benayache, F.; Ameddah, S. Limonium duriusculum (de Girard) kuntze exhibits anti-inflammatory effect via NF-κB pathway modulation. *Braz. Arch. Biol. Techn.* **2021**, *64*, e21200179. [CrossRef]
34. Sanchez-Fidalgo, S.; Villegas, I.; Rosillo, M.A.; Aparicio-Soto, M.; de la Lastra, C.A. Dietary squalene supplementation improves dss-induced acute colitis by downregulating p38 MAPK and NFκB signaling pathways. *Mol. Nutr. Food Res.* **2015**, *59*, 284–292. [CrossRef] [PubMed]
35. Lagha, A.B.; Grenier, D. Tea polyphenols inhibit the activation of NF-κB and the secretion of cytokines and matrix metalloproteinases by macrophages stimulated with *Fusobacterium nucleatum*. *Sci. Rep.* **2016**, *6*, 34520. [CrossRef] [PubMed]
36. Gao, R.; Shu, W.; Shen, Y.; Sun, Q.; Bai, F.; Wang, J.; Li, D.; Li, Y.; Jin, W.; Yuan, L. Sturgeon protein-derived peptides exert anti-inflammatory effects in lps-stimulated RAW264.7 macrophages via the MAPK pathway. *J. Funct. Foods* **2020**, *72*, 104044. [CrossRef]
37. Liu, J.; Huang, H.; Huang, Z.; Ma, Y.; Zhang, L.; He, Y.; Li, D.; Liu, W.; Goodin, S.; Zhang, K.; et al. Eriocitrin in combination with resveratrol ameliorates LPS-induced inflammation in RAW264.7 cells and relieves TPA-induced mouse ear edema. *J. Funct. Foods* **2019**, *56*, 321–332. [CrossRef]
38. Ekstrand-Hammarstrom, B.; Osterlund, C.; Lilliehook, B.; Bucht, A. Vitamin e down-modulates mitogen-activated protein kinases, nuclear factor-κB and inflammatory responses in lung epithelial cells. *Clin. Exp. Immunol.* **2007**, *147*, 359–369. [CrossRef]

antioxidants

MDPI

Review

Critical Review on Molecular Mechanisms for Genistein's Beneficial Effects on Health Through Oxidative Stress Reduction

Ke Zhang †, Jingwen Wang † and Baojun Xu *

Food Science and Technology Program, Department of Life Sciences, Beijing Normal-Hong Kong Baptist University, Zhuhai 519087, China; r130013040@mail.uic.edu.cn (K.Z.); wangjingwen@uic.edu.cn (J.W.)

* Correspondence: baojunxu@uic.edu.cn; Tel.: +86-756-3620636

† These authors contributed equally to this work.

Abstract

Oxidative stress directly or indirectly contributes to the development and progression of various diseases; therefore, regulating oxidative stress is a promising strategy for preventing or treating these conditions. The unique substances in soybeans, soy isoflavones, notably genistein, which have a strong antioxidant capacity, are considered to regulate various signaling pathways, alleviate oxidative stress, and improve gut microbiota imbalance as well as mitochondrial dysfunction. In this literature review, we summarize the latest research on genistein, providing evidence of its development and application as a potential drug for preventing and treating five selected diseases (Parkinson's disease, Alzheimer's disease, diabetes mellitus, cardiovascular disease, and cancers). The literature was searched using keywords that include tripartite combinations of genistein and oxidative stress, along with each of the five selected diseases, from PubMed, Science Direct, and Google Scholar between 2014 and 2024. According to current in vitro, in vivo, and clinical trials, we comprehensively discuss the therapeutic dose used to target various disease entities to achieve optimal efficacy and meet safety requirements. Moreover, considering the poor water solubility and limited bioavailability of genistein, strategies for improving its therapeutic efficacy, such as combining it with exercise, existing medications, and advanced technologies, as well as applying nanotechnology, were assessed. Therefore, this review aims to provide robust evidence for the development and application of genistein as a potential therapeutic agent or functional food for preventing and treating these diseases.

Keywords: genistein; oxidative stress; signaling pathway; antioxidant; anti-inflammatory

1. Introduction

Oxidative stress is an imbalance between the decline in antioxidant mechanisms and the growth of free radicals, such as reactive nitrogen species (RNS) and reactive oxygen species (ROS), leading to the overproduction of oxidative free radicals that damage lipids, proteins, and DNA [1]. Among these, ROS, a highly reactive metabolic byproduct and biomarker of oxidative stress, play a dual role in cell function and reduction–oxidation (redox) homeostasis. At low to intermediate levels, ROS play a positive role in cellular functions (e.g., proliferation, differentiation, migration, apoptosis, and necrosis). However, excessive ROS production leads to oxidative stress and chronic diseases owing to factors such as mitochondrial deficiency and nutritional stress [2,3]. Thus, oxidative stress is considered a causative factor for the onset of numerous diseases such as Parkinson's disease (PD), Alzheimer's disease (AD), diabetes mellitus (DM), cardiovascular diseases (CVDs), and cancer (Figure 1). For instance, plaque aggregation and the high phosphorylation of tau, induced by oxidative stress, are two of the main causes of AD [4]. Moreover, oxidative stress increases the risk of insulin resistance (IR) by impairing insulin signaling transduction [5]. Thus, future drug interventions should maximize the amelioration or prevention of diseases by inhibiting oxidative damage, especially using non-toxic and absorbable natural products.

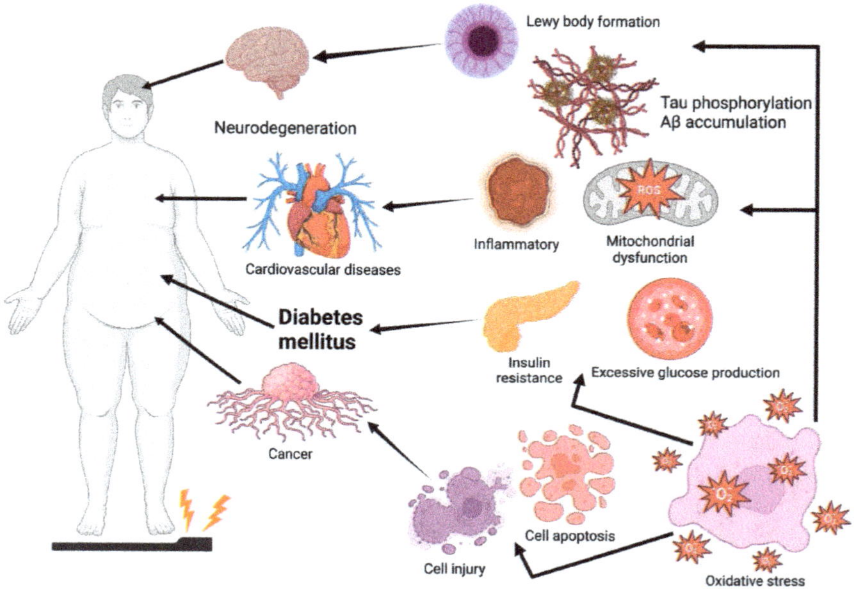

Figure 1. Oxidative stress leads to Lewy body formation, tau phosphorylation, Aβ accumulation, insulin resistance, excessive glucose production, cell injury, and cell apoptosis, which are common causes of Parkinson's disease, Alzheimer's disease, cardiovascular disease, diabetes mellitus, and cancer. Graphical elements were created using BioRender (https://www.biorender.com).

Genistein is a flavonoid commonly found in soy products. It is a phytoestrogen that can combine with estrogen receptors in the body to mimic estrogen and benefit lipid metabolism in postmenopausal women. It has shown excellent anti-aging, anti-inflammatory, and anti-apoptotic abilities, possibly protecting targeting cells and activating critical signaling pathways such as nuclear factor erythroid 2-related factor 2 (Nrf2) or phosphatidylinositol 3-kinase/protein kinase B (PI3K/Akt) against ROS damage. For example, genistein increased the expression of superoxide dismutase (SOD), catalase (CAT), glutathione peroxidase (GPx), and other related enzymes to regulate ROS levels and reduce the incidence of diseases [6]. Numerous preclinical studies have suggested that genistein increases SOD protein expression by upregulating the Nrf2 pathway and the PI3K/Akt pathway as well as downregulating the mitogen-activated protein kinase (MAPK) pathway, especially jun N-terminal kinase (JNK) and p38 MAPK, which decreases the prevalence of aging, AD, PD, CVD, and cancer (e.g., bladder cancer, liver cancer, and breast cancer) induced by ROS [7]. Thus, genistein can have beneficial health effects by increasing the expression of some antioxidant enzymes to reduce the adverse effects caused by unbalanced oxidative stress.

However, the molecular mechanism of genistein treatment has not been thoroughly investigated, and its efficacy in alleviating diseases caused by oxidative stress remains to be determined. Thus, this review aims to summarize the molecular mechanisms by which genistein reduces diseases caused by oxidative stress and further demonstrate the curative effect of genistein on such diseases, thereby assisting in substantiating some hypotheses for the positive therapeutic role of genistein and serving as a reference for selecting different dosage levels in practical research experiments.

Also of note, this article, in the form of a review, serves as a medium to summarize the positive therapeutic effects of genistein on five common diseases that have become increasingly prevalent in recent years. It provides readers interested in the antioxidant effects of genistein with a systematic understanding of its potential in disease treatment through the regulation of oxidative stress, a prominent advantage of the review format. All comments in this review are supported by a substantial body of experimental data, including preclinical experience and actual clinical trials, which highlight the practical potential of plant extracts in therapeutic applications. At the same time, by covering various treatment approaches and dosages for each disease, the review enables readers to compare the relative costs and effects of each approach horizontally. Moreover, given that scientists have expressed reservations about genistein's ability to treat or prevent these diseases due to its limited bioavailability, various strategies for enhancing its therapeutic efficacy have been explored in this article, including its synergistic effects with exercise and existing medications, as well as the research and development of genistein-loaded nanoparticles. This also further provides readers with a new understanding of both the challenges and the latest advancements in enhancing the therapeutic efficacy of genistein.

2. Methodology

This review was conducted by investigating the most relevant literature in the scientific databases PubMed (https://pubmed.ncbi.nlm.nih.gov, assessed on 1 January 2025), Science Direct (https://www.sciencedirect.com, assessed on 3 January 2025), and Google Scholar (https://scholar.google.com, assessed on 5 January 2025). Relevant articles were published from 2014 to 2024, with most published in the past five years. The keywords included tripartite combinations of genistein and oxidative stress with the five selected diseases (PD, AD, DM, CVD, and cancer). Those that did not cover the research topic of interest or that lacked appropriate design descriptions were excluded from this review. Therefore, we have summarized the molecular mechanisms involved in the specific palliative or therapeutic effects of genistein on five common diseases, including the in vivo and in vitro pharmacological effects of genistein and the most recent clinical trials.

3. The Characteristics of Genistein and Its Therapeutic Effect Through Inhibiting Oxidative Stress

3.1. The Chemical Structure of Genistein

Genistein, a 4,5,7-trihydroxy isoflavone, is recognized and used as a phytoestrogen owing to its chemical structure resembling that of mammalian estrogen. This similarity enables genistein to play beneficial roles, providing anxiolytic, neuroprotective, cardioprotective, anticarcinogenic, anti-inflammatory, and antioxidant effects [8–10].

The molecular formula of genistein is $C_{15}H_{10}O_5$ with a molecular weight of 270.241 g/mol. It comprises 15 carbons arranged in two aromatic rings (A and B) and a central pyran ring (ring C). A distinctive feature is the double bond between the second and third positions, which constitutes the fundamental carbon skeleton of genistein [8]. The key functional groups of genistein are the 7-position hydroxyl group of the A ring, 4-position hydroxyl group of the B ring, and 4-position ketone group of the C ring, which are mainly responsible for the biological activity of genistein. The four and seven hydroxyl groups are directly involved in antioxidant activity, binding to estrogen receptors, and regulating key signaling pathways such as PI3K/Akt. The ketone group provides molecular polarity and facilitates the formation of intermolecular hydrogen bonds (Figure 2). Additionally, the three hydroxyl groups of genistein enhance estrogenic activity, increasing its substantial capacity to interact with estrogen receptors and regulate various disorders [11].

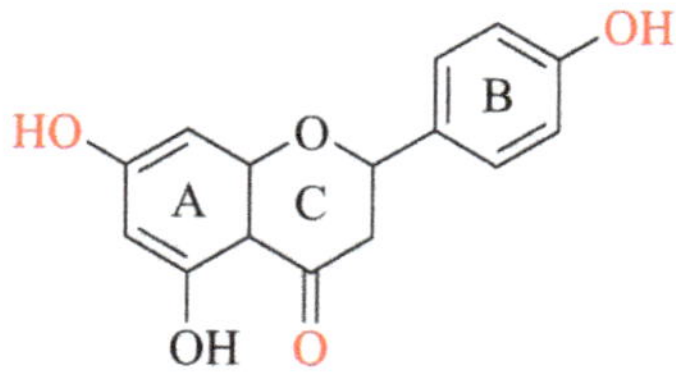

Figure 2. Chemical structures of genistein. Graphical elements were created using Chemdraw (https://www.chemdraw.com.cn).

3.2. The Therapeutic Effect of Genistein Through Inhibiting Oxidative Stress

As for its antioxidant capacity, genistein is a dose-dependent substance with a dual effect against oxidative stress. At relatively low levels, genistein enhances antioxidant enzymes to protect the human body from the destructive damage caused by oxidative stress [12]. When the concentration of genistein is high, although it can also reduce the likelihood of the further progression of cerebral infarction and liver damage in women, it acts as a pro-oxidative stress factor and exacerbates ROS production in the human body [13,14].

On the one hand, at low concentrations, genistein, as a critical upstream stimulatory factor, positively affected many phase II detoxification and antioxidant enzymes through nuclear factor kappa B (NF-κB) to regulate many diseases [15]. Moreover, genistein inhibited the development of neonatal hypoxic-ischemic brain damage (HIBD) by upregulating the NF-κB pathway and altering interleukin-6 (IL-6), tumor necrosis factor-α (TNF-α), and interleukin-1β (IL-1β) [16]. Genistein upregulated the NF-κB pathway and *Nrf2* expression, ultimately inhibiting oxidative stress. Then, it effectively improves oxidative stress and neuroinflammation caused by cerebral ischemia due to NF-κB restraint and antioxidants in the Nrf2 pathway [17]. Similarly, in hypoxic-ischemic encephalopathy therapy, genistein upregulated the erythroid 2-related factor transcription factor/hemoxygenase 1 (Nrf2/HO-1) pathway and inhibited the NF-κB signaling pathway. Genistein diminished brain infarct size and neuronal apoptosis while enhancing neuroprognosis and brain atrophy recovery, leading to the attenuation of oxidative stress and neuroinflammation induced by hypoxic-ischemic brain injury [16]. After genistein treatment, the viability of HIBD cells after oxygen–glucose deprivation/reoxygenation injury increased, whereas neuronal injury and cell apoptosis levels decreased. Notably, no toxic effects can be detected after genistein entered the blood–brain barrier, indicating its potential in brain damage therapy [16].

In contrast, high concentrations of genistein are pro-oxidative stress factors that stimulate ROS production and damage the human body. ROS alter the conformation and detachment of Kelch-like ECH-associated protein 1 (Keap1), along with the intensified ubiquitination and proteolysis of Nrf2 in the cytoplasm, enhancing the interaction between Nrf2 and Keap1 [15,18]. When Nrf2 was translocated into the nucleus and bound to the musculoaponeurotic fibrosarcoma oncogene homolog protein, genistein was attached to antioxidant response elements and activated antioxidant gene expression [19]. Consequently, genistein triggered HO-1, SOD, and CAT activities through Nrf2 release and increased antioxidant response element expression, leading to severe oxidative stress [20]. High concentrations of genistein may lead to hormonal imbalances and endocrine disruption [21,22]. As a phytoestrogen, genistein inhibited natural estrogen activity by competing with endogenous estrogen for receptor binding. In addition, excessive genistein intake potentially increased the risk of hormone-sensitive conditions or the prevalence of cancer, such as breast cancer or endometrial hyperplasia, in susceptible individuals [23].

4. Therapeutic Application of Genistein in Common Oxidative Stress-Induced Diseases

The multiple molecular mechanisms underlying the therapeutic effects of genistein on PD, AD, DM, CVD, and cancer are shown in Figure 3.

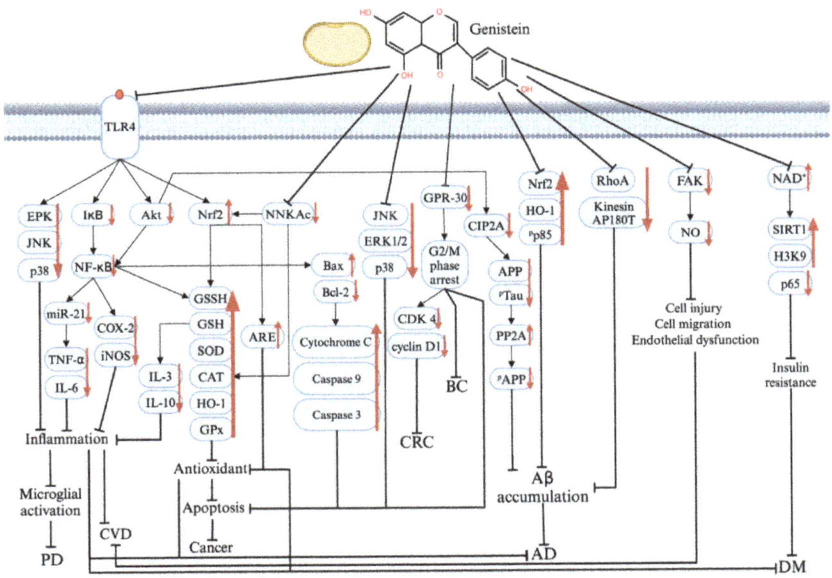

Figure 3. The mechanism of genistein's therapeutic effect on various protein and signaling pathways in PD, AD, DM, CVD, and cancer pathogenesis. Graphical elements were created using BioRender (https://www.biorender.com). ↑ means increase; ↓ means decrease; ⊣ means repressed.

4.1. *Parkinson's Disease (PD)*

PD is a progressive disorder of the nervous system that primarily affects individuals over the age of 60 years. PD is characterized by the unbinding of inflammatory cytokines (e.g., IL-6, TNF-α, and IL-1β) due to microglial activation by the formation of Lewy bodies as well as lipopolysaccharide and dopaminergic neuron loss in the dorsomedial cluster and substantia nigra (SN). Lewy bodies, which contain overexpressed α-synuclein (αS), are the neuropathological hallmark of PD and critical targets for treatment strategies [24].

Genistein, a polyphenol, has the potential to alleviate PD symptoms with few side effects [25]. Anti-inflammatory effects, antioxidant activity, and anti-apoptotic properties help protect dopaminergic neurons and slow the progression of PD. Genistein inhibited MAPK and IκB signaling pathways and reduced cyclooxygenase-2 (COX-2) and inducible nitric oxide synthase (iNOS) expression in the SN of rats through G protein-coupled estrogen receptor and insulin-like growth factor 1 receptor signaling pathway regulation [26]. The

NF-κB signaling pathway was restricted, while the 5′ AMP-activated protein kinase (AMPK) signaling pathway was activated by genistein. Therefore, the inflammatory response in the human synovial membrane of TNF-α-induced MH7A cells was successfully suppressed. In ovariectomized rats induced by amyloid-beta (Aβ), genistein downregulated the expression of MAPK, Toll-like receptor 4 (TLR4), and NF-κB in microglia, offering neuroprotection against inflammation, reducing microglial activation, protecting SN dopaminergic neurons, and lowering the phosphorylation levels of extracellular signal-regulated kinase (ERK), p38, JNK, and IκB [26].

Genistein has a good antioxidant effect, protecting neurons from oxidative damage. An in vitro study investigated the anti-Parkinson effects of genistein (20 μM) on cell viability, apoptosis, and oxidative stress in human SH-SY5Y cells treated with rotenone. The results showed that the genistein intervention increased cell viability, reduced apoptosis, and attenuated the oxidative stress imbalance, primarily through the activation of estrogen receptors and the nuclear factor erythroid 2-related factor 2 signaling pathway [27]. Similarly, genistein treatment at 40 μM in transgenic *Drosophila* was reported to enhance antioxidant capacity, thereby further protecting neurons. Genistein reduced lipid peroxidation by 2.5 times compared to untreated *Drosophila* PD at this concentration. Genistein increased the glutathione (GSH) levels, minimized oxidative stress, and protected neurons by balancing ROS production and oxidative conditions through lower monoamine oxidase activity [28]. However, human αS and Lewy bodies were present in both treated and untreated *Drosophila*, indicating that genistein's therapeutic effect on PD was unable to stem αS expression or Lewy body formation and only enhanced antioxidant ability [24]. Moreover, genistein inhibited the formation of some ROS producers (dopamine, 3,4-Dihydroxyphenylacetic acid, and homovanillic acid) in the treated group, demonstrating its inhibitory effect on oxidative stress [26].

Genistein also controls caspase-8 and caspase-3 to mitigate apoptosis in PD pathological features. Genistein downregulated death receptors, such as tumor necrosis factor receptor superfamily member 6 (*FAS* and TNF, limiting caspase-8 recruitment to the death-inducing signaling complex. It also decreased the levels of the pro-inflammatory cytokine TNF-α, which had a positive relationship with caspase-8 in terms of its anti-inflammatory properties. Genistein prevented oxidative damage to mitochondrial DNA, proteins, and lipids by stabilizing mitochondrial function and reducing ROS levels, thereby avoiding further apoptosis [29]. Moreover, genistein upregulated SOD, GPx, and Bax (Bcl-2-associated X protein) and decreased β-cell lymphoma 2 (Bcl-2) and Bid levels, hindering the growth of both caspase-8 and caspase-3. As mentioned in the review by Singh et al., genistein alleviated the cytotoxic effects of 6-hydroxydopamine (6-OHDA) in the PC12 cell model by inhibiting the activation of caspase-8 and caspase-3 proteins [25].

In conclusion, these in vivo and in vitro experiments elucidated the molecular mechanism of genistein in PD treatment, highlighting its role in attenuating inflammatory responses and oxidative damage, as well as modulating caspase-8 and caspase-3 expression. Notably, although the current experiments found that genistein does not reduce αS expression, future studies should explore its potential in neural stem cell therapy or as an adjuvant therapy to enhance therapeutic strategies for PD by targeting αS reduction in combination with other drugs.

4.2. Alzheimer's Disease (AD)

AD is a common neurodegenerative disorder that is characterized by progressive memory loss and cognitive decline. In the early stages of AD, microglial activation around Aβ deposition sites increases the levels of pro-inflammatory cytokines. ROS cause lysosomal dysfunction and neuroinflammation, contributing to Aβ accumulation and tau phosphorylation, thereby advancing neurodegeneration [30–32].

Genistein, a cell-permeable protein tyrosine kinase inhibitor, regulates various intracellular signaling pathways by the auto-phosphorylation of the epidermal growth factor receptor kinase. It reduced the formation of TNF-α and TLR4, which were typical pro-inflammatory molecules, improving memory and alleviating astrogliosis in SH-SY5Y cell models with AD [33]. Duan et al. have summarized in their literature review that the anti-inflammatory effects of genistein may be associated with the suppression of the TLR4-mediated NF-κB signaling pathway [34].

Genistein has been reported to increase antioxidant effects by blocking Aβ toxicity in cell and animal models. For instance, in the $A\beta_{25-35}$ rat primary hippocampal neuron model, genistein significantly increased cell viability while suppressing ROS production, MDA levels, LDH levels, and apoptosis. Further results suggested that α7nAChR signaling-mediated PI3K/Akt and Nrf2/Keap1 pathways are critical for the neuroprotective effect of genistein [35]. Genistein significantly reduced amyloid plaques in the anterior cingulate gyrus [36]. Genistein reduced amyloid precursor protein (APP) secretion by inhibiting β-site APP-cleaving enzyme 1 and blocking platelet-derived growth factor stimulation [37]. The anti-neurotoxicity ability of genistein, mediated through the Nrf2/HO-1/PI3K pathway, was demonstrated in an $A\beta_{25-35}$-treated SH-SY5Y cell model, showing the inhibitory effect of genistein on Akt and tau phosphorylation, thereby strengthening resistance to Aβ-induced toxicity [34]. Meanwhile, 50 μM genistein upregulated HO-1 expression and strengthened cells' defense in cases of oxidative stress injury, increasing the survival rate of SH-SY5Y cells [33]. It also reduced $A\beta_{42}$-induced neurotoxicity and accumulation by inhibiting kinesin AP180T and Ras homolog family member A (RhoA) [6]. High concentrations of genistein regulate tau phosphorylation and $A\beta_{40}$ and $A\beta_{42}$ levels in the cortex and hippocampus, reducing the abnormal accumulation of harmful proteins in key brain regions and thereby preventing AD development [38].

Notably, alterations in the composition of gut microbiota are highly associated with oxidative stress imbalance, neuroinflammation, and ongoing neuronal loss, causing the onset and progression of brain disorders such as AD [39]. As mentioned in our previous literature review, the consumption of soy isoflavones has been shown to improve gut homeostasis [40]. It increases the population of *Bifidobacteria*, *Eubacteria*, and short-chain fatty acid-producing *Lachnospira*, supporting longevity and improving cognitive behaviors, which may serve as an alternative for AD treatment. Moreover, S-equol (a gut microbiome-derived metabolite of genistein with antioxidant properties) is reported to decrease white matter lesions in the aging brain and lead to the amelioration of AD symptoms due to its anti-amyloid ability [40].

In summary, genistein reduces inflammation and Aβ toxicity to slow down the further development of AD pathology with limited side effects. Viña et al. observed that long-term continuous treatment with genistein (exceeding one year) significantly reduced Aβ accumulation, thereby delaying the onset of AD in patients with prodromal AD. This conclusion was based on the increased results from 18F-flutemetamol in the control group, whereas no significant changes were detected in the genistein-treated group [36]. However, further studies are required to determine the precise duration of genistein treatment required for AD management. At the same time, shortening the duration of genistein treatment in AD and improving its bioavailability are still worth considering in future development.

4.3. Diabetes Mellitus (DM)

DM is a prevalent disease among older people, primarily caused by methylglyoxal (MG) exposure, IR, and mitochondrial dysfunction. Oxidative stress, inflammation, abnormal blood sugar levels, and obesity increase the risk of DM [41]. Genistein is a phytoestrogen that controls insulin sensitivity and glucose homeostasis during DM treatment. It regulates β-cells and promotes efficient glucose regulation by elevating cyclic adenosine monophosphate and protein kinase A [42–44].

Genistein regulates nicotinamide adenine dinucleotide, oxidized form (NAD^+) to prevent IR [45]. In skeletal muscle, the HOMA-IR results showed that genistein increased NAD^+ levels, thereby stimulating insulin sensitivity [46,47]. In the C/EBPβ-expressing 3T3-L1 adipocyte model, genistein increased NAMPT expression to upregulate PHB1 stability. NAD^+ levels then increased, alleviating metabolic dysfunction and decreasing IR [47–49].

Genistein regulates the DM process by mitigating oxidative damage. When *EA. HY926* cells were exposed to MG, the ROS level of the genistein-treated group remained stable, while that of the untreated group showed a significant increase. These results reflected the inhibitory effect of genistein on oxidative damage. Simultaneously, the reduction of Nrf2 translocation in genistein-treated cells proved the mitigation of oxidative stress, which was caused by MG [50]. In addition, genistein reduced MAPK phosphorylation and apoptosis, as well as mitigated endothelial cell oxidative damage [51]. It also reduced chronic inflammation by suppressing NF-κB, IL-1β, and TNF-α and mitigated oxidative stress in pancreatic cells, thereby preserving the function of β-cells [41].

Emerging studies have reported that individuals with T2DM exhibit reduced levels of *Bifidobacterium*, *Firmicutes*, and *Akkermansia* genera in their gut microbiota [52,53]. This gut microbiota dysbiosis causes chronic inflammation and oxidative stress, as well as disorders in glycolipid metabolism and insulin resistance, which are key factors in the pathogenesis of T2DM and contribute to its deterioration. Thus, regulating gut microbiota is considered a promising alternative for protecting against metabolic diseases. According to a recent randomized controlled trial, 2 months of dietary supplementation with genistein effectively increases the expression of fatty acid oxidation-related genes, alters the gut microbiota, and attenuates insulin resistance [45]. Moreover, in a study analyzing causation in genistein interventions, gut microbiota changes, glucose metabolism, and adipose tissue browning, it was found that the genistein treatment not only improved the diversity of gut microbiota, but also attenuated impaired glucose metabolism in mice treated with a high-fat diet, suggesting that healthy gut homeostasis is essential for metabolic health [54]. Similarly, genistein has been reported to significantly attenuate glucose and lipid metabolism dysfunction, inflammation, and insulin resistance by enhancing the abundances of *Bacteroides* and *Prevotella* and reducing those of *Helicobacter* and *Ruminococcus* in T2DM mice [55]. Thus, the alteration in gut microbiota composition and abundance induced by genistein intervention is tightly implicated in DM prevention and treatment.

In summary, genistein hinders DM through IR regulation and the prevention of oxidative damage via various signaling pathways. These do provide a compelling option for the development of novel plant-based functional drugs for DM treatment. However, in animal studies, there was no significant difference in the effect of genistein interventions between 12 and 24 weeks, suggesting that the optimal duration of genistein treatment still needs to be determined through clinical trials [42]. Meanwhile, it is worth considering whether genistein could be combined with existing antidiabetic drugs to optimize therapeutic strategies.

4.4. Cardiovascular Disease (CVD)

CVD has a close relationship with oxidative stress damage, leading to most causes of death worldwide, which is much greater than the cumulative sum of cancer and chronic lung disease [56]. Smoking, a high-salt diet, and oxidative stress generate inflammation, spawning endothelial dysfunction, and eventually trigger vascular lesions, vascular remodeling, thrombosis, and CVD. Inflammatory and cardiovascular system disorders are the two main factors that result in CVD damage in humans.

Genistein reduces inflammation, cell invasion, and migration by regulating the key pathways related to oxidative stress. Genistein prevents inflammation and enlargement in vascular endothelial cells (VECs) through miRNA control. VEC damage, which releases various inflammatory factors and chemokines, causes the local accumulation of lipid responses in the vascular intima. In an LPS-induced C57BL/6 mouse model, genistein downregulated the miR-21/NF-κB p65 pathway, similar to an miR-21 inhibitor, to block the growth of typical anti-inflammatory factors (TNF-α, IL-6, and iNOS), limiting inflammation in VECs [57]. In a C57BL/6J male mouse model, genistein worked as a CB1 antagonist and CB2 agonist to neutralize Δ^9-tetrahydrocannabinol (THC), which was the psychoactive component of cannabis, thus repairing damage to the cardiovascular system. Δ^9-THC not only caused VEC inflammation, but also aggravated endothelial dysfunction. Simultaneously, based on the decrease in NF-κB phosphorylation, genistein decreased IL-3, IL-6, and IL-10 expressions while increasing SOD and GSH expression in mouse serum, reversing the Δ^9-THC-induced effect with little damage to the cardiac structure or function of the heart [58]. In the HU-EST cell model, genistein downregulated the FAK signaling pathway to stimulate cell viability and NO production, while also blocking further focal adhesion, estradiol-induced vascular endothelial injury, cell invasion, and migration [59].

Additionally, genistein hinders hypertension mainly caused by a high-salt diet. Severe oxidative damage and inflammation occur around the hypothalamic paraventricular nucleus due to a high-salt diet. In male Wistar rats on the high-salt model (8% NaCl), after being treated by genistein, their systolic pressure, diastolic pressure, and MAP showed various degrees of reduction, indicating the potential use of genistein in hypertension treatment. Based on the increments of glutathione disulfide (GSSH), SOD, Ac-FOXO1, GSH, SIRT1, Nrf2, HO-1, and NQO-1, genistein inhibited oxidative stress through the SIRT1/Nrf2 pathway, providing a basis for high-salt diet-induced hypertension treatment [60].

As a continuous pathophysiological procedure that is activated by a local redox imbalance to endothelial dysfunction, atherosclerosis (AS) is another serious CVD that can be alleviated by genistein [61]. Before treating AS, it is essential to understand the two hallmarks of AS: leukocyte adhesion and atherosclerotic plaques. Genistein alleviated the formation of these two AS factors to repair the cardiovascular system. It activated NO, boosting cyclic guanosine monophosphate and limiting leukocyte adhesion to the vessel wall [59]. Meanwhile, genistein suppressed platelet aggregation and adhesion to inhibit fibrous plaque formation, preventing the further development of AS.

In other words, genistein exhibits a multifaceted role in alleviating CVD by inhibiting inflammation, promoting the differentiation of abnormal cells, and reducing the NaCl content in the human body. This suggests a promising therapeutic approach for managing vascular inflammation and dysfunction caused by lifestyle and dietary factors. However, it is essential to conduct large-scale clinical trials to validate the efficacy and safety of genistein in the treatment of CVD and to investigate advanced drug morphological designs, such as nanoparticles or liposomes, which enhance absorption and tissue targeting. The role of genistein in treating PD, AD, DM, and CVD, as demonstrated in cell and animal studies, is summarized in Table 1.

Table 1. A summary of preclinical studies of genistein in treating Parkinson's disease (PD), Alzheimer's disease (AD), diabetes mellitus (DM), and cardiovascular disease (CVD).

Targeting Diseases	Experimental Model	Treatment	Outcome Characteristics	References
PD	Human SH-SY5Y cells overexpressing the A53T mutant of α-synuclein	Incubated with 20 μM genistein and rotenone for 24 h.	Rotenone-induced cell death, mitochondrial oxidative stress, and apoptosis ↓ Protein expression level of nuclear NFE2L2, HMOX1, and p-Akt ↑	[27]
	Lipopolysaccharide (LPS)-induced ovariectomized rats	Orally gavage with genistein (10 mg/kg) once daily for 14 consecutive days. Ovariectomy (OVX) was performed to eliminate the endogenous estrogen effect.	Apomorphine-induced rotational behavior ↓ Proinflammatory factor level (TNF-α and IL-1β) ↓ Protein expressions of COX-2 and iNOS ↓ Suppressing LPS-induced activation of MAPK and IκB signaling pathways through GPER and IGF-1R	[26]
	Transgenic *Drosophila* expressing normal human αS panneurally	Treated with 0–40 μM genistein.	Lifespan ↑ Inhibition of oxidative stress damage: GSH ↑ GST activity ↓ LPO level ↓ PC content ↓ Dopamine content ↑ Monoamine oxidase activity ↓	[28]
AD	Amyloid beta (Aβ)$_{25–35}$-treated SH-SY5Y cells	After being treated with 0–50 μM genistein for 90 min, cells were exposed to Aβ$_{25–35}$ at a concentration of 20 mM for 24 h.	Aβ-induced cell death ↓ Protein and mRNA expression of *HO-1* ↑ Activation of Nrf2//HO-1/PI3K signaling	[33]
	Aβ$_{25–35}$-treated Rat primary hippocampal neurons	After incubating with 0–1 μg/mL genistein for 24 h, neurons were continuously exposed to Aβ$_{25–35}$ for 3 days.	Aβ-induced cytotoxicity and necroptosis ↓ Aβ-induced LDH release, ROS accumulation, and MDA production ↓ Activating PI3K/Akt phosphorylation via α7nAChR signaling Activation of endogenous Nrf2/Keap1 transcription factors ↑	[35]
	Streptozotocin (STZ)-induced male Wistar rat model of the sporadic form of AD	Rats were administered 150 mg/kg b.w. via an orogastric probe once a day for 30 or 90 days.	Locomotor activity, memory and cognitive ability ↑ Protein expression level of levels of APP, total Aβ, Aβ$_{40}$, Aβ$_{42}$, and p-tau in the cortex, hippocampus, and the rest of the brain ↓ Autophagy ↑	[38]
DM	Methylglyoxal (MG)-treated *EA.HY926* human endothelial cells	Cells were pre-incubated for 2 h with genistein (0–100 μM) before co-treatment with 250 μM MG for 24 h.	MG-induced toxicity and ROS formation ↓ G0/G1 percentage ↑ Prevented MG-induced apoptosis via Nrf2 activation and MAPK-mediated signaling pathway regulation.	[50]
	Ovariectomized diabetic rats	Animals underwent swimming training (1 h/day) or received genistein (35 mg/kg b.w.) or a combination of both for eight weeks.	Inflammatory protein levels of IL-1β, Nf-κB, and TNF-α ↓ Anti-inflammatory protein levels of SIRT1 ↑	[41]

Table 1. *Cont.*

Targeting Diseases	Experimental Model	Treatment	Outcome Characteristics	References
CVD	High-fat-diet (HFD)- and LPS-induced chronic vascular inflammation in C57BL/6 mice	Intraperitoneal injection of LPS, combined with a HFD, was used to create the chronic vascular inflammation model. Administered orally with genistein (10 mg/kg b.w.) for 20 weeks.	Expression of inflammation-associated factors (mRNA expression of *TNF-α* and *IL-6*, as well as iNOS and NF-κB p65 protein levels) ↓ miR-21 ↓	[57]
	Δ^9-THC-induced C57BL/6J male mice	The mice were randomized into three groups: control group; Δ^9-THC treated group (1 mg/kg b.w./day); Δ^9-THC (1 mg/kg b.w./day)- and genistein (50 mg/kg b.w./day)-treated group.	Reversed Δ^9-THC-induced endothelial dysfunction, oxidative stress, and inflammation NF-κB phosphorylation ↓	[58]
	Estradiol-induced human umbilical vein endothelial cells (HUVECs)	Cells were treated with or without genistein in the presence of estradiol.	Cell viability ↑ NO level ↑ ROS formation ↓ Cell invasion and migration ↓ FAK protein expression ↓	[59]

Notes: NFE2L2, nuclear factor erythroid 2-related factor 2; HMOX1, heme Oxygenase 1; TNF-α, tumor necrosis factor alpha; IL-1β, interleukin-1 beta; IκB, inhibitor of NF-κB; COX-2, cyclooxygenase-2; iNOS, inducible nitric oxide synthase; GPER, G protein-coupled estrogen receptor; IGF-1R, insulin-like growth factor 1 receptor; MAPK, mitogen-activated protein kinase; GSH, glutathione; GST, glutathione S-transferase; LPO, lipid peroxidation; PC, protein carbonyl; Nrf2, nuclear factor erythroid 2-related factor 2; Nf-κB, nuclear factor kappa B; Keap1, Kelch-like ECH-associated protein 1; APP, amyloid precursor protein; ROS, reactive oxygen species; MDA, malondialdehyde; NO, nitric oxide; FAK, focal adhesion kinase; ↓, decrease; ↑, increase.

4.5. Cancer

Cancer remains a leading cause of death worldwide, with a continued increase in incidence and complexity. Chemotherapy, radiation, immunotherapy, and surgery are common cancer treatments. However, adverse effects such as premature ovarian failure resulting from radiotherapy, adverse body reactions after surgical resection, and secondary poisoning infection based on drug accumulation evoke significant side effects in cancer treatment [7,62,63]. As an advanced cancer treatment technology, targeted therapy minimizes side effects and reduces the likelihood of drug resistance, thereby providing new options for patients [62,64,65].

Genistein regulates typical protein expressions to protect against acetaldehyde-mediated cell injury and other liver damage, as shown in Table 2. After being compared with five common flavonoids, genistein had a better ability to reduce TNF-α in the liver [66]. Genistein treatment significantly decreased alanine aminotransferase (ALT), aspartate aminotransferase (AST), TNF-α, Mcp-1, and CXCL1 levels in an alcohol-fed mouse model. HO-1, which repairs the dysfunction caused by acetaldehyde-induced hepatocellular carcinoma and reverses the rise in Nrf2, was stimulated by genistein and induced a decrease in Keap1 downstream, leading to the inhibition of liver cancer [67]. Moreover, hepatocellular carcinoma (HCC) is another common cause of death among patients with cirrhosis. An in vivo study found that after genistein treatment, the levels of MAD and hydrogen peroxide decreased, while the levels of GSH, SOD, and Nrf2 increased in the Sprague Dawley rat model. Compared with rats treated with 25 mg/kg genistein, the survival rate of the rats treated with 75 mg/kg genistein was three times higher, reflecting the dose-dependent property of genistein again. The significant decline in ALT, AST, GGT, alkaline phosphatase, and ERK-1 protein expression demonstrated that genistein was protective against liver damage. In addition, a decline in PDGF protein was observed only in the HCC control group, indicating that genistein restrains the further development of HCC-induced proteins and influences cell repair, which is affected by the activation of downstream signaling pathways, such as the MAPK and PI3K/Akt signaling pathways [68]. Furthermore, genistein downregulated lncRNA TTTY18 and reduced the number of SGK1, AktSer473, p38 MAPK, and Tyr323 pathways to inhibit the ulterior spread of colorectal cancer (CRC) [69].

Additionally, genistein is also widely used for DNA damage repair following ROS exposure [70,71]. Oxidative DNA damage contributes to DNA breakage and has long-term side effects [70]. Similarly, genistein activated the Nrf2/ARE pathway to reduce ROS generation and DNA injury in bronchial epithelial cells [7]. By targeting the DNA injury response and promoting G2/M cell cycle arrest, genistein increased the susceptibility of cancer cells to radiation, stimulated apoptosis, and inhibited the progression of cervical cancer [71].

Table 2. A summary of preclinical studies of genistein in treating cancer.

Targeting Diseases	Experimental Model	Treatment	Outcome Characteristics	References
Liver cancer	Alcohol-fed male ICR mice	Mice in the flavonoid groups were orally administered five different kinds of flavonoids, respectively (quercetin, apigenin, naringenin, epigallocatechin gallate, and genistein; 0.3 mmol/kg b.w.), 1 h before the alcohol consumption for 5 weeks.	Hepatic function ↑ Prevented dyslipidemia Hepatic lipid peroxidation and oxidative stress ↓ Hepatic inflammatory stress ↓ Hepatic fibrosis and apoptosis ↓	[66]
	Chronic alcohol-fed mice	Mice were subjected to a Lieber–DeCarli alcohol liquid diet with or without genistein (1 mg/kg b.w./day mixed into their diet) for 8 weeks.	Liver injury and hepatic steatosis ↓ Hepatic inflammatory cell infiltration ↓ Ameliorated alcohol-induced hepatic oxidative stress, ER stress, and mitochondrial dysfunction Acetaldehyde-induced hepatocyte apoptosis ↓ HO-1 restoration and upregulation of NRF2 are involved in the preventive effect of genistein against ALD	[67]
	Thioacetamide (TAA)-induced Hepatocellular carcinoma (HCC) in Sprague Dawley rats	Rats were randomly divided into five groups: control group; genistein-treated group (75 mg/kg b.w., orally intake); HCC group (200 mg/kg b.w. TAA, i.p., twice a week); HCC + low dosage of genistein-treated group (25 mg/kg b.w., orally intake); and (v) HCC + high dosage of genistein-treated group (75 mg/kg b.w., orally intake). Treatments lasted for 16 weeks.	HCC-induced oxidative stress ↓ (hepatic MDA, and hydrogen peroxide ↓; hepatic Nrf2, GSH, and SOD levels ↑) Liver function ↑ (ALT, AST, alkaline phosphatase, and GGT serum levels ↓; serum albumin levels ↑) Protein expression of PDGF ↑, versican ↑, PKC ↓, and ERK-1 ↓	[68]
Ovarian cancer	γ-radiation to induce POF in female Sprague Dawley rats	Rats were administered a single intraperitoneal injection of genistein (5 mg/kg body weight) for 7 days, followed by exposure to a 3.2 Gy single dose of γ-rays on the 7th day.	Protected the ovarian tissue from hemorrhage and fibrosis Oxidative stress ↓ (GSH level and GPx activity ↑) mRNA expression of *Bax* ↓ and *Bcl-2* ↑ Optical densities of Cytochrome c and Caspase 3 ↓ Ovarian mRNA expression of *ER-β* ↑, *FOXL2* ↑, and *TGF-β* ↓	[62]
Colorectal cancer	Human malignant cell line of SW480	Cells were treated with different concentrations of genistein (0, 25, 50, and 100 μM) for 48 h.	Cell viability ↓ Cell apoptosis ↑ Cellular migration ↓ Protein expression of TTTY18, SGK1, Akt^{Ser473}, p38 $MAPK^{Tyr323}$ ↓	[69]
	Tumor-bearing nude mice	The mice were treated with 0, 20, 30, and 60 mg/kg b.w. genistein for 14 consecutive days.	Body mass ↓ Tumorous TGF-β1 and TTTY18 ↓ Intracellular numbers of SGK1, Akt^{Ser473}, p38 $MAPK^{Tyr323}$ positive cells ↓	[69]

Note: ALT, alanine transaminase; AST, aspartate transferase; GGT, gamma-glutamyl transferase; ER-β, estrogen receptor beta; Bax, Bcl-2-associated X protein; Bcl-2, β-cell lymphoma 2; FOXL2, forkhead box protein L2; TGF-β, transforming growth factor beta; SGK, serum and glucocorticoid-inducible kinase; ↓, decrease; ↑, increase.

Another important mechanism for genistein-targeted therapy is the regulation of the cell cycle and protein synthesis, which prevents the growth of malignant cells. An in vivo study in female Sprague Dawley rats demonstrated that genistein attenuated Bcl-2 expression and stimulated Bax expression, thereby increasing cytochrome C and caspase-3 expression. It upregulated ER-β and FOXL-2, downregulated TGF-β, reversed ovarian apoptosis, and blocked the growth of primordial follicles [62]. Correspondingly, genistein downregulated cyclin B1 and Bcl-2 by activating the NF-κB pathway at concentrations ranging from 5 to 20 μM. It inhibited HER-2 expression and phosphorylation, thereby restraining breast cancer development [72]. Genistein broke down cancer inhibitors of protein phosphatase 2A and increased the cytosolic Ca^{2+} buffering capacity of MCF-7 and T4D7 cells to decrease apoptosis and breast cancer cell proliferation [73]. It prevented the activation of G protein-coupled receptor-30 in breast cancer gene 1, causing phosphorylation of Akt and arresting the G2/M phase [74]. In CRC treatment, genistein similarly arrested the cell cycle at the G2/M phase to downregulate CDK4 and cyclin D1 expression in human salivary adenoid cystic carcinoma cell lines [63,75]. Thus, genistein inhibited the further spread of cancer cells by regulating specific proteins and controlling the cell cycle.

In summary, genistein offers promising therapeutic potential by mitigating the side effects of cancer treatment, exhibiting hepatoprotective properties, repairing oxidative DNA damage, and inhibiting the proliferation of malignant cells. Its mechanisms include protection against liver damage by regulating antioxidant pathways, repairing the DNA damage caused by ROS, and arresting the cell cycle to stunt tumor growth, highlighting its multifaceted role in targeted cancer therapy (Table 2).

5. The Role of Genistein in Recent Treatments

5.1. Pharmacokinetic Profile and Clinical Studies of Genistein

Genistein is present in plant raw materials as glycosides and is converted into an active aglycone for improved absorption by the human body after oral administration [76]. Viña and his team identified that the oral absorption of genistein is stable and results in minimal side effects in the body [36]. Both animal and human studies have demonstrated that the oral bioavailability of free genistein is low. For instance, its absolute bioavailability in rats is only 6.8%, and human pharmacokinetic studies have also reported low plasma and urinary levels of aglycone genistein. This poor oral bioavailability is largely due to the extensive metabolism of genistein in the body [77].

Based on this pharmacokinetic profile, several clinical studies have focused on the anti-inflammatory effects of genistein in preventing CVD, T2DM, and cancer (Table 3). For instance, genistein lowered the risk of hypertension-related complications (e.g., stroke and kidney disease), reduced heart strain, and promoted vascular health due to the decline in SBP, DBP, and MAP [78]. It also reduced the risk of DM in postmenopausal women [41]. Furthermore, in terms of blood sugar control, three months of genistein treatment lowered fasting blood sugar, A1C, and IR in postmenopausal women with DM, downregulated MDA levels, and stimulated thoracic aortic calcium levels to prevent further oxidative damage [42]. Kumar et al. studied the effects of genistein on prostate cancer (PCa), the most prevalent cancer among American men, across both African American men (AAM) and Caucasian men (CM). After a month of treatment, the PSA level in CM and the IGF-1/IGF-BP-3 ratio decreased, suggesting that genistein alleviated PCa to varying degrees in both groups. However, due to the limited number of participants with AAM, the effectiveness of genistein for PCa treatment in AAM requires further confirmation [79].

Table 3. Therapeutic effects of genistein in clinical trials.

Disease	Experiment Model	Treatment	Outcome Characteristics	References
AD	24 prodromal AD patients (54 to 76 years old) in a double-blind, placebo-controlled, and bicentric clinical trial	Randomly received genistein (one capsule/time, 60 mg/capsule) or placebo orally twice per day for up to 12 months. Orally twice per day for up to 12 months	Amyloid-beta deposition uptake in the anterior cingulate gyrus ↓ Individual cognitive behavior ↑ No reported deaths or serious adverse events	[36]
DM	58 postmenopausal women with T2DM (randomized, double-blind, placebo-controlled clinical trial)	Randomly received genistein (two capsules/day, 54 mg/capsule) or placebo orally for 12 weeks.	FBS, A1C, TG, and MDA ↓ TAC, HDL-C, and QUICKI ↑	[42]
Prostate cancer	70 participants, of whom 36 participants (25 CM, 6AAM) were randomized to the isoflavone group and 34 (25 CM, 7AAM) to the placebo group	Administered isoflavones (20 mg BID) or placebo for 3–6 weeks.	No changes in serum steroid hormones PSA levels in CM ↓ IGF-1/IGF-BP-3 ↓ Ki-67 expression in the placebo group ↑	[79]

Note: FBS, fasting blood glucose; A1C, glycated hemoglobin; TG, triglyceride; MDA, malondialdehyde; TAC, total antioxidant capacity; HDL-C, high-density lipoprotein cholesterol; QUICKI, quantitative insulin sensitivity check index; AAM, African American men; CM, Caucasian men; PSA, prostate-specific antigen; ↓, decrease; ↑, increase.

However, based on the lower rate of absorption of genistein, which accounts for only 20–40% of the gastrointestinal tract in enterohepatic cycling, the bioavailability of genistein is at a more fundamental level [80]. Thus, numerous researchers have expressed concerns about its future clinical application and sought alternatives to improve the bioavailability or efficacy of genistein.

5.2. Combination Therapy with Exercise or Existing Agents to Enhance Genistein's Bioavailability and Efficacy

Some scholars have sought to improve the bioavailability and efficacy of genistein by combining it with exercise, existing drugs, or technologies. The synergistic effects of exercise and genistein have been studied in the context of retinal neovascularization, obesity induced by a high-fat, high-sugar (HFHS) diet, and non-alcoholic steatohepatitis (NASH). For instance, a previous study has suggested that retinal neovascularization, a visual disorder that occurs during the postmenopausal period or in patients with T2DM, can be treated with powerful antioxidant supplements or physical activity. Compared with the single application of either swimming or genistein, the co-treatment of genistein and swimming markedly reduces the levels of MDA, IL-6, and TNF-α, as well as the expression of vascular endothelial growth factor, in the retinal tissues of Wistar rats [81]. In addition, genistein and running were combined to mitigate the detrimental effects of an HFHS diet in male C57BL/6 mice with AD. The levels of pGSK-3β, p-IR, caspase 3, CP13, and APP in the combined treatment group were significantly lower than those in the treatment group, indicating a positive change in mice [82]. In previous postmenopausal animal models, exercise has been proven to have an estrogen-like effect in reducing fatty liver disease. Genistein relieved NASH primarily by exerting phytoestrogenic effects. F0 fibrosis observed in the combination treatment group demonstrated an improvement in liver fibrosis during NASH treatment, potentially halting further disease progression and promoting liver function recovery [83].

Likewise, combining genistein with metformin (MET) has been explored as a potential strategy to ameliorate non-alcoholic fatty liver disease (NAFLD). This combination aims to slow NAFLD progression by leveraging the ability of metformin to inhibit hepatic glucose production and increase insulin sensitivity in peripheral tissues, which is beneficial in the treatment of T2DM. Combining these two agents significantly decreased the amelioration of glucose tolerance, ALT, plasma TG, and liver TG in HFD-fed C57BL/6 mice compared to the groups treated with genistein and MET alone. Meanwhile, the combination of genistein and MET inhibited SREBP-1c expression induced by an HFD, a response that was not altered by the separate use of genistein or MET [84]. Additionally, some researchers have tried to supplement genistein with existing drugs for breast cancer treatment to identify their effects on cellular anticancer properties. The research team of Bezerra et al. reported that the combination of exemestane and genistein significantly reduced S-phase and cyclin E levels, enhanced the anti-proliferative and apoptotic effects, and altered the PARP/PARP ratio in MCF-7aro cells [85]. Another in vitro study on breast CSCs found that the combination of genistein and myokines was able to reduce colony and sphere formation in the MCF-7 cell model, highlighting the potential of these treatments to inhibit the self-renewal and stemness of CSCs. Additionally, the expressions of SOX2 and OCT4 (both were CSC markers) were suppressed after the co-treatment, which was more effective than the individual treatments [86].

As a promising approach to enhance cancer treatment outcomes, neutron radiotherapy addresses the limitations of monotherapy, minimizes the risk of drug resistance, and improves patient prognosis. Interestingly, Khamesi and his team discovered that the combination of neutron radiotherapy and genistein, although considered a radiation sensitizer, did not exhibit a better efficacy in killing prostate cancer cells than neutron radiation and genistein alone [87]. Although specific combination therapies with genistein have shown excellent synergistic effects and improved therapeutic outcomes in some cases (Table 4), it is undeniable that not all combinations guarantee superior efficacy, as some may offer no additional benefit compared to monotherapies.

Table 4. Synergistic therapeutic effects of genistein combined with exercise or conventional drugs in preclinical models.

Disease	Experiment Model	Treatment	Outcome Characteristics	References
Obesity and AD	High-fat and high-sugar (HFHS) mouse model of neurodegeneration	Mice were treated with genistein (600 mg/kg in HFHS diet), exercise training, or a combination of both for 12 weeks.	Body weight, adipose mass, and inflammatory marker TNF-α ↓ Key proteins' expression involved in AD (pGSK-3β/GSK, Aβ, ADAM10, Caspase-3, pIR/IR, p-IRS/IRS) ↓	[82]
Non-alcoholic steatohepatitis (NASH)	NASH model of OVX rats fed with high-fat high-fructose (HFHF) diet	Rats were given genistein (16 mg/kg b.w.), engaged in moderate running exercises, or both for 5 weeks.	Did not provide additional benefits for NASH in OVX rats fed with HFHF diet.	[83]
Non-alcoholic fatty liver disease (NAFLD)	HFD-fed mice model of NAFLD	Mice were administered 0.23% metformin (MET, 2.3 g/kg diet) combined with 0.2% genistein (2 g/kg diet), or MET and genistein alone.	Body and liver weights ↓ FBG, fasting plasma insulin, HOMA-IR, glucose tolerance, ALT, AST, plasma TG, and liver TG ↓ Steatosis ↓ Gene expression of *FAS*, pro-inflammatory (*TNFα*, *IL-1β*, and *IL-6*), *PEPCK*, and *G6Pase* ↓ Protein expression of pGSK-3β ↑ Better efficacy than treatment with genistein or metformin alone	[84]
Breast cancer	ER^+ aromatase-overexpressing human breast cancer cell line MCF-7aro	Cells were treated with genistein (0.5–25 μM), with or without exemestane (Exe), anastrozole (Ana), or letrozole (Let) (1, 5, and 10 μM) for 3 days.	Combination of genistein with Ana or Let negatively impacts the therapeutic efficacy of aromatase inhibitors. Genistein enhanced the anticancer properties of Exe. Hormone targets are not affected by this combination treatment.	[85]
	Human breast cancer MCF-7 cells	Cells were treated with various concentrations of myokines, genistein, or their combination for 72 h.	A larger reduction in colony formation than myokines or genistein alone. Higher reduction in sphere formation Higher decrease in *SOX2* and *OCT4* gene expressions	[86]

Note: pGSK-3β, brain phosphorylated glycogen synthase kinase; IR, insulin receptor; IRS, insulin receptor substrate; HOMA-IR, homeostasis model assessment of insulin resistance; ALT, alanine transaminase; AST, aspartate transaminase; TG, triglyceride; TNFα, Tumor necrosis factor alpha; IL-1β, Interleukin-1 beta; IL-6, Interleukin-6; ↓, decrease; ↑, increase.

5.3. Advances in the Use of Nanotechnology for Genistein Delivery to Enhance Bioavailability

Nanotechnology has emerged as a transformative technique in medical research, providing innovative solutions for drug development and improvement [88]. Its advantages, such as enabling the precise targeting of therapeutic agents, facilitating the delivery of large biomolecules, and improving the water solubility and bioavailability of drugs, have attracted extensive research interest from scientists [88,89]. They have developed numerous nanoscale structures, including liposomes, polymer-based nanoparticles, micelles, dendrimers, solid lipid nanoparticles, nanoemulsions, nanocapsules, ceramic particles, and metallic nanoparticles, to further enhance the effectiveness of their drug delivery systems [89]. In recent literature reviews, Khan et al. and Syahputra et al. have highlighted the potential effects of flavonoid nano-formulations (quercetin in eudragit-coated liposomes, red ginger flavonoids encapsulated in nanoemulsion systems, EGCG formulated in PLA-PEG (polylactic acid–polyethylene glycol) nanoparticles, etc.) in inhibiting ROS formation, reducing systolic blood pressure, and preventing or treating cancer [89,90]. Similarly, nanotechnology has significantly advanced the formulation and delivery system of genistein, utilizing lipid nanoparticles, liposomes, tocotrienol-rich nanoemulsions, polymeric nanoparticles, dextran complexes, chitosan complexes, and Fe_3O_4 nanoparticles with carboxymethylated chitosan. These methods enable precise control over drug release and enhance the solubility and stability of genistein [91]. Recently, the novel genistein-loaded solid lipid nanoparticles (SLNs) developed by Obinu et al. have demonstrated excellent bioavailability, stability, and a high drug loading capacity. In their studies, SLNs could be absorbed by the intestinal mucosa and Caco-2 cells, indicating that improved intestinal lymphatic absorption may be achieved by minimizing the first-pass metabolic effect of genistein [92]. Moreover, the development of genistein-loaded nanoparticles not only enhances these physicochemical and pharmacokinetic properties but also enables more precise targeting of specific cells, thereby decreasing the risk of adverse effects on human primary cells or healthy cells. For instance, the genistein–gold nanoparticle conjugates Gen@AuNPs1 and Gen@AuNPs2 developed by Vodnik et al. have been reported to not only effectively suppress cancer cell growth but also enhance the targeting of cancerous phenotypes [93]. In summary, the integration of nanotechnology with genistein holds great promise for overcoming traditional pharmacological limitations, enhancing therapeutic efficacy, and is essential for more precise, safer, and effective treatments.

5.4. Therapeutic Doses of Genistein by Disease Entity

As mentioned above, genistein exhibits dose-dependent, dual effects on regulating oxidative stress. At relatively low concentrations, it acts as an antioxidant by upregulating detoxification and antioxidant enzymes through various signaling pathways, including the PI3K/Akt, NF-κB, and Nrf2/HO-1 pathways, thereby reducing oxidative stress, inflammation, and neuronal toxicity, which highlights its therapeutic potential for brain injury and neurodegenerative diseases. However, exposure to high-dose genistein has a pro-oxidative role, promoting the production of ROS and leading to increased oxidative damage and inflammation. This heightened oxidative stress can decrease the viability of cancer cells, but it also raises the risk of hormonal imbalances and hormone-sensitive diseases due to its phytoestrogenic properties. Overall, genistein also appears to elicit varying responses depending on the treatment concentrations. Thus, this section summarizes the therapeutic doses of genistein reported in recent studies, categorized by different disease entities, to provide practical information for clinical reference and the development of functional foods.

According to the meta-analysis provided by Lei et al. and numerous experimental studies, it is clear that genistein treatment in the AD rat model (treatment concentration: 0.022 to 150 mg/kg b.w./day; time duration: from 6 days to 9 months) led to different outcomes [94]. Moreover, the National Institutes of Health (NIH) in the United States recommends that the daily dietary intake of isoflavones for adults should be in the range of 30 to 50 mg [6]. Certain clinical studies have demonstrated that taking 60 mg of genistein orally twice a day for 12 months significantly enhances cognitive performance in AD patients and exhibits a promising effect by decreasing Aβ plaques within the anterior cingulate gyrus [95]. Unfortunately, this dosage in the trial is much higher than the daily intake recommended by the NIH; as a result, the appropriate dosage for individuals with AD still needs to be established. Unlike the preclinical studies in the AD model, almost all of the genistein experimental groups in the LPS-induced inflammatory PD rat model or the 6-OHDA-induced PD rat model received genistein via i.c.v. injection or i.p injection [26,96,97]. For example, a single, high-dose intraperitoneal injection of genistein (10 mg/kg b.w.) ameliorated rotational behavior and protected neurons in the substantia nigra pars compacta (SNC) in 6-OHDA-induced rats [96,97]. However, there is currently a lack of clinical data in patients with PD, and further research is required to determine the effective dose in humans.

A randomized, double-blind, placebo-controlled clinical trial in postmenopausal women with T2DM has commonly used genistein at a dose of 108 mg/day for 12 weeks, resulting in improved glycemic control and reduced serum lipid levels [42]. Another clinical study investigating the effects of genistein on insulin resistance in pre-diabetic individuals suggested that a genistein supplement (50 mg/day) for 2 months significantly attenuated insulin resistance, which in turn prevented metabolic abnormalities and further progression to diabetes [45]. Thus, in human-based research, the doses of genistein ranging from 50 to 108 mg/day for 2 to 3 months effectively improved glycemic regulation, lipid control, and insulin sensitivity in DM patients, leading to effective DM treatment. Generally, in vivo studies administered a higher level of genistein to determine its anti-DM and antioxidative properties. Gilbert et al. and Yang et al. have mentioned that genistein at 8 mg/kg b.w., the average daily consumption from a soy-rich human diet, exhibited a limited effect on diabetic rats, whereas genistein at 30 mg/kg significantly improved glucose metabolism [98,99]. In summary, both clinical and animal studies suggest that genistein has huge potential to regulate blood glucose and improve metabolic profiles within a specific dosage range; however, the effective dosage in animal experiments is significantly higher than that in the human diet and should be taken into consideration.

Regarding the dose of genistein in cancer treatment, the optimal dose differs depending on the type of cancer and the specific study parameters. In an azoxymethane-induced colon cancer male Sprague Dawley rat model, the genistein intervention (140 mg/kg b.w.) for 13 weeks was reported to prevent early colon neoplasia [100]. In an animal model of liver cancer, the administration of genistein for 15 days (15 mg/kg b.w.) effectively treated liver cancer by inducing apoptosis in HCC rats [101]. Similarly, genistein treatment (75 mg/kg b.w.) for 16 weeks enhances liver function in HCC rats [68]. In a study on ovarian cancer, the administration of genistein (5 mg/kg body weight, administered intraperitoneally for 7 days) showed positive effects on antioxidant activity and ovarian tissue protection [62]. However, it is notable that the impact of exposure to genistein on women with estrogen-dependent breast cancer remains controversial [102]. In a mouse model designed to replicate the low-estrogen environment of postmenopausal women, genistein treatment may present adverse health effects, including promoting estrogen-dependent tumor growth. Thus, the optimal therapeutic dose of genistein should be tailored to the specific disease, patient population, and treatment goals. Its safety and potential adverse effects should be considered carefully. Further animal and clinical trials are needed to determine the recommended dosage for each disease entity.

6. Conclusions

Oxidative stress induces plaque aggregation, insulin signaling, DNA damage, and endothelial dysfunction, which further induce or exacerbate various diseases, including neurodegenerative diseases, metabolic diseases, and cancer. This review systematically elucidates the detailed molecular mechanisms, particularly the roles of the Nrf2/HO-1, PI3K/Akt, and NF-κB pathways, clarifying how genistein exerts its antioxidant, anti-inflammatory, and anti-apoptotic effects in five common diseases (PD, AD, DM, CVD, and cancer) that are prevalent in the population. As a plant-derived compound, genistein is recognized as a promising therapeutic option for symptom relief due to its abundant natural availability, dietary applicability, and favorable side effect profile. In conclusion, a rich body of evidence suggests that it alleviates disease progression by inhibiting oxidative stress, inflammation, apoptosis, and endothelial dysfunction, all of which are closely related to oxidative stress. To improve its bioavailability and therapeutic efficacy, genistein has been combined with exercise, existing drugs, and advanced delivery systems such as nanoparticles, enhancing oral absorption. However, the optimal therapeutic dose of genistein for specific diseases, as well as the efficacy and synergistic effects of some combined therapies, remain obstacles to the use of genistein in disease prevention and the development of functional foods. Further animal and clinical trials, as well as systematic literature reviews, are needed to explore its efficacy and bioavailability.

Author Contributions: K.Z. and J.W.: investigation, writing—original draft preparation. B.X.: software, writing—reviewing and editing. B.X.: conceptualization, methodology, funding, supervision, writing—reviewing and editing. All authors have read and agreed to the published version of the manuscript.

Funding: This work was jointly supported by Guangdong Higher Education Upgrading Plan (2021–2025) (No. UICR0400015-24B and UICR0400016-24B) at Beijing Normal-Hong Kong Baptist University, Zhuhai, PR China.

Conflicts of Interest: The authors declare that there are no conflicts of interest.

Abbreviations

AAM: African American men; AD: Alzheimer's disease; ALT: alanine aminotransferase; AMPK: adenosine 5′-monophosphate-activated protein kinase; APP: amyloid precursor protein; AS: atherosclerotic; AST: aspartate aminotransferase; Bax: Bcl-2-associated X protein; Bcl-2: β-cell lymphoma 2; Bid: Bcl-2-associated X protein; CAT: catalase; CM: Caucasian men; COX-2: cyclooxygenase-2; CRC: colorectal cancer; CVD: cardiovascular disease; DA: dopamine; DM: diabetes mellitus; DOPAC: 3,4-Dihydroxyphenylacetic acid; ERK: extracellular signal-regulated kinase; GPx: glutathione peroxidase; GSH: glutathione; GSSH: glutathione disulfide; HCC: hepatocellular carcinoma; HDAC3: histone deacetylase 3; HER: human epidermal growth factor receptor 2; HFD: high-fat diet; HFHS: high-fat high-sugar; HIBD: hypoxic-ischemic brain damage; HO-1: hemoxygenase 1; HOMA-IR: homeostatic model assessment of insulin resistance; HVA: homovanillic acid; IGF-1R: insulin-like growth factor 1 receptor; IL-1β: interleukin-1β; IL-6: interleukin-6; iNOS: inducible nitric oxide synthase; IR: insulin resistance; IκB: inhibitor of NF-κB; JNK: jun N-terminal kinase; Keap1: Kelch-like ECH-associated protein 1; MAD: multiple ascending dose; MAPK: mitogen-activated protein kinase; mCRC: metastatic colorectal cancer; MET: metformin; MG: methylglyoxal; NAFLD: non-alcoholic fatty liver disease; NAMPT: nicotinamide phosphoribosyltransferase; NASH: non-alcoholic steatohepatitis; NF-κB: nuclear factor κ B; NNK: 4-(methylnitrosamino)-1-(3-pyridyl)-1-butanone; NIH: National Institutes of Health; Nrf2: nuclear factor erythroid 2-related factor 2; PCa: prostate cancer; PD: Parkinson's disease; PHB1: prohibitin 1; PI3K/Akt: phosphatidylinositol 3-kinase/protein kinase 3; RhoA: Ras homolog family member A; RNS: reactive nitrogen species; ROS: reactive oxygen species; SIRI1: Sirtuin 1; SN: substantia nigra; SOD: superoxide dismutase; T2DM: type 2 diabetes mellitus; TG: triglyceride; TLR4: toll-like receptor 4; TNF: tumor necrosis factor; TNF-α: tumour necrosis factor-α; VECs: vascular endothelial cells;; αS: α-synuclein; 6-OHDA: 6-hydroxydopamine; SLNs: solid lipid nanoparticles; NAD^+, nicotinamide adenine dinucleotide, oxidized form.

References

1. Preiser, J.C. Oxidative stress. *J. Parenter. Enter. Nutr.* **2012**, *36*, 147–154. [CrossRef] [PubMed]
2. Rani, V.; Deep, G.; Singh, R.K.; Palle, K.; Yadav, U.C. Oxidative stress and metabolic disorders: Pathogenesis and therapeutic strategies. *Life Sci.* **2016**, *148*, 183–193. [CrossRef] [PubMed]
3. González, P.; Lozano, P.; Ros, G.; Solano, F. Hyperglycemia and oxidative stress: An integral, updated and critical overview of their metabolic interconnections. *Int. J. Mol. Sci.* **2023**, *24*, 9352. [CrossRef] [PubMed]
4. Cioffi, F.; Adam, R.H.I.; Broersen, K. Molecular mechanisms and genetics of oxidative stress in Alzheimer's disease. *J. Alzheimers Dis.* **2019**, *72*, 981–1017. [CrossRef] [PubMed]
5. Yaribeygi, H.; Sathyapalan, T.; Atkin, S.L.; Sahebkar, A. Molecular mechanisms linking oxidative stress and diabetes mellitus. *Oxid. Med. Cell Longev.* **2020**, *2020*, 8609213. [CrossRef] [PubMed]
6. Yu, L.; Rios, E.; Castro, L.; Liu, J.; Yan, Y.; Dixon, D. Genistein: Dual role in women's health. *Nutrients* **2021**, *13*, 3048. [CrossRef] [PubMed]
7. Suraweera, T.L.; Merlin, J.P.J.; Dellaire, G.; Xu, Z.; Rupasinghe, H.P.V. Genistein and procyanidin B2 reduce carcinogen-induced reactive oxygen species and DNA damage through the activation of Nrf2/ARE cell signaling in bronchial epithelial cells in vitro. *Int. J. Mol. Sci.* **2023**, *24*, 3676. [CrossRef] [PubMed]
8. Goh, Y.X.; Jalil, J.; Lam, K.W.; Husain, K.; Premakumar, C.M. Genistein: A review on its anti-inflammatory properties. *Front. Pharmacol.* **2022**, *13*, 820969. [CrossRef] [PubMed]
9. Rajput, M.S.; Sarkar, P.D.; Nirmal, N.P. Inhibition of DPP-4 activity and neuronal atrophy with genistein attenuates neurological deficits induced by transient global cerebral ischemia and reperfusion in streptozotocin-induced diabetic mice. *Inflammation* **2017**, *40*, 623–635. [CrossRef] [PubMed]
10. Sarao, L.; Kaur, S.; Malik, T.; Singh, A. Genistein and daidzein. In *Nutraceuticals and Health Care*, 2022nd ed.; Kour, J., Nayik, G.A., Eds.; Academic Press: Cambridge, MA, USA, 2022; pp. 331–341.
11. Kumar, V.; Chauhan, S.S. Daidzein induces intrinsic pathway of apoptosis along with ER α/β ratio alteration and ROS production. *Asian Pac. J. Cancer Prev.* **2021**, *22*, 603–610. [CrossRef] [PubMed]
12. Alorda-Clara, M.; Torrens-Mas, M.; Morla-Barcelo, P.M.; Roca, P.; Sastre-Serra, J.; Pons, D.G.; Oliver, J. High concentrations of genistein decrease cell viability depending on oxidative stress and inflammation in colon cancer cell lines. *Int. J. Mol. Sci.* **2022**, *23*, 7526. [CrossRef] [PubMed]
13. Li, Y.; Ou, S.; Liu, Q.; Gan, L.; Zhang, L.; Wang, Y.; Qin, J.; Liu, J.; Wu, W. Genistein improves mitochondrial function and inflammatory in rats with diabetic nephropathy via inhibiting MAPK/NF-κB pathway. *Acta Cir. Bras.* **2022**, *37*, e370601. [CrossRef] [PubMed]
14. Wang, F.; Han, L.; Wang, X.; Li, Y.; Zhu, Y.; Wang, J.; Xue, C. Sialoglycoprotein isolated from eggs of *Carassius auratus* promotes fracture healing in osteoporotic mice. *J. Food Drug Anal.* **2018**, *26*, 716–724. [CrossRef] [PubMed]
15. Wu, C.; Zhou, S.; Ma, S.; Suzuki, K. Effect of genistein supplementation on exercise-induced inflammation and oxidative stress in mice liver and skeletal muscle. *Medicina* **2021**, *57*, 1028. [CrossRef] [PubMed]
16. Li, Y.; Zhang, J.J.; Chen, R.J.; Chen, L.; Chen, S.; Yang, X.F.; Min, J.W. Genistein mitigates oxidative stress and inflammation by regulating Nrf2/HO-1 and NF-κB signaling pathways in hypoxic-ischemic brain damage in neonatal mice. *Ann. Transl. Med.* **2022**, *10*, 32. [CrossRef] [PubMed]
17. Schreihofer, D.A.; Oppong-Gyebi, A. Genistein: Mechanisms of action for a pleiotropic neuroprotective agent in stroke. *Nutr. Neurosci.* **2019**, *22*, 375–391. [CrossRef] [PubMed]
18. Lignitto, L.; LeBoeuf, S.E.; Homer, H.; Jiang, S.; Askenazi, M.; Karakousi, T.R.; Pass, H.I.; Bhutkar, A.J.; Tsirigos, A.; Ueberheide, B.; et al. Nrf2 activation promotes lung cancer metastasis by inhibiting the degradation of Bach1. *Cell* **2019**, *178*, 316–329. [CrossRef] [PubMed]
19. Done, A.J.; Traustadóttir, T. Nrf2 mediates redox adaptations to exercise. *Redox Biol.* **2016**, *10*, 191–199. [CrossRef] [PubMed]

20. Yang, L.; Shen, L.; Li, Y.; Li, Y.; Yu, S.; Wang, S. Hyperoside attenuates dextran sulfate sodium-induced colitis in mice possibly via activation of the Nrf2 signalling pathway. *J. Inflamm.* **2017**, *14*, 25. [CrossRef] [PubMed]
21. Caceres, S.; Crespo, B.; Alonso-Diez, A.; de Andrés, P.J.; Millan, P.; Silván, G.; Illera, M.J.; Illera, J.C. Long-term exposure to isoflavones alters the hormonal steroid homeostasis-impairing reproductive function in adult male Wistar rats. *Nutrients* **2023**, *15*, 1261. [CrossRef] [PubMed]
22. Mondal, S.; Bandyopadhyay, A. Antioxidants in mitigating phthalate-induced male reproductive toxicity: A comprehensive review. *Chemosphere* **2024**, *364*, 143297. [CrossRef] [PubMed]
23. Singh, S. Review on natural agents as aromatase inhibitors: Management of breast cancer. *Comb. Chem. High Throughput Screen.* **2024**, *27*, 2623–2638. [CrossRef] [PubMed]
24. Li, D.W.; Zhou, F.Z.; Sun, X.C.; Li, S.C.; Yang, J.B.; Sun, H.H.; Wang, A.H. Ginsenoside Rb1 protects dopaminergic neurons from inflammatory injury induced by intranigral lipopolysaccharide injection. *Neural Regen. Res.* **2019**, *14*, 1814–1822. [CrossRef] [PubMed]
25. Singh, A.; Tripathi, P.; Yadawa, A.K.; Singh, S. Promising polyphenols in Parkinson's disease therapeutics. *Neurochem. Res.* **2020**, *45*, 1731–1745. [CrossRef] [PubMed]
26. Du, Z.R.; Gu, Y.; Xie, X.M.; Zhang, M.; Jiang, G.Y.; Chen, W.F. GPER and IGF-1R mediate the anti-inflammatory effect of genistein against lipopolysaccharide (LPS)-induced nigrostriatal injury in rats. *J. Steroid Biochem. Mol. Biol.* **2021**, *214*, 105989. [CrossRef] [PubMed]
27. Wu, H.C.; Hu, Q.L.; Zhang, S.J.; Wang, Y.M.; Jin, Z.K.; Lv, L.F.; Zhang, S.; Liu, Z.L.; Wu, H.L.; Cheng, O.M. Neuroprotective effects of genistein on SH-SY5Y cells overexpressing A53T mutant α-synuclein. *Neural Regen. Res.* **2018**, *13*, 1375–1383. [PubMed]
28. Siddique, Y.H.; Naz, F.; Jyoti, S.; Ali, F.; Rahul. Effect of genistein on the transgenic *Drosophila* model of Parkinson's disease. *J. Diet. Suppl.* **2019**, *16*, 550–563. [CrossRef] [PubMed]
29. Khan, A.; Jahan, S.; Imtiyaz, Z.; Alshahrani, S.; Antar Makeen, H.; Mohammed Alshehri, B.; Kumar, A.; Arafah, A.; Rehman, M.U. Neuroprotection: Targeting multiple pathways by naturally occurring phytochemicals. *Biomedicines* **2020**, *8*, 284. [CrossRef] [PubMed]
30. Almeida, M.F.; Bahr, B.A.; Kinsey, S.T. Endosomal-lysosomal dysfunction in metabolic diseases and Alzheimer's disease. *Int. Rev. Neurobiol.* **2020**, *154*, 303–324. [PubMed]
31. Brand, A.L.; Lawler, P.E.; Bollinger, J.G.; Li, Y.; Schindler, S.E.; Li, M.; Lopez, S.; Ovod, V.; Nakamura, A.; Shaw, L.M.; et al. The performance of plasma amyloid beta measurements in identifying amyloid plaques in Alzheimer's disease: A literature review. *Alzheimers Res. Ther.* **2022**, *14*, 195. [CrossRef] [PubMed]
32. Rahman, M.M.; Lendel, C. Extracellular protein components of amyloid plaques and their roles in Alzheimer's disease pathology. *Mol. Neurodegener.* **2021**, *16*, 59. [CrossRef] [PubMed]
33. Yi, S.; Chen, S.; Xiang, J.; Tan, J.; Huang, K.; Zhang, H.; Wang, Y.; Wu, H. Genistein exerts a cell-protective effect via Nrf2/HO-1/ /PI3K signaling in Ab25-35-induced Alzheimer's disease models in vitro. *Folia Histochem. Cytobiol.* **2021**, *59*, 49–56. [CrossRef] [PubMed]
34. Duan, X.; Li, Y.; Xu, F.; Ding, H. Study on the neuroprotective effects of genistein on Alzheimer's disease. *Brain Behav.* **2021**, *11*, e02100. [CrossRef] [PubMed]
35. Guo, J.; Yang, G.; He, Y.; Xu, H.; Fan, H.; An, J.; Zhang, L.; Zhang, R.; Cao, G.; Hao, D.; et al. Involvement of α7nAChR in the protective effects of genistein against β-amyloid-induced oxidative stress in neurons via a PI3K/Akt/Nrf2 pathway-related mechanism. *Cell Mol. Neurobiol.* **2021**, *41*, 377–393. [CrossRef] [PubMed]
36. Viña, J.; Escudero, J.; Baquero, M.; Cebrián, M.; Carbonell-Asíns, J.A.; Muñoz, J.E.; Satorres, E.; Meléndez, J.C.; Ferrer-Rebolleda, J.; Cózar-Santiago, M.D.P.; et al. Genistein effect on cognition in prodromal Alzheimer's disease patients. The GENIAL clinical trial. *Alzheimers Res. Ther.* **2022**, *14*, 164. [CrossRef] [PubMed]
37. Aimaier, S.; Tao, Y.; Lei, F.; Yupeng, Z.; Wenhui, S.; Aikemu, A.; Maimaitiyiming, D. Protective effects of the *Terminalia bellirica* tannin-induced Nrf2/HO-1 signaling pathway in rats with high-altitude pulmonary hypertension. *BMC Complement. Altern. Med.* **2023**, *23*, 150. [CrossRef] [PubMed]
38. Pierzynowska, K.; Podlacha, M.; Gaffke, L.; Majkutewicz, I.; Mantej, J.; Węgrzyn, A.; Osiadły, M.; Myślińska, D.; Węgrzyn, G. Autophagy-dependent mechanism of genistein-mediated elimination of behavioral and biochemical defects in the rat model of sporadic Alzheimer's disease. *Neuropharmacology* **2019**, *148*, 332–346. [CrossRef] [PubMed]
39. Liu, S.; Gao, J.; Zhu, M.; Liu, K.; Zhang, H.L. Gut microbiota and dysbiosis in Alzheimer's disease: Implications for pathogenesis and treatment. *Mol. Neurobiol.* **2020**, *57*, 5026–5043. [CrossRef] [PubMed]
40. Wang, J.W.; Zhang, J.; Yu, Z.L.; Chung, S.K.; Xu, B.J. The roles of dietary polyphenols at crosstalk between type 2 diabetes and Alzheimer's disease in ameliorating oxidative stress and mitochondrial dysfunction via PI3K/Akt signaling pathways. *Ageing Res. Rev.* **2024**, *99*, 102416. [CrossRef] [PubMed]

41. Khalilzadeh, B.; Zayer, M.; Yousefi, H. Anti-inflammatory effect of swimming exercise and genistein in combination in the pancreas of ovariectomized diabetic rats. *Med. J. Tabriz Univ. Med. Sci. Health Serv.* **2024**, *46*, 38–47. [CrossRef]
42. Braxas, H.; Rafraf, M.; Hasanabad, S.K.; Jafarabadi, M.A. Effectiveness of genistein supplementation on metabolic factors and antioxidant status in postmenopausal women with type 2 diabetes mellitus. *Can. J. Diabetes.* **2019**, *43*, 490–497. [CrossRef] [PubMed]
43. Goswami, K.; Badruddeen; Arif, M.; Akhtar, J.; Khan, M.I.; Ahmad, M. Flavonoids, isoflavonoids and others bioactives for insulin sensitizations. *Curr. Diabetes Rev.* **2024**, *20*, e270423216247. [CrossRef] [PubMed]
44. Ram Makena, M.; Gatla, H.; Verlekar, D.; Sukhavasi, S.; Pandey, M.K.; Pramanik, K.C. Wnt/β-catenin signaling: The culprit in pancreatic carcinogenesis and therapeutic resistance. *Int. J. Mol. Sci.* **2019**, *20*, 4242. [CrossRef] [PubMed]
45. Guevara-Cruz, M.; Godinez-Salas, E.T.; Sanchez-Tapia, M.; Torres-Villalobos, G.; Pichardo-Ontiveros, E.; Guizar-Heredia, R.; Arteaga-Sanchez, L.; Gamba, G.; Mojica-Espinosa, R.; Schcolnik-Cabrera, A.; et al. Genistein stimulates insulin sensitivity through gut microbiota reshaping and skeletal muscle AMPK activation in obese subjects. *BMJ Open Diabetes Res.* **2020**, *8*, e000948. [CrossRef]
46. Braxas, H.; Musazadeh, V.; Zarezadeh, M.; Ostadrahimi, A. Genistein effectiveness in improvement of glucose and lipid metabolism and homocysteine levels: A systematic review and meta-analysis. *J. Funct. Foods* **2023**, *102*, 105433. [CrossRef]
47. Watanabe, S.; Haruyama, R.; Umezawa, K.; Tomioka, I.; Nakamura, S.; Katayama, S.; Mitani, T. Genistein enhances NAD(+) biosynthesis by upregulating nicotinamide phosphoribosyltransferase in adipocytes. *J. Nutr. Biochem.* **2023**, *121*, 109433. [CrossRef] [PubMed]
48. Mitani, T.; Watanabe, S.; Wada, K.; Fujii, H.; Nakamura, S.; Katayama, S. Intracellular cAMP contents regulate NAMPT expression via induction of C/EBPβ in adipocytes. *Biochem. Biophys. Res. Commun.* **2020**, *522*, 770–775. [CrossRef] [PubMed]
49. Stromsdorfer, K.L.; Yamaguchi, S.; Yoon, M.J.; Moseley, A.C.; Franczyk, M.P.; Kelly, S.C.; Qi, N.; Imai, S.; Yoshino, J. NAMPT-Mediated NAD+ biosynthesis in adipocytes regulates adipose tissue function and multi-organ insulin sensitivity in mice. *Cell Rep.* **2016**, *16*, 1851–1860. [CrossRef] [PubMed]
50. Liccardo, M.; Sapio, L.; Perrella, S.; Sirangelo, I.; Iannuzzi, C. Genistein prevents apoptosis and oxidative stress induced by methylglyoxal in endothelial cells. *Molecules* **2024**, *29*, 1712. [CrossRef] [PubMed]
51. Laddha, A.P.; Kulkarni, Y.A. Tannins and vascular complications of diabetes: An update. *Phytomedicine* **2019**, *56*, 229–245. [CrossRef] [PubMed]
52. Gurung, M.; Li, Z.; You, H.; Rodrigues, R.; Jump, D.B.; Morgun, A.; Shulzhenko, N. Role of gut microbiota in type 2 diabetes pathophysiology. *EBioMedicine* **2020**, *51*, 102590. [CrossRef] [PubMed]
53. Larsen, N.; Vogensen, F.K.; van den Berg, F.W.; Nielsen, D.S.; Andreasen, A.S.; Pedersen, B.K.; Al-Soud, W.A.; Sørensen, S.J.; Hansen, L.H.; Jakobsen, M. Gut microbiota in human adults with type 2 diabetes differs from non-diabetic adults. *PLoS ONE* **2010**, *5*, e9085. [CrossRef] [PubMed]
54. Li, S.; Zhou, L.; Zhang, Q.; Yu, M.; Xiao, X. Genistein improves glucose metabolism and promotes adipose tissue browning through modulating gut microbiota in mice. *Food Funct.* **2022**, *13*, 11715–11732. [CrossRef] [PubMed]
55. Yang, R.; Jia, Q.; Mehmood, S.; Ma, S.; Liu, X. Genistein ameliorates inflammation and insulin resistance through mediation of gut microbiota composition in type 2 diabetic mice. *Eur. J. Nutr.* **2021**, *60*, 2155–2168. [CrossRef] [PubMed]
56. Senoner, T.; Dichtl, W. Oxidative stress in cardiovascular diseases: Still a therapeutic target? *Nutrients* **2019**, *11*, 2090. [CrossRef] [PubMed]
57. Xie, X.; Cong, L.; Liu, S.; Xiang, L.; Fu, X. Genistein alleviates chronic vascular inflammatory response via the miR-21/NF-κB p65 axis in lipopolysaccharide-treated mice. *Mol. Med. Rep.* **2021**, *23*, 192. [CrossRef] [PubMed]
58. Wei, T.T.; Chandy, M.; Nishiga, M.; Zhang, A.; Kumar, K.K.; Thomas, D.; Manhas, A.; Rhee, S.; Justesen, J.M.; Chen, I.Y.; et al. Cannabinoid receptor 1 antagonist genistein attenuates marijuana-induced vascular inflammation. *Cell* **2022**, *185*, 1676–1693.e1623. [CrossRef] [PubMed]
59. Liu, B.; Xu, L.; Yu, X.; Jiao, X.; Yan, J.; Li, W.; Guo, M. Genistein inhibited estradiol-induced vascular endothelial cell injury by downregulating the FAK/focal adhesion pathway. *Cell Physiol. Biochem.* **2018**, *49*, 2277–2292. [CrossRef] [PubMed]
60. Niu, L.G.; Sun, N.; Liu, K.L.; Su, Q.; Qi, J.; Fu, L.Y.; Xin, G.R.; Kang, Y.M. Genistein alleviates oxidative stress and inflammation in the hypothalamic paraventricular nucleus by activating the SIRT1/Nrf2 pathway in high salt-induced hypertension. *Cardiovasc. Toxicol.* **2022**, *22*, 898–909. [CrossRef]
61. Zhang, H.; Zhao, Z.; Pang, X.; Yang, J.; Yu, H.; Zhang, Y.; Zhou, H.; Zhao, J. MiR-34a/sirtuin-1/foxo3a is involved in genistein protecting against ox-LDL-induced oxidative damage in HUVECs. *Toxicol. Lett.* **2017**, *277*, 115–122. [CrossRef] [PubMed]
62. Haddad, Y.H.; Said, R.S.; Kamel, R.; Morsy, E.M.E.; El-Demerdash, E. Phytoestrogen genistein hinders ovarian oxidative damage and apoptotic cell death-induced by ionizing radiation: Co-operative role of ER-β, TGF-β, and FOXL-2. *Sci. Rep.* **2020**, *10*, 13551. [CrossRef] [PubMed]

63. Cheng, Y.; Tang, Y.; Tan, Y.; Li, J.; Zhang, X. KCNK9 mediates the inhibitory effects of genistein on hepatic metastasis from colon cancer. *Clinics* **2023**, *78*, 100141. [CrossRef] [PubMed]
64. Chu, X.; Tian, W.; Ning, J.; Xiao, G.; Zhou, Y.; Wang, Z.; Zhai, Z.; Tanzhu, G.; Yang, J.; Zhou, R. Cancer stem cells: Advances in knowledge and implications for cancer therapy. *Sig. Transduct Target Ther.* **2024**, *9*, 170.f. [CrossRef] [PubMed]
65. Victoir, B.; Croix, C.; Gouilleux, F.; Prié, G. Targeted therapeutic strategies for the treatment of cancer. *Cancers* **2024**, *16*, 461. [CrossRef] [PubMed]
66. Zhao, L.; Zhang, N.; Yang, D.; Yang, M.; Guo, X.; He, J.; Wu, W.; Ji, B.; Cheng, Q.; Zhou, F. Protective effects of five structurally diverse flavonoid subgroups against chronic alcohol-induced hepatic damage in a mouse model. *Nutrients* **2018**, *10*, 1754. [CrossRef] [PubMed]
67. Ding, Q.; Pi, A.; Hao, L.; Xu, T.; Zhu, Q.; Shu, L.; Yu, X.; Wang, W.; Si, C.; Li, S. Genistein protects against acetaldehyde-induced oxidative stress and hepatocyte injury in chronic alcohol-fed mice. *J. Agric. Food Chem.* **2023**, *71*, 1930–1943. [CrossRef] [PubMed]
68. El-Far, Y.M.; Khodir, A.E.; Emarah, Z.A.; Ebrahim, M.A.; Al-Gayyar, M.M.H. Chemopreventive and hepatoprotective effects of genistein via inhibition of oxidative stress and the versican/PDGF/PKC signaling pathway in experimentally induced hepatocellular carcinoma in rats by thioacetamide. *Redox Rep.* **2022**, *27*, 9–20. [CrossRef] [PubMed]
69. Chen, X.; Wu, Y.; Gu, J.; Liang, P.; Shen, M.; Xi, J.; Qin, J. Anti-invasive effect and pharmacological mechanism of genistein against colorectal cancer. *Biofactors* **2020**, *46*, 620–628. [CrossRef] [PubMed]
70. Poetsch, A.R. The genomics of oxidative DNA damage, repair, and resulting mutagenesis. *Comput. Struct. Biotechnol. J.* **2020**, *18*, 207–219. [CrossRef]
71. Salini, D.; Debanjan, T.; Debomita, S.; Sutapa, M. Restoration of radiosensitivity by soya isoflavone genistein is accomplished by facilitating DNA damage response in radioresistant cervical cancer in vitro. *J. Radiat. Cancer Res.* **2024**, *15*, 200–210. [CrossRef]
72. Sohel, M.; Biswas, P.; Al Amin, M.; Hossain, M.A.; Sultana, H.; Dey, D.; Aktar, S.; Setu, A.; Khan, M.S.; Paul, P. Genistein, a potential phytochemical against breast cancer treatment-insight into the molecular mechanisms. *Processes* **2022**, *10*, 415. [CrossRef]
73. Bhat, S.S.; Prasad, S.K.; Shivamallu, C.; Prasad, K.S.; Syed, A.; Reddy, P.; Cull, C.A.; Amachawadi, R.G. Genistein: A potent anti-breast cancer agent. *Curr. Issues Mol. Biol.* **2021**, *43*, 1502–1517. [CrossRef] [PubMed]
74. Kim, G.Y.; Suh, J.; Jang, J.H.; Kim, D.H.; Park, O.J.; Park, S.K.; Surh, Y.J. Genistein inhibits proliferation of brca1 mutated breast cancer cells: The GPR30-Akt axis as a potential target. *J Cancer Prev.* **2019**, *24*, 197–207. [CrossRef] [PubMed]
75. Lin, X.; Wu, J.F.; Wang, D.M.; Zhang, J.; Zhang, W.J.; Xue, G. The correlation and role analysis of KCNK2/4/5/15 in human papillary thyroid carcinoma microenvironment. *J. Cancer.* **2020**, *11*, 5162–5176. [CrossRef] [PubMed]
76. Garbiec, E.; Cielecka-Piontek, J.; Kowalówka, M.; Hołubiec, M.; Zalewski, P. Genistein-opportunities related to an interesting molecule of natural origin. *Molecules* **2022**, *27*, 815. [CrossRef] [PubMed]
77. Yang, Z.; Kulkarni, K.; Zhu, W.; Hu, M. Bioavailability and pharmacokinetics of genistein: Mechanistic studies on its ADME. *Anticancer Agents Med. Chem.* **2012**, *12*, 1264–1280. [CrossRef] [PubMed]
78. Braxas, H.; Rafraf, M.; Karimi Hasanabad, S.; Asghari Jafarabadi, M. Genistein supplementation improves some cardiovascular risk factors in postmenopausal women with Type 2 diabetes mellitus. *Nutr. Food Sci.* **2021**, *51*, 125–136. [CrossRef]
79. Kumar, N.B.; Pow-Sang, J.; Spiess, P.; Dickinson, S.; Schell, M.J. A phase II randomized clinical trial using aglycone isoflavones to treat patients with localized prostate cancer in the pre-surgical period prior to radical prostatectomy. *Oncotarget* **2020**, *11*, 1218–1234. [CrossRef] [PubMed]
80. Jaiswal, N.; Akhtar, J.; Singh, S.P.; Ahsan, F. An overview on genistein and its various formulations. *Drug Res.* **2019**, *69*, 305–313. [CrossRef] [PubMed]
81. Sadeghian, R.; Shahidi, S.; Komaki, A.; Habibi, P.; Ahmadiasl, N.; Yousefi, H.; Daghigh, F. Synergism effect of swimming exercise and genistein on the inflammation, oxidative stress, and VEGF expression in the retina of diabetic-ovariectomized rats. *Life Sci.* **2021**, *284*, 119931. [CrossRef] [PubMed]
82. Shah, J.; Orosz, T.; Singh, A.; Laxma, S.P.; Gross, R.E.; Smith, N.; Vroegop, S.; Sudler, S.; Porter, J.T.; Colon, M.; et al. Influence of exercise and genistein to mitigate the deleterious effects of high-fat high-sugar diet on Alzheimer's disease-related markers in male mice. *Int. J. Mol. Sci.* **2024**, *25*, 9019. [CrossRef] [PubMed]
83. Witayavanitkul, N.; Werawatganon, D.; Chayanupatkul, M.; Klaikeaw, N.; Siriviriyakul, P. Genistein and exercise treatment reduced NASH related HDAC3, IL-13 and MMP-12 expressions in ovariectomized rats fed with high fat high fructose diet. *J. Tradit. Complement. Med.* **2021**, *11*, 503–512. [CrossRef] [PubMed]
84. Zamani-Garmsiri, F.; Hashemnia, S.M.R.; Shabani, M.; Bagherieh, M.; Emamgholipour, S.; Meshkani, R. Combination of metformin and genistein alleviates non-alcoholic fatty liver disease in high-fat diet-fed mice. *J. Nutr. Biochem.* **2021**, *87*, 108505. [CrossRef] [PubMed]

85. Bezerra, P.H.A.; Amaral, C.; Almeida, C.F.; Correia-da-Silva, G.; Torqueti, M.R.; Teixeira, N. *In vitro* effects of combining genistein with aromatase inhibitors: Concerns regarding its consumption during breast cancer treatment. *Molecules* **2023**, *28*, 4893. [CrossRef] [PubMed]
86. Kwon, H.; Kim, Y.; Kim, J.H. A combination of myokines and genistein suppresses cancer stemness in MCF-7 human breast cancer cells. *Nutr. Res. Pract.* **2024**, *18*, 436–445. [CrossRef] [PubMed]
87. Khamesi, S.M.; Barough, M.S.; Zargan, J.; Shayesteh, M.; Banaee, N.; Noormohammadi, A.H.; Mousavi, M.; Alikhani, H.K. Combined anticancer effects of neutron radiation and genistein on prostate cancer cells. *J. Radiat. Res. Appl. Sci.* **2023**, *16*, 100731. [CrossRef]
88. Haleem, A.; Javaid, M.; Singh, R.P.; Shanay Rab, S.R.; Suman, R. Applications of nanotechnology in medical field: A brief review. *Glob. Health J.* **2023**, *7*, 70–77. [CrossRef]
89. Khan, H.; Ullah, H.; Martorell, M.; Valdes, S.E.; Belwal, T.; Tejada, S.; Sureda, A.; Kamal, M.A. Flavonoids nanoparticles in cancer: Treatment, prevention and clinical prospects. *Semin Cancer Biol.* **2021**, *69*, 200–211. [CrossRef] [PubMed]
90. Syahputra, R.A.; Dalimunthe, A.; Utari, Z.D.; Halim, P.; Sukarno, M.A.; Zainalabidin, S.; Salim, E. Nanotechnology and flavonoids: Current research and future perspectives on cardiovascular health. *J. Funct. Foods.* **2024**, *120*, 106355. [CrossRef]
91. Coutinho, A.J.; Pinheiro, M.; Neves, A.R.; Pinto, M.M.M. Therapeutic potential of genistein: Preclinical studies, clinical evidence, and nanotechnology application. *Curr. Med. Chem.* **2023**, *30*, 2480–2517. [CrossRef] [PubMed]
92. Obinu, A.; Burrai, G.P.; Cavalli, R.; Galleri, G.; Migheli, R.; Antuofermo, E.; Rassu, G.; Gavini, E.; Giunchedi, P. Transmucosal solid lipid nanoparticles to improve genistein absorption via intestinal lymphatic transport. *Pharmaceutics.* **2021**, *13*, 267. [CrossRef] [PubMed]
93. Vodnik, V.V.; Mojić, M.; Stamenović, U.; Otoničar, M.; Ajdžanović, V.; Maksimović-Ivanić, D.; Mijatović, S. Development of genistein-loaded gold nanoparticles and their antitumor potential against prostate cancer cell lines. *Mater. Sci. Eng. C Mater. Biol. Appl.* **2021**, *124*, 112078. [CrossRef] [PubMed]
94. Lei, H.; Dong, X.Y.; Gao, Y. Genistein in the treatment of Alzheimer's disease: A systematic review and meta-analysis of preclinical studies. *J. Agric. Food Chem.* **2024**, *72*, 13500–13512. [CrossRef]
95. Wang, J.W.; Yu, Z.L.; Peng, Y.; Xu, B.J. Insights into prevention mechanisms of bioactive components from healthy diets against Alzheimer's disease. *J. Nutr. Biochem.* **2023**, *119*, 109397. [CrossRef] [PubMed]
96. Baluchnejadmojarad, T.; Roghani, M.; Nadoushan, M.R.; Bagheri, M. Neuroprotective effect of genistein in 6-hydroxydopamine hemi-parkinsonian rat model. *Phytother. Res.* **2009**, *23*, 132–135. [CrossRef] [PubMed]
97. Kyuhou, S. Preventive effects of genistein on motor dysfunction following 6-hydroxydopamine injection in ovariectomized rats. *Neurosci. Lett.* **2008**, *448*, 10–14. [CrossRef] [PubMed]
98. Gilbert, E.R.; Liu, D. Anti-diabetic functions of soy isoflavone genistein: Mechanisms underlying its effects on pancreatic β-cell function. *Food Funct.* **2013**, *4*, 200–212. [CrossRef] [PubMed]
99. Yang, W.; Wang, S.; Li, L.; Liang, Z.; Wang, L. Genistein reduces hyperglycemia and islet cell loss in a high-dosage manner in rats with alloxan-induced pancreatic damage. *Pancreas.* **2011**, *40*, 396–402. [CrossRef] [PubMed]
100. Zhang, Y.; Li, Q.; Zhou, D.; Chen, H. Genistein, a soya isoflavone, prevents azoxymethane-induced up-regulation of WNT/β-catenin signalling and reduces colon pre-neoplasia in rats. *Br. J. Nutr.* **2013**, *109*, 33–42. [CrossRef] [PubMed]
101. Chodon, D.; Banu, S.M.; Padmavathi, R.; Sakthisekaran, D. Inhibition of cell proliferation and induction of apoptosis by genistein in experimental hepatocellular carcinoma. *Mol. Cell Biochem.* **2007**, *297*, 73–80. [CrossRef] [PubMed]
102. Spagnuolo, C.; Russo, G.L.; Orhan, I.E.; Habtemariam, S.; Daglia, M.; Sureda, A.; Nabavi, S.F. Genistein and cancer: Current status, challenges, and future directions. *Adv. Nutr.* **2015**, *6*, 408–419. [CrossRef] [PubMed]

antioxidants

Review

Oxidative Stress and Dietary Antioxidants in Head and Neck Cancer

A Jeong You [1,2], Jaehyung Park [1], Jae-Min Shin [1,2] and Tae Hoon Kim [1,2,*]

1 Department of Otorhinolaryngology-Head and Neck Surgery, College of Medicine, Korea University, Seoul 02841, Republic of Korea; ajungyi@naver.com (A.J.Y.); soonsliy131@gmail.com (J.P.); shinjm0601@korea.ac.kr (J.-M.S.)
2 Mucosal Immunology Institute, College of Medicine, Korea University, Seoul 02841, Republic of Korea
* Correspondence: doctorth@korea.ac.kr; Tel.: +82-02-920-5486

Abstract: Oxidative stress serves as both a driver and result of redox metabolism across diverse physiological and pathological states, including cancer. Head and neck squamous cell carcinoma (HNSCC), the sixth most prevalent malignancy worldwide, is no exception. HNSCC is strongly linked to modifiable external risk factors such as tobacco smoking, alcohol consumption, and high-risk human papilloma (HR-HPV) infection. These risk factors are associated with elevated oxidative stress, which contributes to carcinogenesis through DNA damage, chronic inflammation, and dysregulation of cell signaling pathways. Current treatment options for HNSCC have limitations and burden of side effects. Studies have been conducted on potent dietary antioxidants for the prevention and adjunctive treatment of HNSCC. This review aims to explore the contribution of oxidative stress to carcinogenesis in general and the three major risk factors for HNSCC. We evaluate latest evidence for nine dietary antioxidants such as vitamin C, vitamin E, carotenoids, epigallocatechin-3-gallate (EGCG), and curcumin, that have shown promise in preclinical and clinical studies. We discuss how these compounds mitigate ROS, influence cancer-related signaling pathways, and modulate tumor microenvironment. Despite encouraging findings, current clinical data remain limited and inconclusive, highlighting the need for further research on possible dietary antioxidants for HNSCC.

Keywords: oxidative stress; reactive oxygen species (ROS); antioxidants; dietary antioxidants; head and neck squamous cell carcinoma (HNSCC)

1. Introduction

Head and neck cancer (HNC) is a heterogeneous group of malignancies arising from the mucosal epithelium of the oral cavity, pharynx, and larynx. Among HNCs, head and neck squamous cell carcinoma (HNSCC) accounts for more than 90% of cases and represents a significant global health burden. HNSCC is the sixth most common cancer worldwide, with approximately 870,000 new cases and 440,000 deaths annually. The disease primarily affects individuals over the age of 50 and is more prevalent in men, with geographic variation influenced by cultural and environmental exposures. Compared to many other cancers, HNSCC has relatively well-established risk factors. Alcohol consumption, tobacco smoking, and high-risk human papilloma virus (HR-HPV) infection—particularly HPV type 16—are recognized as the primary etiological factors [1]. These factors not only increase the likelihood of developing HNSCC but also influence the molecular and clinical characteristics. While tobacco- and alcohol-associated HNSCCs are often linked to extensive genetic alterations and poorer outcomes, HPV-related HNSCC tends to occur in younger patients and is generally associated with a more favorable response to therapy [2]. Interestingly, these risk factors converge on a shared biological mechanism: the induction of oxidative stress. Oxidative stress results from a disrupted redox equilibrium between reactive oxygen/nitrogen species (ROS/RNS) and the antioxidant protection mechanism against them [3]. This imbalance, whether caused by increased ROS/RNS production or diminished antioxidant capacity, contributes to a wide range of pathological states [4], including the development and progression of HNSCC. Excessive ROS levels can promote DNA damage, chronic inflammation, and aberrant cell signaling—processes that are key drivers of carcinogenesis. Given this metastatic link, there is growing interest in the role of dietary antioxidants as potential modulators of oxidative stress and supportive agents in cancer prevention and therapy. As advanced HNSCC often carries a poor prognosis and current treatment options remain limited, further investigation into preventive and adjunctive strategies is needed—particularly in relation to oxidative stress.

Although existing clinical data on the role of dietary antioxidants in HNSCC remain limited and largely inconclusive, the biological plausibility and public health relevance of this topic warrant a critical and integrative review. By acknowledging both the promise and the current limitations of available evidence, this review seeks to provide a balanced perspective that may inform future research and clinical considerations.

In this review, we aim to investigate how oxidative stress contributes to the initiation and progression of HNSCC and to evaluate the current evidence on the potential dietary antioxidants in its prevention and treatment. To date, few reviews have comprehensively examined the intersection of oxidative stress and HNSCC through the lens of its primary external risk factors—tobacco, alcohol, and HR-HPV infection. This review offers a mechanistic perspective on how these etiological agents may promote redox imbalance and drive carcinogenesis. Moreover, unlike previous reviews that narrowly focus on individual antioxidants, our approach provides a broader yet focused synthesis, specifically tailored to HNSCC subsites classified by ICD codes. We critically evaluate nine dietary antioxidants based on both preclinical and clinical evidence, offering both mechanistic insights as well as practical relevance for future research and clinical strategies in HNSCC management.

2. Oxidative Stress in Cancer

2.1. Overview of Oxidative Stress

Oxidative stress, a term first defined by Helmut Sies in 1985, refers to "an imbalance between oxidants and antioxidants in favor of oxidants, leading to the disruption of redox signaling and control and/or molecular damage" [3]. Oxidants are an inevitable byproduct of oxidation–reduction (redox) reactions occurring in living cells, and aerobic metabolism for energy production generates diverse ROSs. Some ROSs are free radicals, which are atoms or molecules possessing one or more unpaired electrons within an atomic or molecular orbital. Other ROSs are nonradicals that have all their electrons paired [5]. The two primary endogenous sources of ROSs are mitochondria and nicotinamide adenine dinucleotide phosphate oxidases (NOXs). Mitochondria generate superoxide as a byproduct of aerobic metabolism, and NOXs generate superoxide in response to various intrinsic and extrinsic stimuli [6]. ROSs are highly reactive molecules capable of damaging nucleic acids, lipids, and proteins, leading to alterations in their functions [7]. Therefore, cells possess a defense system called the antioxidant network, which maintains ROSs at physiological levels [8]. Antioxidants can be endogenously produced within the body, acquired through diet, or artificially synthesized. Endogenous antioxidants include enzymatic antioxidants, such as superoxide dismutase (SOD), glutathione peroxidase (GPx), catalase, and peroxiredoxin (Prx). Nonenzymatic antioxidants such as reduced glutathione (GSH) and coenzyme Q10 (CoQ) are endogenous antioxidants. Dietary antioxidants include vitamins C and E, phenolic compounds, ergothionein, and carotenoids. Clinically used synthetic antioxidants include ebselen (2-phenyl-1,2-benzoisoselenazol-3(2H)-one), edaravone, and N-acetylcysteine [5]. The damage caused by ROS has been established as a contributing factor to various chronic diseases such as neurodegenerative diseases, emphysema, cardiovascular and inflammatory diseases, cataracts, and malignancy [4].

2.2. Role of Oxidative Stress in Cancer

Oxidative stress plays a critical role in driving and sustaining key cancer hallmarks, such as deoxyribonucleic acid (DNA) damage, angiogenesis, cell cycle progression, and metastatic ability. ROSs interact with DNA molecules, causing damage such as single-strand breaks, double-strand breaks, and base modifications. ROSs, such as nitric oxide, interfere with the repair mechanisms of DNA, such as direct reversal and base excision repair [9]. Uncontrolled DNA damage leads to genomic instability in healthy cells and cancer [10]. During the early stages of tumor initiation, autophagy removes damaged organelles and cells, thereby decreasing ROS levels and exerting tumor-suppressive effect [11,12]. Excessive ROS levels interfere with autophagy [13]. At this stage, deletion of autophagy-related genes (*ATG*) 5 and 7 results in autophagy inhibition, leading to the accumulation of oxidative stress, damaged tissues, and inflammation, all of which favor tumor initiation [12,14]. Mutual interactions between chronic inflammation and cancer have been clearly identified. Chronic inflammation results in a significant number of immunosuppressive cells and release of cytokines, leading to ROS generation This is associated with signaling pathways like nuclear factor kappa-light-chain-enhancer of activated B cells (NF-κB), and overexpression of ROS in turn worsens chronic inflammation. Excessive ROS activate the mitogen-activated protein kinase (MAPK) signaling pathway, which phosphorylates and activates activator protein-1 (AP-1) components such as c-Jun and c-Fos. AP-1 and oxidative stress have a mutual relationship, and dysregulated AP-1 expression contributes to various diseases, including lung cancer [15]. Researchers have found that NF-κB may contribute to tumor cell survival by increasing the expression of antiapoptotic genes, such as B-cell lymphoma2 (*BCL2*), which suppresses apoptosis. Moreover, NF-κB enhance the expression of hypoxia-inducible factor-1α (HIF-1α), contributing to cellular resistance against stress and hypoxic conditions [16]. Apoptosis dysregulation is a defining characteristic of cancer. Apoptosis in carcinogenesis is inhibited through the activation of NF-κB, which upregulates the production of antiapoptotic proteins such as BCL-XL (B-cell lymphoma XL), BFL1 (a BCL-2-related protein), and GADD45β (growth arrest and DNA-damage-inducible 45β). These proteins promote cell survival by counteracting pro-apoptotic signals. Increased ROS levels inactivate phosphatase and tension homolog (PTEN) and upregulate the phosphoinositide 3-kinase/protein kinase B (PI3K/Akt) signaling pathway. The Akt pathway inactivates proapoptotic transcription factors BCL2-associated agonist of cell death (BAD) and BCL2-associated X protein (BAX), resulting in cell proliferation. Elevated ROS generation is also associated with the protein kinase D (PKD) signaling pathway. Specifically, PKD1 enhances cell viability by downregulating the pro-apoptotic c-Jun N terminal protein kinase (JNK) pathway and upregulating the pro-survival transcription factor NF-κB. This dual regulatory action promotes the resistance of cancer cells to oxidative stress and apoptosis [13].

As described above, ROS may play a role in preventing damaged cells from undergoing apoptosis after DNA damage, thereby promoting their transformation into cancer cells. After developing into cancer cells, ROS drive cancer cell proliferation through multiple pathways. As mentioned earlier, the PI3K/Akt signaling pathway is activated by elevated ROS levels. PI3K/Akt activate downstream mammalian target of rapamycin (mTOR) and HIF-1α, promoting angiogenesis, which is important for cancer cell survival in a hypoxic environment [17]. In the advanced phases of cancer progression, cancer cells induce oxidative stress, activating transcription factors HIF-1α and NF-κB. This promotes autophagy of stromal cells in the tumor microenvironment, supporting tumor growth by supplying nutrients-rich metabolites such as lactate and ketones [12]. Increased ROS levels induce the inhibition of the cell division cycle 14B (CDC14B) phosphatase, leading to the upregulation of cyclin-dependent kinase1 (CDK1). CDK1 contribute to the transition from G2 to M phase of the cell cycle. Unchecked activity can drive the cell cycle forward even in the presence of DNA damage, thereby contributing to cancer cell progression [13,18]. ROS also promote metastasis and the epithelial-to-mesenchymal transition (EMT) process. EMT contributes to cancer evolution by enhancing tumor invasion, proliferation, and metastasis. During the EMT, epithelial cells undergo a loss of adhesion properties and gain mesenchymal traits, thereby increasing their motility and invasiveness [19]. ROSs, such as H_2O_2, elevate the expression or activity of proteins, such as matrix metalloproteinases (MMPs), vascular endothelial growth factor (VEGF), epidermal growth factor (EGF), and EGF receptor (EGFR), which are crucial in tumor metastasis [20]. In summary, oxidative stress is thought to be involved in multiple stages of cancer development, from the accumulation of DNA damage in healthy cells that may lead to malignant transformation to the subsequent proliferation and metastasis of cancer cells.

3. Oxidative Stress and Dietary Antioxidants in HNSCC

3.1. Overview of HNSCC

HNC ranks as the sixth most common malignancy globally, with more than 870,000 new diagnoses and 440,000 associated deaths reported in 2020. HNSCC constitutes approximately 90% of HNC cases [1]. HNSCC presents significant treatment challenges and requires a multidisciplinary approach. Surgery, radiotherapy, and systemic therapy constitute the cornerstone of treatment for locally advanced disease. The advent of immunotherapy in the conventional treatment of recurrent or metastatic HNSCCs has significantly revolutionized their management. However, much is yet to be understood, and HNSCC remains a highly complex disease. Patients with stage III or higher HNSCC generally have a poor prognosis. For instance, the 5-year survival rate of locoregionally advanced laryngeal cancer is approximately 40% [21], and over 60% of HNSCC patients are diagnosed at stage III or IV disease [22]. Moreover, HNSCC imposes a substantial socioeconomic burden due to the high costs associated with its diagnosis, treatment, and long-term management, as well as the loss of productivity due to high morbidity and mortality [23].

The major risk factors for HNSCC are tobacco smoking, alcohol intake, and high-risk human papillomavirus (HR-HPV) infection [1]. HNSCC, which is traditionally linked to heavy tobacco and alcohol use, is typically diagnosed in older individuals and is gradually declining worldwide owing to reduced tobacco consumption. In contrast, the incidence of HPV-associated oropharyngeal cancer, mainly caused by HPV type 16, is increasing, particularly among young populations in North America and northern Europe [22]. These risk factors associated with HNSCC are difficult to eliminate. A pooled analysis of 30 case–control studies related to drinking or smoking cessation revealed that it may take up to 20 years or more for the risk of HNSCC to reach that of never drinkers or smokers [24]. Moreover, HPV infections are irreversible. Therefore, more chemopreventive measures are needed rather than simply trying to avoid these risk factors. Changes in redox metabolism are thought to be involved in all stages of HNSCC through cancer etiology, progression, therapy, and quality of life after treatment [25]. Therefore, we reviewed the involvement of oxidative stress in the three major risk factors of HNSCC and the possible dietary antioxidants that patients can easily access. For this review, we classified HNSCCs into cancer of oral cavity, oropharynx, hypopharynx, oral cavity or pharynx not otherwise specified, or larynx or HNSCC unspecified. Cancers of the salivary glands, nasal cavity, ear, paranasal sinuses, and esophagus were not included owing to differences in epidemiology and pathology. The subsites were classified based on disease codes from the International Classification of Diseases (ICD)-10 [26].

3.2. Oxidative Stress in HNSCC

The three primary risk factors for HNSCC are tobacco smoking, alcohol consumption, and HR-HPV infection [1]. When these risk factors are combined, there are devastating results. For instance, a pooled analysis of 17 case–control studies (11,221 cases and 16,168 controls) demonstrated that the combined effect of alcohol and tobacco use was greater than a multiplicative effect on HNSCC risk [27]. HNSCC has relatively distinct external risk factors compared with other cancers. These risk factors contribute to cancer development through various mechanisms; however, they share the common feature of being associated with oxidative stress, which is crucial in carcinogenesis. Therefore, we focused on how alcohol consumption, tobacco smoking, and HR-HPV infection contribute to the development of HNSCCs related to oxidative stress.

Ethanol is metabolized to the carcinogen acetaldehyde by NADPH-dependent cytochrome P450 2E1 (CYP2E1), alcohol dehydrogenase (ADH), and catalase. Induction of CYP2E1 affects the cellular oxidative balance by generating ROS, leading to the oxidation of proteins, lipids, and DNAs [28]. Tobacco smoking is attributed to 42% of HNSCC-related deaths [25]. Tobacco smoke can be classified into two phases, namely gas and tar (particulates). The gas phase occurs during the combustion of tobacco, and consists of chemicals smaller than 0.1 μm. The tar phase contains compounds with diameters ranging from 0.1–1 μm, averaging 0.2 μm. Both phases produce a significant number of free radicals [29]. A recent preclinical study on HPV-negative HNSCC cell lines revealed the molecular mechanism by which ROS produced by alcohol and tobacco smoking contribute to the development of HNSCC. Oxidative stress induced by tobacco and alcohol induces the dimerization of transmembrane 4 L6 family member 19 (TM4SF19) in the endoplasmic reticulum. This dimerization prevents guanine and adenine-binding protein β1 (GABPβ1) from proteasomal degradation, and GABP transcription factor complex promotes Yes-associated protein 1 (YAP1) transcription. YAP, a key effector of the Hippo-YAP signaling pathway, is frequently dysregulated in various human cancers. Increased YAP activity is strongly associated with dismal prognosis and therapeutic resistance in multiple cancers, including HNSCC [30].

HPV infection and oxidative stress have been suggested to influence each other in a bidirectional manner, potentially acting as both causes and consequences depending on the context. Studies have revealed that oxidative stress affects early stages of viral infection by altering the local redox environment and facilitating viral integration by enhancing DNA damage and weakening repair mechanisms via HPV oncoproteins such as E6 and E7. Moreover, high levels of oxidative stress markers, such as ferritin, are associated with reduced clearance of HPV infection [31]. HPV induces ROS production via multiple pathways. HPV infection drives chronic inflammation through various mechanisms, leading to oxidative stress. Compared to HPV-negative HNSCC, HPV-positive HNSCC demonstrates increased levels of β-oxidation associated genes. β-oxidation is a fatty acid oxidation in mitochondria, which can generate ATP energy more efficiently than glycolysis (108 ATP molecules from oxidation of palmitoyl-CoA compared with 32 ATP molecules from oxidation of glucose). β-oxidation produces H_2O_2, which consequently induces more ROS, such as superoxide radical ($O_2^{\bullet-}$). $O_2^{\bullet-}$ diffuses into the cytosol and produces hydroxyl radicals (•OH) by Fenton and Haber-Weiss reactions. In turn, •OH trigger oxidative damage by inducing malondialdehye (MDA) and 4-hydroxynonenal (4-HNE) [32]. MDA is a marker of oxidative stress and its elevation is associated with tumor extent and unfavorable prognosis [33].

In addition, the viral proteins E6 and E7 induce nicotinamide adenine dinucleotide phosphate oxidase 2 (NOX2)-dependent ROS generation, leading to oxidative DNA damage. ROS-mediated genomic instability is more apparent in HPV-positive HNSCC than in HPV-negative HNSCC [6]. Oncogenic protein E6 also inactivates antioxidants such as SOD2 and GPx, leading to increased oxidative DNA damage [34]. E6 and E7 oncoproteins promote p53 and retinoblastoma protein (pRb) degradation, which is thought to be one of the key mechanisms by which HPV oncoproteins induce genomic instability [35]. In summary, HR-HPV infection leads to ROS generation by multiple pathways such as inducing chronic inflammation, β-oxidation, NOX2 activation, and decreasing antioxidants. ROSs, in turn, cause oxidative DNA damage, which can be detected by measuring the level of 8-oxo-7,8-dihydro-2′-deoxyguanosine (8-oxo-dG). Guanine has the lowest redox potential among the four DNA bases, making it the most prone to oxidation. Its oxidized form, 8-oxo-dG, has an even lower redox potential, rendering it highly susceptible to further oxidation. Therefore, 8-oxo-dG is commonly recognized as a biomarker of oxidative stress [9,29]. In summary, the three major risk factors for HNSCC are closely related to oxidative stress, highlighting the importance of maintaining the redox balance for the prevention and treatment of HNSCC.

However, numerous studies have revealed that components of the antioxidant defense system, such as paraoxonase-2 (PON2) and the Nuclear Factor Erythroid 2-Related Factor 2/Kelch Like ECH Associated Protein 1 (NRF2/KEAP1) signaling axis, can paradoxically support tumor progression and therapy resistance in HNSCC. PON2, a membrane-bound enzyme with antioxidant properties, is overexpressed in HNSCC tissues [36] and has been shown to promote tumor cell survival by mitigating mitochondrial ROS production and preventing apoptosis [37]. High PON2 expression was associated with decreased overall survival (Hazard Ratio = 1.53, 95% CI: 1.16–2.23, $p = 0.0025$) based on The Cancer Genome Atlas data [36], and has also been linked to resistance to both radiotherapy and chemotherapy [38]. Similarly, NRF2, a regulator of cellular redox homeostasis, is known to be constitutively activated in several cancers, including HNSCC. Under normal physiological conditions, NRF2 is tightly regulated by its cytoplasmic inhibitor KEAP1. However, dysregulation of this pathway can lead to persistent NRF2 activation, which facilitates tumor cell proliferation, metabolic rewiring, and evasion of oxidative-stress-induced apoptosis [39]. Furthermore, NRF2/KEAP1 signaling has been linked to cisplatin and radiotherapy resistance in HNSCC models [40,41]. These findings highlight that while antioxidant pathways play certain roles in protecting normal tissues, their dysregulation in cancer cells may confer selective advantages, contributing to malignant behavior and treatment resistance. Given these complex and sometimes opposing roles, the impact of oxidative stress on HNSCC remains debated and context-dependent. Some antioxidant mechanisms may suppress carcinogenesis under certain conditions, whereas others—particularly when overactivated—may paradoxically promote tumor survival and progression. Despite these discrepancies, a growing number of studies suggest that specific dietary antioxidants may provide chemopreventive or therapeutic benefit in subsets of HNSCC patients. In light of this, our review aims to provide an overview of nine dietary antioxidants that have shown mechanistic or clinical relevance, while also acknowledging the limitations and duality of antioxidant biology in cancer.

3.3. Dietary Antioxidants in HNSCC

As described above, impaired redox balance contributes to the development of HNSCC, from etiology to treatment [25]. In addition, it is very costly to treat HNSCC, and there are clear limitations to current therapies. Hence, prevention and increasing the efficiency of treatment are important. Dietary antioxidants that may have chemopreventive or adjuvant therapeutic roles in HNSCC have been investigated. Currently, no antioxidants are routinely used as therapeutic agents for HNSCC. In addition, research to date has not proven that any antioxidants are effective in preventing second primary malignancies in patients with HNSCC [42]. However, there have been many studies on the prevention of primary cancer and the alleviation of the side effects of chemotherapy and radiation therapy. One of the most referenced large-scale prospective cohort studies on diet and cancer is the NIH-AARP (National Institutes of Health–American Association of Retired Persons) Diet and Health Study. This study collected dietary and health data from over 500,000 participants through annual questionnaires, cancer registries, and national death indexes. It has provided valuable epidemiological insights, including analyses on dietary pattern in relation to HNC incidence. However, the NIH-AARP study relies heavily on self-reported dietary data, which introduces recall bias and limits the ability to establish causal relationships. Additionally, due to its study design, it cannot assess the effects of specific antioxidant compounds but instead categorizes dietary patterns broadly, such as "fruit and vegetable" [43] or "dietary fiber and grain" [44] intake. Therefore, while its findings are hypothesis-generating, they are not conclusive. Despite the limitations of large cohort studies, their findings support further investigation of individual antioxidant compounds, which we explore in detail in the following sections. We consider nine dietary antioxidants that have promising cancer preventive and/or therapeutic potential (Table 1).

Vitamin C is a potent aqueous-soluble electron donor in humans. Two pooled analysis studies by the INHANCE (International Head and Neck Cancer Epidemiology) Consortium, of supplement intake [45] and natural food intake [46], suggested an negative association between vitamin C intake and HNSCC incidence. A recent case–control study of 101 patients also reported corresponding results [47]. A large prospective cohort study in the Netherlands confirmed these results, especially for oral-cavity cancer, compared with oro/hypopharyngeal and laryngeal cancers. As daily vitamin C intake increased, the relative risk of HNSCC gradually declined. The highest vitamin C dose in the study was 144.8–153.3 mg/day, which was associated with 61% reduced risk for HNSCC, compared to 55.2–63.5 mg intake per day [48]. However, none of these studies offered a specific tolerable dose of vitamin C, and no statistically clear dose–response relationship was observed [45]. Moreover, most studies rely on observational or case–control designs with self-reported dietary data, introducing potential bias and limiting causal inference. For instance, one study reported that high citrus fruit intake was linked to an elevated risk of HPV16-associated HNSCC. In addition, the risk was greatly reduced in subjects with low citrus fruit exposure and polymorphisms in the *SLC23A2* allele, which encodes the sodium-dependent vitamin C transporter (SVCT2). This report suggested that high dietary consumption of vitamin C may exhibit paradoxical impacts on HNSCC in certain situations and genetic backgrounds [49]. Moreover, a meta-analysis conducted by Patini et al., following PRISMA guidelines, found no statistically significant association between vitamin C intake and the risk of oral cavity carcinoma [50]. Therefore, although some studies suggest that vitamin C may play a preventive effect against HNSCC, further research is needed to determine its optimal use, accounting for individual variability and study design limitations.

Vitamin E is a fat-soluble antioxidant including four tocopherols and four tocotrienols: α-, β-, γ-, and δ-tocopherol and α-, β-, γ-, and δ-tocotrienol. In a pooled analysis of 10 case–control studies in the INHANCE consortium, vitamin E intake showed an inverse association with HNSCC subtypes [51]. However, in a double-blind, placebo-controlled randomized controlled trial (RCT) involving 540 patients with HNSCC, patients receiving long-term supplementation with 400 IU of vitamin E per day had higher all-cause mortality [52] and a higher rate of recurrence or second primary malignancy [53]. These results may be due to the paradoxical pro-oxidant effect at high doses [54]; therefore, long-term use of high doses requires caution. Vitamin E also has positive effects on patients undergoing radiation therapy. α-Tocopherol failed to alleviate severe radiation-induced side effects when administered as a supplement [55]. However, when combined with pentoxifylline, which blocks the production of the inflammatory marker TGF-β1, vitamin E reduced oral mucositis (OM) and dysphagia [56]. Moreover, when combined with vitamin C, it also reduced radiotherapy-induced xerostomia [57] Besides oral administration, when patients were assigned to rinse the oral cavity with a vitamin E-containing oil solution, it reduced radiation-induced OM [58]. Hence, vitamin E can be an effective measure for attenuating the side effects induced by radiotherapy, with appropriate usage and in combination with other nutrients. α-Tocopherol is also known to reduce the toxicity of vitamin A analogs. Vitamin A analogs have been actively studied over the last few decades. However, their use is limited owing to the short duration of response, resistance, and toxicities such as cheilitis, dry skin, and conjunctivitis [59,60]. One vitamin A analog is 13-cis-retinoic acid (13-cRA), which has shown promising results in HNSCC treatment when combined with α-tocopherol and interferon-α (IFN-α). In a phase-II bioadjuvant trial of patients with locally advanced HNSCC, 86% of patients successfully completed the 12-month treatment with a 13-cRA, IFN-α, and α-tocopherol combination, and it was generally well tolerated with promising results [61]. In the long-term follow-up, the combination demonstrated significant efficacy in preventing both recurrence and development of second primary tumors. The 5-year overall survival was 81.3% (95% confidence interval (CI), 63.7–90%), which was significantly higher than the historical 5-year overall rate for advanced HNSCC (approximately 40%) [62].

Carotenoids (CTDs) are a broad group of tetraterpenoids that are abundant in yellow to red fruits and vegetables. In human plasma, α-carotene, β-carotene, lycopene, lutein, zeaxanthin, and β-cryptoxanthin constitute more than 95% of the total CTDs. Nutrients like α-, β-, and γ-carotene and β-zeacarotene is easily converted into vitamin A, while nonprovitamin A CTDs include lycopene, lutein, and zeaxanthine. CTDs effectively prevent ROS formation because of the long polyene chain of 8–13 conjugated double-bond structures [63]. In addition to their antioxidant properties, many in vitro studies have revealed chemopreventive properties of CTDs through their action in key intracellular tumor signaling processes, such as the ERK-MAPK and PI3K/AKT/mTOR pathways [64]. An important study with a considerable sample size on the relationship between CTDs and HNSCC is a pooled analysis of 10 case–control studies of 18,207 subjects in the INHANCE consortium by Leoncini et al. The authors showed a 39% risk reduction of oropharyngeal and laryngeal cancer in the highest quintile of CTDs intake compared to the lowest quintile of CTDs intake [65]. A meta-analysis of 15 case–control studies and one prospective cohort study found that intake of β-carotene equivalents was associated with a risk reduction of 46% (95% CI 20–63%) for oral cancer, and 57% (95% CI 23–76%) for cancer of the larynx. Lycopene and β-cryptoxanthin also reduced the risk of laryngeal cancer. Specifically, subjects in the highest category of carotenoid consumption experienced 50% (95% CI 11–72%) risk reduction for laryngeal cancer with lycopene and 59% (95% CI 49–67%) reduction with β-cryptoxanthin. Additionally, lycopene, α-carotene, and β-cryptoxanthin were linked to at least a 26% reduction (95% CI 2–4%) in the risk of oropharyngeal cancers, emphasizing their protective role in HNSCC prevention [66]. Edefonti et al. conducted a pooled analysis of five case–control studies of 2452 cases, and identified three dietary patterns. The pattern referred to as 'antioxidant vitamins and fiber' showed the highest levels for vitamin C, total carotene, and lutein. This dietary pattern showed an inverse association with oropharyngeal cancer (odds ratio (OR) = 0.57, 95% CI 0.43–0.76) when comparing the highest to the lowest score quintile [67]. De Vito et al. conducted a multi-study factor analysis of five studies analyzed by Edefonti et al. The authors reported corresponding results with a high intake of antioxidant vitamins, with the greatest loading on vitamin C and total carotene [68]. Although not specifically targeting CTDs, numerous studies with large sample sizes have also proven that large consumption of fruits and vegetables, which are rich in CTDs, is associated with lower risk for HNSCC [69–71]. Beyond cancer prevention, carotenoids have also been found to improve treatment outcomes in patients with HNSCC. An RCT of 540 patients demonstrated that carotenoids reduced the risk of local recurrence and alleviated the severe adverse effects of radiation therapy, indicating their dual roles in prevention and therapeutic support [55]. However, results across studies are inconsistent. For instance, a large cohort study in the Netherlands showed no notable association between intake of α-carotene, β-carotene, lutein plus zeaxanthin, or lycophene with overall risk of HNSCC [48]. This may be accounted for by many factors, such as heterogeneity in dietary sources, variations in study design, varying degrees of exposure to smoking and alcohol use, and differences in HNSCC subtypes.

Green tea is effective in various cancer types. Its efficacy is primarily associated with its polyphenol content, which includes epicatechin, epigallocatechin, epicatechin-3-gallate, and epigallocatechin-3-gallate (EGCG). Of these, EGCG is the predominant and has been studied for its chemopreventive role [72]. Polyphenols are also potent antioxidants and radical scavengers. The role of green tea in HNSCC prevention was supported by a case–control study of 147 patients in Iran, which used a standardized questionnaire. Compared to individuals who never consumed green tea, the risk of developing oral cancer for those using green tea was significantly reduced. Individuals who consumed less than a cup of green tea daily had an OR of 0.29 (0.16–0.52), while those who consumed one or more cups of green tea daily had an OR of 0.38 (0.17–0.86) [73]. Another case–control study involving 396 patients with HNSCC in Taiwan reported similar results. There was a 6% reduction in HNSCC risk in subjects who had a cup of green tea daily (OR = 0.94, 95% Cl, 0.90–0.98), while oolong and black tea consumption had no statistical association with HNSCC risk [74]. One mechanism underlying the anticancer activity of EGCG is the targeting of tyrosine kinase receptors, such as EGFR and VEGFR [75]. In a mouse xenograft model of HNSCC, EGCG augmented the cell-growth inhibition by the EGFR-tyrosine kinase inhibitor, erlotinib [76]. A phase Ib study of green tea polyphenon E and erlotinib in advanced premalignant lesions (APLs) of the oral cavity and larynx showed similar results. This combination resulted in pathologic responses in 17/19 patients, with 66.3% cancer-free survival and 93% overall survival [77]. When not administered with erlotinib, green tea extract showed positive results in patients with high-risk oral premalignant lesions. High doses of green tea extract resulted in clinical responses in 58.8% (750 and 1000 mg/m^2) and 36.4% (500 mg/m^2) of patients, whereas the placebo group had a clinical response of 18.2% [78]. In patients diagnosed with oral cancer, mouth rinsing with green tea improved oral health status [79]. In a xenograft model of HNSCC stem cells, EGCG inhibited the expression of stem cell markers by inhibiting the Notch pathway and augmented cisplatin-mediated chemosensitivity, demonstrating that EGCG is a promising dietary antioxidant with both chemopreventive and therapeutic potential [80]. However, these studies are based on small sample sizes or preclinical models; thus, their translation into clinical outcomes requires further validation in larger human trials.

Curcumin is a natural compound found in the rhizomes of *Curcuma longa* L. (turmeric) plants. Numerous preclinical studies have revealed the therapeutic potential of curcumin in the treatment of HNSCC. Curcumin exerts broad anti-HNSCC effects by targeting multiple cancer-driving pathways such as NF-κB, JAK/STAT, and EGFR. Curcumin modulates the tumor microenvironment not only by promoting cell cycle arrest, apoptosis, and cytotoxicity, but also targeting cancer-associated fibroblasts, immune responses, and lymphovascular niches [81]. These mechanisms contribute to its observed antiproliferative and antimetastatic activity and support its development as an adjunct to conventional therapies in HNSCC [82,83]. Despite its safety being recognized by the Food and Drug Administration (FDA), the clinical use of curcumin is hindered by its low bioavailability. Researchers have attempted to overcome this problem by developing curcumin analogs and novel drug delivery systems [81]. A clinical trial involving 15 patients demonstrated enhanced bioavailability of curcumin following the transmucosal administration of microgranular curcumin. Higher serum levels of curcumin decreased the serum levels of fibroblast growth factor-2, granulocyte macrophage colony-stimulating factor, and interleukin (IL)-17 in patients with HNSCC [84]. APG-157 is a botanical drug composed of multiple polyphenols, including curcumin. Its oral delivery showed systemic absorption, and decreased Bacteroidetes species and inflammatory markers such as IL-1β, IL-6, and IL-8 in the saliva of subjects with oral cancer. Immune T cells also increased in tumor tissue, highlighting the therapeutic potential of curcumin by improving its bioavailability [85]. Kim et al. also conducted a pilot study of 39 subjects (21 HNSCC patients, 13 patients with dental caries, and 5 healthy controls) and demonstrated that curcumin treatment significantly reduced the activity of IκB kinase β (IKKβ), which is known to promote cancer progression through activation of NF-κB pathway [86]. There has been little clinical research on curcumin in patients with HNSCC, but existing research has shown promising results. Kuriakose et al. conducted a phase IIb double-blind, placebo-controlled RCT with 223 patients with oral leukoplakia. Patients either had 3.6 g curcumin/day or a placebo for 6 months. The clinical response was better in the curcumin group (67.5%, 95% CI 58.4–75.6%) than that in the control group (55.3%, 95% CI 46.1–64.2%). The combined clinical and histologic response was also better in curcumin group (hazard ratio 0.50, 95% CI 0.27–0.92), without any safety concern [87]. Zhang et al. conducted a meta-analysis of six RCTs on the preventive and therapeutic effects of curcumin on treatment-related OM in patients with HNSCC. Curcumin did not lower the incidence of OM but lowered the incidence of severe OM and decreased its mean severity. It also considerably reduced weight loss [88].

Quercetin is a plant flavonol belonging to the flavonoid polyphenol group. Quercetin has been actively studied, especially regarding its antitumor effect in targeting the apoptotic pathway in oral squamous cell carinoma (OSCC) cell lines. A recent in vitro study has shown that quercetin induces ferroptosis by inactivating the mTOR/S6KP70 cascade and inhibiting cell growth in OSCC [89]. Quercetin induces G1 cell cycle arrest and apoptosis in OSCC cells by activating the p38 MAPK signaling pathway regardless of *TP53* mutation status [90]. Moreover, apoptosis is also induced by JNK-activation-regulated ERK1/2 and GSK3-α/β-mediated mitochondria-dependent apoptosis signaling pathway in tongue squamous cell carcinoma [91]. In addition to the apoptotic pathway, quercetin targets other pathways associated with cancer cell proliferation. In studies using the CAL-27 OSCC cell line, quercetin inhibited glycolysis and proliferation of the cells by inactivating the G3BP1/YWHAZ axis [92] and suppressed cell invasion by activating the miR-1254/CD36 signaling pathway [93]. Another in vitro study demonstrated that quercetin selectively induced cell cycle arrest in OSCC cell lines but had no effect on human keratinocytes. Moreover, quercetin suppressed cancer cell metastasis via an EMT-mediated pathway [94]. Although promising laboratory results have demonstrated the therapeutic potential of quercetin, there are no clinical trials to date; hence, its relevance to patient outcomes remains speculative.

Isothiocyanates (ITCs) are produced by enzymatic conversion of metabolites called glucosinolates. Glucosinolates are rich in cruciferous vegetables, such as broccoli, cabbage, cauliflower, and kale. Phenethyl isothiocyanate (PEITC), allyl isothiocyanate (AITC), benzyl isothiocyanate (BITC), and sulforaphane (SFN) have been studied for their chemopreventive role [72,95]. A blinded, randomized, placebo-controlled trial analyzed the safety and efficacy of PEITC, in the form of nutri-PEITC jelly, in oral or oropharyngeal cancer. The group administered 20 mg of PEITC (in 200 g of nutrient jelly) for 3 months showed improved health-related quality of life, stable disease, and longer progression-free survival than the control group ($p < 0.001$). Severe adverse events were not noticed [96]. Several in vivo studies have revealed underlying molecular mechanisms. In a mouse xenograft model of oral cancer, PEITC increased the levels of the oxidative DNA damage markers 8-oxo-dG and p53. The researchers also revealed that PEITC induced ROS formation and cell cycle arrest in the G1/S phase in vitro, suggesting that PEITC triggers ROS-mediated cell cycle arrest [97]. Another in vivo study found that PEITC induced apoptosis by upregulating tumor necrosis factor-related apoptosis-inducing ligand (TRAIL), death receptor (DR) 4, and DR5 [98]. Multiple studies have highlighted the potential of AITC and BITC in HNSCC treatment through the regulation of apoptosis, cell cycle arrest, cell migration, and chemotherapy sensitivity [99]. For example, AITC reduced cell viability by activating transient receptor potential ankyrin1 (TRPA1) receptor in human OSCC [100]. Moreover, in human CAL-27 cisplatin-resistant oral cancer cells, AITC suppressed the Akt/mTOR proliferative signaling pathway and induced apoptosis via caspase-9 and caspase-3 [101]. An in vitro study of multiple HNSCC cell lines indicated that BITC inhibited the expression of the EMT marker vimentin, decreased cell migration, and increased cisplatin toxicity [102]. Lan et al. demonstrated that SFN activated the NRF2 pathway, thereby inhibiting oxidative stress-associated carcinogenesis in oral cancer cells [103]. In an HNSCC human cell line, treatment with broccoli extract, which is rich in SFN, increased the cytotoxicity of cisplatin two-fold, and that of 5-fluorouracil ten-fold, while the combinations had little effect on noncancerous cells [104]. In a pilot study of 10 healthy subjects, SFN-rich broccoli sprout extracts showed high bioavailability [105], supporting the potential of SFN as a safe and efficient therapeutic compound for HNSCC.

Resveratrol is a phytoalexin with antioxidant, antimicrobial, and antiinflammatory activities and is abundant in grapes, berries, and peanuts [106]. Its chemopreventive and therapeutic potential in HNSCC is mediated by diverse molecular pathways, as demonstrated by several preclinical studies. A recent meta-analysis of five in vivo studies of oral cancer concluded that resveratrol has the potential to suppress cancer cell proliferation and drive neoplastic apoptosis. A statistically significant reduction was demonstrated in neoplastic parameters, with an overall effect size of 0.85 (95% CI 0.74–0.98) [107]. Fukuda et al. showed that resveratrol induced selective autophagy in human OSCCs by blocking sterol regulatory element binding protein 1 (*SREBP1*) gene expression and suppressing lipid metabolism [108]. One preclinical study found that resveratrol targeted tumor-initiating stem-like and EMT properties. The expression of stemness gene signatures (Oct4, Nanog, and Nestin) and EMT markers (Slug, ZEB1, N-cadherin, and vimentin) decreased [109]. Furthermore, Tyagi et al. demonstrated that resveratrol selectively induced DNA damage, cell cycle inhibition, and apoptosis, independent of Smad4 status. Loss or alterations in Smad4 signaling leads to genomic instability in HNSCC, broadening its applicability in diverse HNSCC cases [110]. Resveratrol has also demonstrated the ability to enhance the efficacy of standard cancer therapies. Resveratrol enhanced cisplatin sensitivity in HNSCC cell lines, which was linked to increased *MYC* and *TP53* expression and decreased *BCL-2* expression. It also reduced cisplatin toxicity in normal adherent cells [111]. Similarly, another study reported enhanced cisplatin and radiation sensitivity in HNSCC xenograft nude mouse models through upregulation of regenerating gene III (*REGIII*) expression, further emphasizing its potential as an adjunct therapy [112]. Resveratrol has also been studied in combination with other dietary antioxidants. When administered with curcumin, resveratrol had a synergistic antitumor effect, compared to when each was administered separately. The combination increased PARP cleavage, Bax/Bcl-2 ratio, cytoplasmic NF-κB accumulation, and autophagy vacuoles. This combination treatment inhibited ERK1 and ERK2 phosphorylation [113]. Moreover, combination treatment with EGCG synergistically induced apoptosis in HNSCC cell lines via caspase-3 and PARP cleavage in vitro. In addition, the inhibition of the Akt/mTOR signaling pathway was observed both in vitro and in vivo in HNSCC-xenografted nude mice. Xenografted tumor volume and weight were significantly reduced [114]. Further clinical studies are necessary to validate the efficacy of resveratrol and to establish optimal dosing strategies for its integration into standard HNSCC care.

Luteolin is a flavonoid that has been extensively studied in various types of cancer treatments. Several studies have found the antitumor effect of luteolin relates to integrin β1. Integrin β1 is a key molecule involved in cancer progression, angiogenesis, invasion, and therapeutic resistance. Integrin β1 enhances VEGF signaling, which is essential for angiogenesis [115], and interacts with CD147 [116] and PLOD2 [117] in laryngeal cancer proliferation. Integrin β1 was also revealed to regulate perineural invasion and radio resistance of OSCC [118]. A recent preclinical study has demonstrated that luteolin enhanced tumor suppression and angiogenesis of laryngeal cancer cells during radiotherapy by downregulating integrin β1 [115]. Luteolin-enhanced radiation sensitivity has also been observed in an in vitro study of oral cancer stem cells. In this study, luteolin was found to inhibit IL-6/STAT3 signaling, which is associated with cancer stem-like properties and progression [119]. An in vivo study using an HNSCC xenograft mouse model showed that luteolin treatment reduced tumor growth. Luteolin treatment disrupted gene expression and microRNA profiles, thereby promoting tumor suppression [120]. Another in vivo study used a water-soluble polymer-encapsulated nano-luteolin to overcome low bioavailability and systemic delivery. Nano-luteolin had a lower half maximal inhibitory concentration than luteolin and had a greater inhibitory effect on tumor growth [121]. These findings demonstrate the promise of luteolin as a protective anticancer agent. However, most findings are based on preclinical models; clinical evidence is scarce and novel approaches to improve its bioavailability in clinical settings are important. Therefore, further clinical studies are warranted.

Table 1. Possible preventive or therapeutic dietary antioxidants in HNSCC.

Dietary Antioxidants	Diet Source	Study Design	Mechanism or Effect	Authors (Years)
Vitamin C (ascorbate)	Citrus fruits, broccoli, bell peppers, strawberries	Prospective cohort study (120,852 subjects)	Lower the risk of HNSCC subtypes	Munter et al. (2015) [48]
		Pooled analysis of 12 case–control studies (7002 cases)	Lower the risk of HNSCC subtypes	Li et al. (2012) [45]
		Pooled analysis of 10 case–control studies (5959 cases)	Lower the risk of HNSCC subtypes	Edefonti et al. (2015) [46]
		Case–control study (101 patients)	Lower the risk of HNSCC subtypes	Saka-Herrán et al. (2023) [47]
Vitamin E (tocopherols and tocotrienols)	Peanuts, sunflower seeds, almonds, olive oil, spinach	Pooled analysis of 10 case–control studies (5959 cases)	Lower the risk of HNSCC subtypes	Edefonti et al. (2015) [51]
Vitamin E (tocopherols and tocotrienols)	Peanuts, sunflower seeds, almonds, olive oil, spinach	Randomized controlled trial (60 patients)	Reduce the severity and duration of radiation-induced oral mucositis and dysphagia when given with pentoxifylline	Sayed et al. (2019) [56]
		Double-blinded randomized placebo-controlled trial (45 patients)	Protective effect against radiotherapy-induced xerostomia when given with vitamin C	Chung et al. (2016) [57]
		Double-blinded randomized controlled trial (54 patients)	Decrease the incidence of radiation-induced oral mucositis	Ferreira et al. (2004) [58]
Carotenoids	Fruits and vegetables, mainly yellow to red (pumpkin, carrot, spinach, tomato, tangerine)	Pooled analysis of 10 case–control studies (5959 cases)	Lower the risk of HNSCC subtypes	Leoncini et al. (2016) [65]
		Meta-analysis of 15 case–control studies and one prospective cohort study	Lower the risk of HNSCC subtypes	Leoncini et al. (2015) [66]
		Pooled analysis of five case–control studies (2452 cases)	Lower the risk of HNSCC subtypes	Edefonti et al. (2012) [67]
		Randomized controlled trial (540 patients)	Decrease local recurrence and severe adverse effects of radiation therapy	Meyer et al. (2007) [55]

Table 1. *Cont.*

Dietary Antioxidants	Diet Source	Study Design	Mechanism or Effect	Authors (Years)
Epigallocatechin-3-gallate	Green tea	Case–control study (147 patients)	Reduce the risk of HNSCC	Rafieian et al. (2019) [73]
		Case–control study (396 patients)	Reduce the risk of HNSCC	Huang et al. (2014) [74]
		Phase Ib clinical trial (19 patients)	Response in advanced premalignant lesions when given with erlotinib	Shin et al. (2020) [77]
		Phase II randomized, placebo-controlled trial (39 patients)	Response in oral premalignant lesions	Tsao et al. (2009) [78]
		Single-blind, randomized, controlled trial (61 patients)	Mouthwash improves oral health in patients with oral cancer	Liao et al. (2021) [79]
		In vivo study of HNSCC stem cell xenograft mouse model	Suppress stem cell markers by inhibiting Notch pathway and augment cisplatin sensitivity	Lee et al. (2013) [80]
Curcumin	Turmeric	Meta-analysis of six randomized controlled trials (266 patients)	Prevent and ameliorate therapy-induced oral mucositis and weight loss	Zhang et al. (2021) [88]
		Randomized double blind placebo-controlled phase IIb (223 patients)	Clinical and histological response in oral leukoplakia	Kuriakose et al. (2016) [87]
		Randomized double-blind placebo-controlled phase I trial (12 patients)	Decrease inflammatory markers and *Bacteroides* species in saliva and increase immune T cells	Basak et al. (2020) [85]
		Pilot clinical trial (15 patients)	Decrease factors involved in angiogenesis and cell invasion, such as FGF-2, GM-CSF, and IL-7	Latimer et al. (2015) [84]
		Pilot clinical trial (39 patients)	Suppress tumor progression by reducing IKKβ activity	Kim et al. (2011) [122]
		Systematic review of 30 in vitro and in vivo studies of HNSCC cell lines	Induce cytotoxicity, apoptosis, cell cycle arrest	Borges et al. (2017) [82]

Table 1. *Cont.*

Dietary Antioxidants	Diet Source	Study Design	Mechanism or Effect	Authors (Years)
Quercetin	Onion, kale, caper	In vitro study of OSCC line	Induces ferroptosis by inactivating mTOR/S6KP70 pathway	Zhu et al. (2024) [89]
		In vitro study of OSCC line	Induce cell cycle arrest and apoptosis by activating p38 pathway	Son et al. (2023) [90]
		In vitro study of tongue SCC cell line	Induce apoptosis via the JNK activation-regulated ERK/GSK-3α/β-mediated mitochondria-dependent apoptotic signaling pathway	Huang et al. (2022) [91]
		In vitro study of OSCC line	Inhibit glycolysis and cell proliferation by inhibiting G3BP1/YWHAZ axis	Hu et al. (2023) [92]
		In vitro study of OSCC line	Inhibit cell survival and invasion via miR-1254/CD36 cascade	Chen et al. (2021) [93]
		In vitro study of OSCC line	Inhibit cell survival and metastasis by inhibiting TGF-β1 inducing EMT	Kim et al. (2020) [94]
Isothiocyanates	Cruciferous vegetables	Randomized blinded placebo-controlled trial (72 patients)	Stabilize disease, improve QoL and PFS in patients with oral and oropharyngeal cancer	Lam-Ubol et al. (2023) [96]
		In vitro study of OSCC line and in vivo study of OSCC xenograft mice model	Induce ROS-mediated cell cycle arrest	Lam-Ubol et al. (2018) [97]
		In vivo study of OSCC xenograft mice model	Induced apoptosis by enhancing TRAIL and upregulating DR4 and DR5	Yeh et al. (2015) [98]
		In vitro study of OSCC line	Reduce cell viability by activating TRPA1 receptor	Kiss et al. (2022) [100]
		In vitro study of OSCC line	Induce apoptosis by inhibiting Akt/mTOR pathway and enhancing caspase-3 and caspase-9 in cisplatin-resistant OSCC	Chang et al. (2021) [101]
		In vitro study of HNSCC cell line	Inhibit cell migration and increase cisplatin sensitivity	Wolf et al. (2014) [102]
		In vitro study of human keratinocytes and in vivo study of mouse tongue	Inhibit oxidative stress-associated oral carcinogenesis by activating NRF2 pathway	Lan et al. (2016) [103]
		In vitro study of HNSCC cell line	Increase cisplatin and 5-FU cytotoxicity	Elkashty et al. (2018) [104]

Table 1. *Cont.*

Dietary Antioxidants	Diet Source	Study Design	Mechanism or Effect	Authors (Years)
Resveratrol	Grapes, red wine, berries, peanuts	Systematic review and meta-analysis of five in vivo studies of oral cancer cells	Suppress tumor growth and induce apoptosis by activating various pathways	Alam et al. (2024) [107]
		In vitro study of OSCC cell line and in vivo study of xenograft mouse model	Induce autophagy by blocking *SREBP1* expression	Fukuda et al. (2022) [108]
		In vitro study of HNSCC line	Increase cisplatin sensitivity by inducing apoptosis and *TP53*, *BCL-2*, and *MYC* modulation	Bostan et al. (2021) [111]
		In vivo study of HNSCC xenograft mouse model	Increase cisplatin and radiation sensitivity by enhancing REGIII expression	Mikami et al. (2019) [112]
Resveratrol	Grapes, red wine, berries, peanuts	In vitro study of HNSCC cell line and in vivo study of xenograft mouse model	Reduce tumor initiating stem-like and EMT properties	Hu et al. (2012) [109]
		In vitro study of HNSCC cell line and in vivo study of xenograft mouse model	Induce selective DNA damage, cell cycle arrest, and apoptosis independent of Smad4 status	Tyagi et al. (2011) [110]
Luteolin	Chamomile tea, celery, parsley	In vitro study of laryngeal cancer cells and in vivo study of xenograft mouse model	Enhance radiation sensitivity by downregulating integrin β1	Li et al. (2023) [115]
		In vitro study of oral cancer stem cells	Enhance radiation sensitivity and inactivate IL-6/STAT3 signaling	Tu et al. (2016) [119]
		In vivo study of HNSCC xenograft mouse model	Reduce tumor growth by inhibiting histone acetylation	Selvi et al. (2015) [120]
		In vitro and in vivo study of HNSCC cell line	Inhibit tumor growth	Majumdar et al. (2014) [121]

Abbreviations: HNSCC, head and neck squamous cell carcinoma; FGF-2, fibroblast growth factor-2; GM-CSF, granulocyte-macrophage colony stimulating factor; IL, interleukin; IKKβ, IκB kinase β; OSCC, oral squamous cell carcinoma; mTOR, mammalian target of rapamycin; JNK, c-Jun N-terminal kinase pathway; ERK, extracellular signal-regulated kinase; GSK, glycogen synthase kinase; G3BP1, ras-GTPase-activating protein SH3 domain-binding protein 1; YWHAZ, tyrosine 3-monooxygenase/tryptophan 5-monooxygenase activation protein zeta (14-3-3ζ); TGF-β1, transforming growth factor-β1; QoL, quality of life; PFS, progression-free survival; TRAIL, tumor necrosis factor-related apoptosis-inducing ligand; DR, death receptor; TRPA1, transient receptor potential ankyrin 1; Akt, protein kinase B; NRF2, nuclear factor erythroid 2-related factor 2; 5-FU, 5-fluorouracil; *SREBP1*, sterol regulatory element-binding protein1; *TP53*, tumor protein p53; *BCL-2*, B-cell leukemia/lymphoma 2 protein; *REGIII*, regenerating islet-derived III; EMT, epithelial-mesenchymal transition; DNA, deoxyribonucleic acid; Stat-3, signal transducer and activator of transcription-3.

4. Conclusions

Dietary antioxidants have been proposed as potential modulators of oxidative stress and may offer chemopreventive or therapeutic benefits by neutralizing reactive oxygen species (ROS), influencing redox-sensitive signaling pathways, and modulating the tumor microenvironment.

In this review, we discussed nine dietary antioxidants—some of which, including vitamins C and E, carotenoids, epigallocatechin-3-gallate, and curcumin, have shown possible benefits in preclinical and early clinical studies. Others, such as quercetin, isothiocyanates, resveratrol, and luteolin, remain primarily supported by experimental evidence. However, despite this growing body of research, the current clinical application of dietary antioxidants in HNSCC remains limited. Notably, many studies in this field rely on observational designs, small sample sizes, or self-reported dietary data, often lacking standardized protocols or long-term follow-up. While some large cohort studies and meta-analyses have suggested potential associations between antioxidant intake and reduced cancer risk, these findings are often undermined by methodological limitations and inconsistencies. As such, the relationship between dietary antioxidants and clinical outcomes in HNSCC should be regarded as plausible but not yet definitive. Moreover, challenges, such as inconsistent clinical trial outcomes, variations in bioavailability, and the need for tailored dosing regimens, must be addressed to fully realize their therapeutic potential.

It is also important to acknowledge that the role of antioxidants in cancer remains complex and, at times, contradictory, as certain antioxidant mechanisms may exhibit both tumor-suppressive and tumor-promoting effects depending on the context. Given these uncertainties, our review does not intend to assert conclusive therapeutic recommendations but rather to provide a mechanistic and evidence-based overview of dietary antioxidants that have demonstrated potential in modulating oxidative stress. Future research should prioritize well-designed randomized clinical trials, mechanistic studies in human models, and strategies to enhance the bioavailability and delivery of promising compounds. By integrating these efforts into a multidisciplinary approach, dietary antioxidants could potentially offer safe, accessible, and effective strategies for the prevention and management of HNSCC.

Author Contributions: Conceptualization, A.J.Y. and T.H.K.; methodology, A.J.Y.; software, J.P.; validation, A.J.Y., J.P. and J.-M.S.; formal analysis, J.P. and J.-M.S.; investigation, A.J.Y. and T.H.K.; resources, J.P.; data curation, A.J.Y. and J.P.; writing—original draft preparation, A.J.Y.; writing—review and editing, A.J.Y. and T.H.K.; visualization, A.J.Y. and T.H.K.; supervision, J.-M.S. and T.H.K.; project administration, T.H.K.; funding acquisition, T.H.K. All authors have read and agreed to the published version of the manuscript.

Funding: This work was supported by the Basic Science Research Program of the National Research Foundation of Korea and funded by the Ministry of Science and Technology (RS-2024-00441029 and RS-2025-00513676), ICT Creative Consilience Program through the Institute of Information & Communications Technology Planning & Evaluation (IITP) grant funded by the Korea government (MSIT) (IITP-2025-RS-2020-II201819, 25%), Korea Health Technology R&D Project through the Korea Health Industry Development Institute (KHIDI), funded by the Ministry of Health & Welfare, Republic of Korea (RS-2022-KH129266). Additionally, this research was supported by a grant from the Korea University College of Medicine and Anam Hospital in Seoul, Republic of Korea.

Conflicts of Interest: The authors declare no conflicts of interest.

References

1. Li, Q.; Tie, Y.; Alu, A.; Ma, X.; Shi, H. Targeted therapy for head and neck cancer: Signaling pathways and clinical studies. *Signal Transduct. Target Ther.* **2023**, *8*, 31. [CrossRef] [PubMed]
2. Barsouk, A.; Aluru, J.S.; Rawla, P.; Saginala, K.; Barsouk, A. Epidemiology, Risk Factors, and Prevention of Head and Neck Squamous Cell Carcinoma. *Med. Sci.* **2023**, *11*, 42. [CrossRef] [PubMed]
3. Sies, H. Oxidative stress: A concept in redox biology and medicine. *Redox Biol.* **2015**, *4*, 180–183. [CrossRef] [PubMed]
4. Pisoschi, A.M.; Pop, A. The role of antioxidants in the chemistry of oxidative stress: A review. *Eur. J. Med. Chem.* **2015**, *97*, 55–74. [CrossRef]

5. Halliwell, B. Understanding mechanisms of antioxidant action in health and disease. *Nat. Rev. Mol. Cell Biol.* **2024**, *25*, 13–33. [CrossRef]
6. Marullo, R.; Werner, E.; Zhang, H.; Chen, G.Z.; Shin, D.M.; Doetsch, P.W. HPV16 E6 and E7 proteins induce a chronic oxidative stress response via NOX2 that causes genomic instability and increased susceptibility to DNA damage in head and neck cancer cells. *Carcinogenesis* **2015**, *36*, 1397–1406. [CrossRef]
7. Iqbal, M.J.; Kabeer, A.; Abbas, Z.; Siddiqui, H.A.; Calina, D.; Sharifi-Rad, J.; Cho, W.C. Interplay of oxidative stress, cellular communication and signaling pathways in cancer. *Cell Commun. Signal.* **2024**, *22*, 7. [CrossRef]
8. Cadenas, E.; Sies, H. Lester Packer: On His Life and His Legacy. *Antioxid. Redox Signal.* **2023**, *38*, 768–774. [CrossRef]
9. Kay, J.; Thadhani, E.; Samson, L.; Engelward, B. Inflammation-induced DNA damage, mutations and cancer. *DNA Repair.* **2019**, *83*, 102673. [CrossRef]
10. Rezatabar, S.; Karimian, A.; Rameshknia, V.; Parsian, H.; Majidinia, M.; Kopi, T.A.; Bishayee, A.; Sadeghinia, A.; Yousefi, M.; Monirialamdari, M.; et al. RAS/MAPK signaling functions in oxidative stress, DNA damage response and cancer progression. *J. Cell Physiol.* **2019**, *234*, 14951–14965. [CrossRef]
11. Yun, H.R.; Jo, Y.H.; Kim, J.; Shin, Y.; Kim, S.S.; Choi, T.G. Roles of Autophagy in Oxidative Stress. *Int. J. Mol. Sci.* **2020**, *21*, 3289. [CrossRef] [PubMed]
12. Poillet-Perez, L.; Despouy, G.; Delage-Mourroux, R.; Boyer-Guittaut, M. Interplay between ROS and autophagy in cancer cells, from tumor initiation to cancer therapy. *Redox Biol.* **2015**, *4*, 184–192. [CrossRef] [PubMed]
13. Arfin, S.; Jha, N.K.; Jha, S.K.; Kesari, K.K.; Ruokolainen, J.; Roychoudhury, S.; Rathi, B.; Kumar, D. Oxidative Stress in Cancer Cell Metabolism. *Antioxidants* **2021**, *10*, 642. [CrossRef]
14. Karantza-Wadsworth, V.; Patel, S.; Kravchuk, O.; Chen, G.; Mathew, R.; Jin, S.; White, E. Autophagy mitigates metabolic stress and genome damage in mammary tumorigenesis. *Genes Dev.* **2007**, *21*, 1621–1635. [CrossRef]
15. Ashique, S.; Mishra, N.; Mantry, S.; Garg, A.; Kumar, N.; Gupta, M.; Kar, S.K.; Islam, A.; Mohanto, S.; Subramaniyan, V. Crosstalk between ROS-inflammatory gene expression axis in the progression of lung disorders. *Naunyn-Schmiedeberg's Arch. Pharmacol.* **2024**, *398*, 417–448. [CrossRef]
16. Zhao, H.; Wu, L.; Yan, G.; Chen, Y.; Zhou, M.; Wu, Y.; Li, Y. Inflammation and tumor progression: Signaling pathways and targeted intervention. *Signal Transduct. Target Ther.* **2021**, *6*, 263. [CrossRef]
17. Korbecki, J.; Simińska, D.; Gąssowska-Dobrowolska, M.; Listos, J.; Gutowska, I.; Chlubek, D.; Baranowska-Bosiacka, I. Chronic and Cycling Hypoxia: Drivers of Cancer Chronic Inflammation through HIF-1 and NF-κB Activation: A Review of the Molecular Mechanisms. *Int. J. Mol. Sci.* **2021**, *22*, 701. [CrossRef]
18. Manzano-López, J.; Monje-Casas, F. The Multiple Roles of the Cdc14 Phosphatase in Cell Cycle Control. *Int. J. Mol. Sci.* **2020**, *21*, 709. [CrossRef]
19. Huang, Z.; Zhou, L.; Duan, J.; Qin, S.; Jiang, J.; Chen, H.; Wang, K.; Liu, R.; Yuan, M.; Tang, X.; et al. Oxidative Stress Promotes Liver Cancer Metastasis via RNF25-Mediated E-Cadherin Protein Degradation. *Adv. Sci.* **2024**, *11*, e2306929. [CrossRef]
20. Belkhiri, A.; Richards, C.; Whaley, M.; McQueen, S.A.; Orr, F.W. Increased expression of activated matrix metalloproteinase-2 by human endothelial cells after sublethal H2O2 exposure. *Lab. Investig.* **1997**, *77*, 533–539.
21. Mody, M.D.; Rocco, J.W.; Yom, S.S.; Haddad, R.I.; Saba, N.F. Head and neck cancer. *Lancet* **2021**, *398*, 2289–2299. [CrossRef] [PubMed]
22. Chow, L.Q.M. Head and Neck Cancer. *N. Engl. J. Med.* **2020**, *382*, 60–72. [CrossRef] [PubMed]
23. Silfverschiöld, M.; Jarl, J.; Hafström, A.; Greiff, L.; Sjövall, J. Cost of Illness of Head and Neck Cancer in Sweden. *Value Health* **2024**, *27*, 425–432. [CrossRef]
24. Marron, M.; Boffetta, P.; Zhang, Z.F.; Zaridze, D.; Wünsch-Filho, V.; Winn, D.M.; Wei, Q.; Talamini, R.; Szeszenia-Dabrowska, N.; Sturgis, E.M.; et al. Cessation of alcohol drinking, tobacco smoking and the reversal of head and neck cancer risk. *Int. J. Epidemiol.* **2010**, *39*, 182–196. [CrossRef]
25. Chen, X.; Mims, J.; Huang, X.; Singh, N.; Motea, E.; Planchon, S.M.; Beg, M.; Tsang, A.W.; Porosnicu, M.; Kemp, M.L.; et al. Modulators of Redox Metabolism in Head and Neck Cancer. *Antioxid. Redox Signal.* **2018**, *29*, 1660–1690. [CrossRef]
26. Hashibe, M.; Brennan, P.; Benhamou, S.; Castellsague, X.; Chen, C.; Curado, M.P.; Dal Maso, L.; Daudt, A.W.; Fabianova, E.; Fernandez, L.; et al. Alcohol drinking in never users of tobacco, cigarette smoking in never drinkers, and the risk of head and neck cancer: Pooled analysis in the International Head and Neck Cancer Epidemiology Consortium. *J. Natl. Cancer Inst.* **2007**, *99*, 777–789. [CrossRef]
27. Hashibe, M.; Brennan, P.; Chuang, S.C.; Boccia, S.; Castellsague, X.; Chen, C.; Curado, M.P.; Dal Maso, L.; Daudt, A.W.; Fabianova, E.; et al. Interaction between tobacco and alcohol use and the risk of head and neck cancer: Pooled analysis in the International Head and Neck Cancer Epidemiology Consortium. *Cancer Epidemiol. Biomark. Prev.* **2009**, *18*, 541–550. [CrossRef]

28. Ferraguti, G.; Terracina, S.; Petrella, C.; Greco, A.; Minni, A.; Lucarelli, M.; Agostinelli, E.; Ralli, M.; de Vincentiis, M.; Raponi, G.; et al. Alcohol and Head and Neck Cancer: Updates on the Role of Oxidative Stress, Genetic, Epigenetics, Oral Microbiota, Antioxidants, and Alkylating Agents. *Antioxidants* **2022**, *11*, 145. [CrossRef]
29. Caliri, A.W.; Tommasi, S.; Besaratinia, A. Relationships among smoking, oxidative stress, inflammation, macromolecular damage, and cancer. *Mutat. Res. Rev. Mutat. Res.* **2021**, *787*, 108365. [CrossRef]
30. Shin, E.; Kwon, Y.; Jung, E.; Kim, Y.J.; Kim, C.; Hong, S.; Kim, J. TM4SF19 controls GABP-dependent YAP transcription in head and neck cancer under oxidative stress conditions. *Proc. Natl. Acad. Sci. USA* **2024**, *121*, e2314346121. [CrossRef]
31. De Marco, F. Oxidative stress and HPV carcinogenesis. *Viruses* **2013**, *5*, 708–731. [CrossRef] [PubMed]
32. Cruz-Gregorio, A.; Aranda-Rivera, A.K.; Ortega-Lozano, A.J.; Pedraza-Chaverri, J.; Mendoza-Hoffmann, F. Lipid metabolism and oxidative stress in HPV-related cancers. *Free Radic. Biol. Med.* **2021**, *172*, 226–236. [CrossRef] [PubMed]
33. Salzman, R.; Pácal, L.; Tomandl, J.; Kanková, K.; Tóthová, E.; Gál, B.; Kostrica, R.; Salzman, P. Elevated malondialdehyde correlates with the extent of primary tumor and predicts poor prognosis of oropharyngeal cancer. *Anticancer Res.* **2009**, *29*, 4227–4231. [PubMed]
34. Williams, V.M.; Filippova, M.; Filippov, V.; Payne, K.J.; Duerksen-Hughes, P. Human papillomavirus type 16 E6* induces oxidative stress and DNA damage. *J. Virol.* **2014**, *88*, 6751–6761. [CrossRef]
35. Hoppe-Seyler, K.; Bossler, F.; Braun, J.A.; Herrmann, A.L.; Hoppe-Seyler, F. The HPV E6/E7 Oncogenes: Key Factors for Viral Carcinogenesis and Therapeutic Targets. *Trends Microbiol.* **2018**, *26*, 158–168. [CrossRef]
36. Campagna, R.; Pozzi, V.; Salvucci, A.; Togni, L.; Mascitti, M.; Sartini, D.; Salvolini, E.; Santarelli, A.; Lo Muzio, L.; Emanuelli, M. Paraoxonase-2 expression in oral squamous cell carcinoma. *Hum. Cell* **2023**, *36*, 1211–1213. [CrossRef]
37. Krüger, M.; Pabst, A.M.; Al-Nawas, B.; Horke, S.; Moergel, M. Paraoxonase-2 (PON2) protects oral squamous cell cancer cells against irradiation-induced apoptosis. *J. Cancer Res. Clin. Oncol.* **2015**, *141*, 1757–1766. [CrossRef]
38. Campagna, R.; Belloni, A.; Pozzi, V.; Salvucci, A.; Notarstefano, V.; Togni, L.; Mascitti, M.; Sartini, D.; Giorgini, E.; Salvolini, E.; et al. Role Played by Paraoxonase-2 Enzyme in Cell Viability, Proliferation and Sensitivity to Chemotherapy of Oral Squamous Cell Carcinoma Cell Lines. *Int. J. Mol. Sci.* **2022**, *24*, 338. [CrossRef]
39. Fantone, S.; Marzioni, D.; Tossetta, G. NRF2/KEAP1 signaling inhibitors in gynecologic cancers. *Expert Rev. Anticancer Ther.* **2024**, *24*, 1191–1194. [CrossRef]
40. Osman, A.A.; Arslan, E.; Bartels, M.; Michikawa, C.; Lindemann, A.; Tomczak, K.; Yu, W.; Sandulache, V.; Ma, W.; Shen, L.; et al. Dysregulation and Epigenetic Reprogramming of NRF2 Signaling Axis Promote Acquisition of Cisplatin Resistance and Metastasis in Head and Neck Squamous Cell Carcinoma. *Clin. Cancer Res.* **2023**, *29*, 1344–1359. [CrossRef]
41. Guan, L.; Nambiar, D.K.; Cao, H.; Viswanathan, V.; Kwok, S.; Hui, A.B.; Hou, Y.; Hildebrand, R.; von Eyben, R.; Holmes, B.J.; et al. NFE2L2 Mutations Enhance Radioresistance in Head and Neck Cancer by Modulating Intratumoral Myeloid Cells. *Cancer Res.* **2023**, *83*, 861–874. [CrossRef] [PubMed]
42. Meliante, P.G.; Petrella, C.; Fiore, M.; Minni, A.; Barbato, C. Antioxidant Use after Diagnosis of Head and Neck Squamous Cell Carcinoma (HNSCC): A Systematic Review of Application during Radiotherapy and in Second Primary Cancer Prevention. *Antioxidants* **2023**, *12*, 1753. [CrossRef] [PubMed]
43. Freedman, N.D.; Park, Y.; Subar, A.F.; Hollenbeck, A.R.; Leitzmann, M.F.; Schatzkin, A.; Abnet, C.C. Fruit and vegetable intake and head and neck cancer risk in a large United States prospective cohort study. *Int. J. Cancer* **2008**, *122*, 2330–2336. [CrossRef] [PubMed]
44. Lam, T.K.; Cross, A.J.; Freedman, N.; Park, Y.; Hollenbeck, A.R.; Schatzkin, A.; Abnet, C. Dietary fiber and grain consumption in relation to head and neck cancer in the NIH-AARP Diet and Health Study. *Cancer Causes Control* **2011**, *22*, 1405–1414. [CrossRef]
45. Li, Q.; Chuang, S.C.; Eluf-Neto, J.; Menezes, A.; Matos, E.; Koifman, S.; Wünsch-Filho, V.; Fernandez, L.; Daudt, A.W.; Curado, M.P.; et al. Vitamin or mineral supplement intake and the risk of head and neck cancer: Pooled analysis in the INHANCE consortium. *Int. J. Cancer* **2012**, *131*, 1686–1699. [CrossRef]
46. Edefonti, V.; Hashibe, M.; Parpinel, M.; Turati, F.; Serraino, D.; Matsuo, K.; Olshan, A.F.; Zevallos, J.P.; Winn, D.M.; Moysich, K.; et al. Natural vitamin C intake and the risk of head and neck cancer: A pooled analysis in the International Head and Neck Cancer Epidemiology Consortium. *Int. J. Cancer* **2015**, *137*, 448–462. [CrossRef]
47. Saka-Herrán, C.; Pereira-Riveros, T.; Jané-Salas, E.; López-López, J. Association between the Mediterranean Diet and Vitamin C and the Risk of Head and Neck Cancer. *Nutrients* **2023**, *15*, 2846. [CrossRef]
48. de Munter, L.; Maasland, D.H.; van den Brandt, P.A.; Kremer, B.; Schouten, L.J. Vitamin and carotenoid intake and risk of head-neck cancer subtypes in the Netherlands Cohort Study. *Am. J. Clin. Nutr.* **2015**, *102*, 420–432. [CrossRef]
49. Chen, A.A.; Marsit, C.J.; Christensen, B.C.; Houseman, E.A.; McClean, M.D.; Smith, J.F.; Bryan, J.T.; Posner, M.R.; Nelson, H.H.; Kelsey, K.T. Genetic variation in the vitamin C transporter, SLC23A2, modifies the risk of HPV16-associated head and neck cancer. *Carcinogenesis* **2009**, *30*, 977–981. [CrossRef]

50. Patini, R.; Favetti Giaquinto, E.; Gioco, G.; Castagnola, R.; Perrotti, V.; Rupe, C.; Di Gennaro, L.; Nocca, G.; Lajolo, C. Malnutrition as a Risk Factor in the Development of Oral Cancer: A Systematic Literature Review and Meta-Analyses. *Nutrients* **2024**, *16*, 360. [CrossRef]
51. Edefonti, V.; Hashibe, M.; Parpinel, M.; Ferraroni, M.; Turati, F.; Serraino, D.; Matsuo, K.; Olshan, A.F.; Zevallos, J.P.; Winn, D.M.; et al. Vitamin E intake from natural sources and head and neck cancer risk: A pooled analysis in the International Head and Neck Cancer Epidemiology consortium. *Br. J. Cancer* **2015**, *113*, 182–192. [CrossRef]
52. Bairati, I.; Meyer, F.; Jobin, E.; Gélinas, M.; Fortin, A.; Nabid, A.; Brochet, F.; Têtu, B. Antioxidant vitamins supplementation and mortality: A randomized trial in head and neck cancer patients. *Int. J. Cancer* **2006**, *119*, 2221–2224. [CrossRef] [PubMed]
53. Bairati, I.; Meyer, F.; Gélinas, M.; Fortin, A.; Nabid, A.; Brochet, F.; Mercier, J.P.; Têtu, B.; Harel, F.; Mâsse, B.; et al. A randomized trial of antioxidant vitamins to prevent second primary cancers in head and neck cancer patients. *J. Natl. Cancer Inst.* **2005**, *97*, 481–488. [CrossRef] [PubMed]
54. Roberts, H.J. Perspective on vitamin E as therapy. *Jama* **1981**, *246*, 129–131. [CrossRef] [PubMed]
55. Meyer, F.; Bairati, I.; Jobin, E.; Gélinas, M.; Fortin, A.; Nabid, A.; Têtu, B. Acute adverse effects of radiation therapy and local recurrence in relation to dietary and plasma beta carotene and alpha tocopherol in head and neck cancer patients. *Nutr. Cancer* **2007**, *59*, 29–35. [CrossRef]
56. Sayed, R.; El Wakeel, L.; Saad, A.S.; Kelany, M.; El-Hamamsy, M. Pentoxifylline and vitamin E reduce the severity of radiotherapy-induced oral mucositis and dysphagia in head and neck cancer patients: A randomized, controlled study. *Med. Oncol.* **2019**, *37*, 8. [CrossRef]
57. Chung, M.K.; Kim, D.H.; Ahn, Y.C.; Choi, J.Y.; Kim, E.H.; Son, Y.I. Randomized Trial of Vitamin C/E Complex for Prevention of Radiation-Induced Xerostomia in Patients with Head and Neck Cancer. *Otolaryngol. Head Neck Surg.* **2016**, *155*, 423–430. [CrossRef]
58. Ferreira, P.R.; Fleck, J.F.; Diehl, A.; Barletta, D.; Braga-Filho, A.; Barletta, A.; Ilha, L. Protective effect of alpha-tocopherol in head and neck cancer radiation-induced mucositis: A double-blind randomized trial. *Head Neck* **2004**, *26*, 313–321. [CrossRef]
59. Smith, W.; Saba, N. Retinoids as chemoprevention for head and neck cancer: Where do we go from here? *Crit. Rev. Oncol. Hematol.* **2005**, *55*, 143–152. [CrossRef]
60. Brown, K.S.; Kane, M.A. Chemoprevention of squamous cell carcinoma of the oral cavity. *Otolaryngol. Clin. N. Am.* **2006**, *39*, 349–363. [CrossRef]
61. Shin, D.M.; Khuri, F.R.; Murphy, B.; Garden, A.S.; Clayman, G.; Francisco, M.; Liu, D.; Glisson, B.S.; Ginsberg, L.; Papadimitrakopoulou, V.; et al. Combined interferon-alfa, 13-cis-retinoic acid, and alpha-tocopherol in locally advanced head and neck squamous cell carcinoma: Novel bioadjuvant phase II trial. *J. Clin. Oncol.* **2001**, *19*, 3010–3017. [CrossRef] [PubMed]
62. Seixas-Silva, J.A., Jr.; Richards, T.; Khuri, F.R.; Wieand, H.S.; Kim, E.; Murphy, B.; Francisco, M.; Hong, W.K.; Shin, D.M. Phase 2 bioadjuvant study of interferon alfa-2a, isotretinoin, and vitamin E in locally advanced squamous cell carcinoma of the head and neck: Long-term follow-up. *Arch. Otolaryngol. Head Neck Surg.* **2005**, *131*, 304–307. [CrossRef] [PubMed]
63. Starska-Kowarska, K. Dietary Carotenoids in Head and Neck Cancer-Molecular and Clinical Implications. *Nutrients* **2022**, *14*, 531. [CrossRef]
64. Varghese, R.; Efferth, T.; Ramamoorthy, S. Carotenoids for lung cancer chemoprevention and chemotherapy: Promises and controversies. *Phytomedicine* **2023**, *116*, 154850. [CrossRef]
65. Leoncini, E.; Edefonti, V.; Hashibe, M.; Parpinel, M.; Cadoni, G.; Ferraroni, M.; Serraino, D.; Matsuo, K.; Olshan, A.F.; Zevallos, J.P.; et al. Carotenoid intake and head and neck cancer: A pooled analysis in the International Head and Neck Cancer Epidemiology Consortium. *Eur. J. Epidemiol.* **2016**, *31*, 369–383. [CrossRef]
66. Leoncini, E.; Nedovic, D.; Panic, N.; Pastorino, R.; Edefonti, V.; Boccia, S. Carotenoid Intake from Natural Sources and Head and Neck Cancer: A Systematic Review and Meta-analysis of Epidemiological Studies. *Cancer Epidemiol. Biomark. Prev.* **2015**, *24*, 1003–1011. [CrossRef]
67. Edefonti, V.; Hashibe, M.; Ambrogi, F.; Parpinel, M.; Bravi, F.; Talamini, R.; Levi, F.; Yu, G.; Morgenstern, H.; Kelsey, K.; et al. Nutrient-based dietary patterns and the risk of head and neck cancer: A pooled analysis in the International Head and Neck Cancer Epidemiology consortium. *Ann. Oncol.* **2012**, *23*, 1869–1880. [CrossRef]
68. De Vito, R.; Lee, Y.C.A.; Parpinel, M.; Serraino, D.; Olshan, A.F.; Zevallos, J.P.; Levi, F.; Zhang, Z.F.; Morgenstern, H.; Garavello, W.; et al. Shared and Study-specific Dietary Patterns and Head and Neck Cancer Risk in an International Consortium. *Epidemiology* **2019**, *30*, 93–102. [CrossRef]
69. Chuang, S.C.; Jenab, M.; Heck, J.E.; Bosetti, C.; Talamini, R.; Matsuo, K.; Castellsague, X.; Franceschi, S.; Herrero, R.; Winn, D.M.; et al. Diet and the risk of head and neck cancer: A pooled analysis in the INHANCE consortium. *Cancer Causes Control* **2012**, *23*, 69–88. [CrossRef]

70. Galvão De Podestá, O.P.; Peres, S.V.; Salaroli, L.B.; Cattafesta, M.; De Podestá, J.R.V.; von Zeidler, S.L.V.; de Oliveira, J.C.; Kowalski, L.P.; Ikeda, M.K.; Brennan, P.; et al. Consumption of minimally processed foods as protective factors in the genesis of squamous cell carcinoma of the head and neck in Brazil. *PLoS ONE* **2019**, *14*, e0220067. [CrossRef]
71. Chang, C.C.; Lee, W.T.; Lee, Y.C.; Huang, C.C.; Ou, C.Y.; Lin, Y.H.; Huang, J.S.; Wong, T.Y.; Chen, K.C.; Hsiao, J.R.; et al. Investigating the association between diet and risk of head and neck cancer in Taiwan. *Oncotarget* **2017**, *8*, 98865–98875. [CrossRef] [PubMed]
72. Crooker, K.; Aliani, R.; Ananth, M.; Arnold, L.; Anant, S.; Thomas, S.M. A Review of Promising Natural Chemopreventive Agents for Head and Neck Cancer. *Cancer Prev. Res.* **2018**, *11*, 441–450. [CrossRef] [PubMed]
73. Rafieian, N.; Azimi, S.; Manifar, S.; Julideh, H.; ShirKhoda, M. Is there any association between green tea consumption and the risk of head and neck squamous cell carcinoma: Finding from a case-control study. *Arch. Oral. Biol.* **2019**, *98*, 280–284. [CrossRef]
74. Huang, C.C.; Lee, W.T.; Tsai, S.T.; Ou, C.Y.; Lo, H.I.; Wong, T.Y.; Fang, S.Y.; Chen, K.C.; Huang, J.S.; Wu, J.L.; et al. Tea consumption and risk of head and neck cancer. *PLoS ONE* **2014**, *9*, e96507. [CrossRef]
75. Kim, J.W.; Amin, A.R.; Shin, D.M. Chemoprevention of head and neck cancer with green tea polyphenols. *Cancer Prev. Res.* **2010**, *3*, 900–909. [CrossRef]
76. Zhang, X.; Zhang, H.; Tighiouart, M.; Lee, J.E.; Shin, H.J.; Khuri, F.R.; Yang, C.S.; Chen, Z.; Shin, D.M. Synergistic inhibition of head and neck tumor growth by green tea (-)-epigallocatechin-3-gallate and EGFR tyrosine kinase inhibitor. *Int. J. Cancer* **2008**, *123*, 1005–1014. [CrossRef]
77. Shin, D.M.; Nannapaneni, S.; Patel, M.R.; Shi, Q.; Liu, Y.; Chen, Z.; Chen, A.Y.; El-Deiry, M.W.; Beitler, J.J.; Steuer, C.E.; et al. Phase Ib Study of Chemoprevention with Green Tea Polyphenon E and Erlotinib in Patients with Advanced Premalignant Lesions (APL) of the Head and Neck. *Clin. Cancer Res.* **2020**, *26*, 5860–5868. [CrossRef]
78. Tsao, A.S.; Liu, D.; Martin, J.; Tang, X.M.; Lee, J.J.; El-Naggar, A.K.; Wistuba, I.; Culotta, K.S.; Mao, L.; Gillenwater, A.; et al. Phase II randomized, placebo-controlled trial of green tea extract in patients with high-risk oral premalignant lesions. *Cancer Prev. Res.* **2009**, *2*, 931–941. [CrossRef]
79. Liao, Y.C.; Hsu, L.F.; Hsieh, L.Y.; Luo, Y.Y. Effectiveness of green tea mouthwash for improving oral health status in oral cancer patients: A single-blind randomized controlled trial. *Int. J. Nurs. Stud.* **2021**, *121*, 103985. [CrossRef] [PubMed]
80. Lee, S.H.; Nam, H.J.; Kang, H.J.; Kwon, H.W.; Lim, Y.C. Epigallocatechin-3-gallate attenuates head and neck cancer stem cell traits through suppression of Notch pathway. *Eur. J. Cancer* **2013**, *49*, 3210–3218. [CrossRef]
81. Zhao, C.; Zhou, X.; Cao, Z.; Ye, L.; Cao, Y.; Pan, J. Curcumin and analogues against head and neck cancer: From drug delivery to molecular mechanisms. *Phytomedicine* **2023**, *119*, 154986. [CrossRef] [PubMed]
82. Borges, G.; Rêgo, D.F.; Assad, D.X.; Coletta, R.D.; De Luca Canto, G.; Guerra, E.N. In vivo and in vitro effects of curcumin on head and neck carcinoma: A systematic review. *J. Oral. Pathol. Med.* **2017**, *46*, 3–20. [CrossRef] [PubMed]
83. Veselá, K.; Kejík, Z.; Masařík, M.; Babula, P.; Dytrych, P.; Martásek, P.; Jakubek, M. Curcumin: A Potential Weapon in the Prevention and Treatment of Head and Neck Cancer. *ACS Pharmacol. Transl. Sci.* **2024**, *7*, 3394–3418. [CrossRef] [PubMed]
84. Latimer, B.; Ekshyyan, O.; Nathan, N.; Moore-Medlin, T.; Rong, X.; Ma, X.; Khandelwal, A.; Christy, H.T.; Abreo, F.; McClure, G.; et al. Enhanced Systemic Bioavailability of Curcumin Through Transmucosal Administration of a Novel Microgranular Formulation. *Anticancer Res.* **2015**, *35*, 6411–6418.
85. Basak, S.K.; Bera, A.; Yoon, A.J.; Morselli, M.; Jeong, C.; Tosevska, A.; Dong, T.S.; Eklund, M.; Russ, E.; Nasser, H.; et al. A randomized, phase 1, placebo-controlled trial of APG-157 in oral cancer demonstrates systemic absorption and an inhibitory effect on cytokines and tumor-associated microbes. *Cancer* **2020**, *126*, 1668–1682. [CrossRef]
86. Schiavoni, V.; Emanuelli, M.; Sartini, D.; Salvolini, E.; Pozzi, V.; Campagna, R. Curcumin and its Analogues in Oral Squamous Cell Carcinoma: State-of-the-art and Therapeutic Potential. *Anticancer Agents Med. Chem.* **2025**, *25*, 313–329. [CrossRef]
87. Kuriakose, M.A.; Ramdas, K.; Dey, B.; Iyer, S.; Rajan, G.; Elango, K.K.; Suresh, A.; Ravindran, D.; Kumar, R.R.; Ramachandran, S.; et al. A Randomized Double-Blind Placebo-Controlled Phase IIB Trial of Curcumin in Oral Leukoplakia. *Cancer Prev. Res.* **2016**, *9*, 683–691. [CrossRef]
88. Zhang, L.; Tang, G.; Wei, Z. Prophylactic and Therapeutic Effects of Curcumin on Treatment-Induced Oral Mucositis in Patients with Head and Neck Cancer: A Meta-Analysis of Randomized Controlled Trials. *Nutr. Cancer* **2021**, *73*, 740–749. [CrossRef]
89. Zhu, Y.W.; Liu, C.L.; Li, X.M.; Shang, Y. Quercetin induces ferroptosis by inactivating mTOR/S6KP70 pathway in oral squamous cell carcinoma. *Toxicol. Mech. Methods* **2024**, *34*, 669–675. [CrossRef]
90. Son, H.K.; Kim, D. Quercetin Induces Cell Cycle Arrest and Apoptosis in YD10B and YD38 Oral Squamous Cell Carcinoma Cells. *Asian Pac. J. Cancer Prev.* **2023**, *24*, 283–289. [CrossRef]
91. Huang, C.F.; Liu, S.H.; Ho, T.J.; Lee, K.I.; Fang, K.M.; Lo, W.C.; Liu, J.M.; Wu, C.C.; Su, C.C. Quercetin induces tongue squamous cell carcinoma cell apoptosis via the JNK activation-regulated ERK/GSK-3α/β-mediated mitochondria-dependent apoptotic signaling pathway. *Oncol. Lett.* **2022**, *23*, 78. [CrossRef] [PubMed]

92. Hu, M.; Song, H.Y.; Chen, L. Quercetin acts via the G3BP1/YWHAZ axis to inhibit glycolysis and proliferation in oral squamous cell carcinoma. *Toxicol. Mech. Methods* **2023**, *33*, 141–150. [CrossRef] [PubMed]
93. Chen, L.; Xia, J.S.; Wu, J.H.; Chen, Y.G.; Qiu, C.J. Quercetin suppresses cell survival and invasion in oral squamous cell carcinoma via the miR-1254/CD36 cascade in vitro. *Hum. Exp. Toxicol.* **2021**, *40*, 1413–1421. [CrossRef]
94. Kim, S.R.; Lee, E.Y.; Kim, D.J.; Kim, H.J.; Park, H.R. Quercetin Inhibits Cell Survival and Metastatic Ability via the EMT-mediated Pathway in Oral Squamous Cell Carcinoma. *Molecules* **2020**, *25*, 757. [CrossRef] [PubMed]
95. Hoch, C.C.; Shoykhet, M.; Weiser, T.; Griesbaum, L.; Petry, J.; Hachani, K.; Multhoff, G.; Bashiri Dezfouli, A.; Wollenberg, B. Isothiocyanates in medicine: A comprehensive review on phenylethyl-, allyl-, and benzyl-isothiocyanates. *Pharmacol. Res.* **2024**, *201*, 107107. [CrossRef]
96. Lam-Ubol, A.; Sukhaboon, J.; Rasio, W.; Tupwongse, P.; Tangshewinsirikul, T.; Trachootham, D. Nutri-PEITC Jelly Significantly Improves Progression-Free Survival and Quality of Life in Patients with Advanced Oral and Oropharyngeal Cancer: A Blinded Randomized Placebo-Controlled Trial. *Int. J. Mol. Sci.* **2023**, *24*, 7824. [CrossRef]
97. Lam-Ubol, A.; Fitzgerald, A.L.; Ritdej, A.; Phonyiam, T.; Zhang, H.; Myers, J.N.; Huang, P.; Trachootham, D. Sensory acceptable equivalent doses of β-phenylethyl isothiocyanate (PEITC) induce cell cycle arrest and retard the growth of p53 mutated oral cancer in vitro and in vivo. *Food Funct.* **2018**, *9*, 3640–3656. [CrossRef]
98. Yeh, C.C.; Ko, H.H.; Hsieh, Y.P.; Wu, K.J.; Kuo, M.Y.; Deng, Y.T. Phenethyl isothiocyanate enhances TRAIL-induced apoptosis in oral cancer cells and xenografts. *Clin. Oral. Investig.* **2016**, *20*, 2343–2352. [CrossRef]
99. Rekha, K.; Venkidasamy, B.; Govindasamy, R.; Neralla, M.; Thiruvengadam, M. Isothiocyanates (AITC & BITC) bioactive molecules: Therapeutic potential for oral cancer. *Oral. Oncol.* **2022**, *133*, 106060. [CrossRef]
100. Kiss, F.; Kormos, V.; Szőke, É.; Kecskés, A.; Tóth, N.; Steib, A.; Szállási, Á.; Scheich, B.; Gaszner, B.; Kun, J.; et al. Functional Transient Receptor Potential Ankyrin 1 and Vanilloid 1 Ion Channels Are Overexpressed in Human Oral Squamous Cell Carcinoma. *Int. J. Mol. Sci.* **2022**, *23*, 1921. [CrossRef]
101. Chang, P.Y.; Tsai, F.J.; Bau, D.T.; Hsu, Y.M.; Yang, J.S.; Tu, M.G.; Chiang, S.L. Potential effects of allyl isothiocyanate on inhibiting cellular proliferation and inducing apoptotic pathway in human cisplatin-resistant oral cancer cells. *J. Formos. Med. Assoc.* **2021**, *120*, 515–523. [CrossRef] [PubMed]
102. Wolf, M.A.; Claudio, P.P. Benzyl isothiocyanate inhibits HNSCC cell migration and invasion, and sensitizes HNSCC cells to cisplatin. *Nutr. Cancer* **2014**, *66*, 285–294. [CrossRef] [PubMed]
103. Lan, A.; Li, W.; Liu, Y.; Xiong, Z.; Zhang, X.; Zhou, S.; Palko, O.; Chen, H.; Kapita, M.; Prigge, J.R.; et al. Chemoprevention of oxidative stress-associated oral carcinogenesis by sulforaphane depends on NRF2 and the isothiocyanate moiety. *Oncotarget* **2016**, *7*, 53502–53514. [CrossRef] [PubMed]
104. Elkashty, O.A.; Ashry, R.; Elghanam, G.A.; Pham, H.M.; Su, X.; Stegen, C.; Tran, S.D. Broccoli extract improves chemotherapeutic drug efficacy against head-neck squamous cell carcinomas. *Med. Oncol.* **2018**, *35*, 124. [CrossRef]
105. Bauman, J.E.; Zang, Y.; Sen, M.; Li, C.; Wang, L.; Egner, P.A.; Fahey, J.W.; Normolle, D.P.; Grandis, J.R.; Kensler, T.W.; et al. Prevention of Carcinogen-Induced Oral Cancer by Sulforaphane. *Cancer Prev. Res.* **2016**, *9*, 547–557. [CrossRef]
106. Rauf, A.; Imran, M.; Butt, M.S.; Nadeem, M.; Peters, D.G.; Mubarak, M.S. Resveratrol as an anti-cancer agent: A review. *Crit. Rev. Food Sci. Nutr.* **2018**, *58*, 1428–1447. [CrossRef]
107. Alam, M.K.; Alqhtani, N.R.; Alnufaiy, B.; Alqahtani, A.S.; Elsahn, N.A.; Russo, D.; Di Blasio, M.; Cicciù, M.; Minervini, G. A systematic review and meta-analysis of the impact of resveratrol on oral cancer: Potential therapeutic implications. *BMC Oral. Health* **2024**, *24*, 412. [CrossRef]
108. Fukuda, M.; Ogasawara, Y.; Hayashi, H.; Inoue, K.; Sakashita, H. Resveratrol Inhibits Proliferation and Induces Autophagy by Blocking SREBP1 Expression in Oral Cancer Cells. *Molecules* **2022**, *27*, 8250. [CrossRef]
109. Hu, F.W.; Tsai, L.L.; Yu, C.H.; Chen, P.N.; Chou, M.Y.; Yu, C.C. Impairment of tumor-initiating stem-like property and reversal of epithelial-mesenchymal transdifferentiation in head and neck cancer by resveratrol treatment. *Mol. Nutr. Food Res.* **2012**, *56*, 1247–1258. [CrossRef]
110. Tyagi, A.; Gu, M.; Takahata, T.; Frederick, B.; Agarwal, C.; Siriwardana, S.; Agarwal, R.; Sclafani, R.A. Resveratrol selectively induces DNA Damage, independent of Smad4 expression, in its efficacy against human head and neck squamous cell carcinoma. *Clin. Cancer Res.* **2011**, *17*, 5402–5411. [CrossRef]
111. Bostan, M.; Mihaila, M.; Petrica-Matei, G.G.; Radu, N.; Hainarosie, R.; Stefanescu, C.D.; Roman, V.; Diaconu, C.C. Resveratrol Modulation of Apoptosis and Cell Cycle Response to Cisplatin in Head and Neck Cancer Cell Lines. *Int. J. Mol. Sci.* **2021**, *22*, 6322. [CrossRef] [PubMed]
112. Mikami, S.; Ota, I.; Masui, T.; Uchiyama, T.; Okamoto, H.; Kimura, T.; Takasawa, S.; Kitahara, T. Resveratrol-induced REG III expression enhances chemo- and radiosensitivity in head and neck cancer in xenograft mice. *Oncol. Rep.* **2019**, *42*, 436–442. [CrossRef] [PubMed]

113. Masuelli, L.; Di Stefano, E.; Fantini, M.; Mattera, R.; Benvenuto, M.; Marzocchella, L.; Sacchetti, P.; Focaccetti, C.; Bernardini, R.; Tresoldi, I.; et al. Resveratrol potentiates the in vitro and in vivo anti-tumoral effects of curcumin in head and neck carcinomas. *Oncotarget* **2014**, *5*, 10745–10762. [CrossRef] [PubMed]
114. Amin, A.; Wang, D.; Nannapaneni, S.; Lamichhane, R.; Chen, Z.G.; Shin, D.M. Combination of resveratrol and green tea epigallocatechin gallate induces synergistic apoptosis and inhibits tumor growth in vivo in head and neck cancer models. *Oncol. Rep.* **2021**, *45*, 87. [CrossRef]
115. Li, Z.; Ge, H.; Xie, Y.; Zhang, Y.; Zhao, X.; Sun, W.; Song, M. Luteolin inhibits angiogenesis and enhances radiotherapy sensitivity of laryngeal cancer via downregulating Integrin β1. *Tissue Cell* **2023**, *85*, 102235. [CrossRef]
116. Li, L.; Dong, X.; Peng, F.; Shen, L. Integrin β1 regulates the invasion and radioresistance of laryngeal cancer cells by targeting CD147. *Cancer Cell. Int.* **2018**, *18*, 80. [CrossRef]
117. Song, M.; Liu, X.; Li, T.; Zhang, Y.; Zhao, X.; Sun, W.; Li, Z. Silencing PLOD2 attenuates cancer stem cell-like characteristics and cisplatin-resistant through Integrin β1 in laryngeal cancer. *Transl. Oncol.* **2022**, *22*, 101460. [CrossRef]
118. Park, S.J.; Min, H.J.; Yoon, C.; Kim, S.H.; Kim, J.H.; Lee, S.Y. Integrin β1 regulates the perineural invasion and radioresistance of oral squamous carcinoma cells by modulating cancer cell stemness. *Cell Signal.* **2023**, *110*, 110808. [CrossRef]
119. Tu, D.G.; Lin, W.T.; Yu, C.C.; Lee, S.S.; Peng, C.Y.; Lin, T.; Yu, C.H. Chemotherapeutic effects of luteolin on radio-sensitivity enhancement and interleukin-6/signal transducer and activator of transcription 3 signaling repression of oral cancer stem cells. *J. Formos. Med. Assoc.* **2016**, *115*, 1032–1038. [CrossRef]
120. Selvi, R.B.; Swaminathan, A.; Chatterjee, S.; Shanmugam, M.K.; Li, F.; Ramakrishnan, G.B.; Siveen, K.S.; Chinnathambi, A.; Zayed, M.E.; Alharbi, S.A.; et al. Inhibition of p300 lysine acetyltransferase activity by luteolin reduces tumor growth in head and neck squamous cell carcinoma (HNSCC) xenograft mouse model. *Oncotarget* **2015**, *6*, 43806–43818. [CrossRef]
121. Majumdar, D.; Jung, K.H.; Zhang, H.; Nannapaneni, S.; Wang, X.; Amin, A.R.; Chen, Z.; Chen, Z.G.; Shin, D.M. Luteolin nanoparticle in chemoprevention: In vitro and in vivo anticancer activity. *Cancer Prev. Res.* **2014**, *7*, 65–73. [CrossRef]
122. Kim, S.G.; Veena, M.S.; Basak, S.K.; Han, E.; Tajima, T.; Gjertson, D.W.; Starr, J.; Eidelman, O.; Pollard, H.B.; Srivastava, M.; et al. Curcumin treatment suppresses IKKβ kinase activity of salivary cells of patients with head and neck cancer: A pilot study. *Clin. Cancer Res.* **2011**, *17*, 5953–5961. [CrossRef]

MDPI AG
Grosspeteranlage 5
4052 Basel
Switzerland
Tel.: +41 61 683 77 34

Antioxidants Editorial Office
E-mail: antioxidants@mdpi.com
www.mdpi.com/journal/antioxidants

www.ingramcontent.com/pod-product-compliance
Lightning Source LLC
Chambersburg PA
CBHW061606210926
51850CB00009B/2235